Causal Inference in R

Decipher complex relationships with advanced R techniques
for data-driven decision-making

Subhajit Das

Causal Inference in R

Group Product Manager: Niranjan Naikwadi

Publishing Product Manager: Nitin Nainani

Book Project Manager: Aparna Nair

Senior Content Development Editor: Shreya Moharir

Technical Editor: Seemanjay Ameriya

Copy Editor: Safis Editing

Proofreader: Shreya Moharir

Indexer: Hemangini Bari

Production Designer: Shankar Kalbhor

DevRel Marketing Coordinator: Vinishka Kalra

First published: November 2024

Production reference: 1311024

Published by Packt Publishing Ltd.

Grosvenor House

11 St Paul's Square

Birmingham

B3 1RB, UK

ISBN 978-1-83763-902-1

www.packtpub.com

I dedicate this book to my mother, Tapasi, and my father, Rabi Sankar, whose immense belief and steadfast faith kept me going even when the journey seemed impossible. To my wife, Florina, for her incredible confidence in me, especially on my darkest days. To my brother, Biswajit, whose smiles and joy in life's little moments have always inspired and encouraged me.

– Subhajit Das

Contributors

About the author

Subhajit Das holds a PhD in computer science from Georgia Institute of Technology, USA, specializing in machine learning (ML) and visual analytics. With 10+ years of experience, he is an expert in causal inference, revealing complex relationships and data-driven decision-making. His work has influenced millions in AI, e-commerce, logistics, and 3D software sectors. He has collaborated with leading companies, such as Amazon, Microsoft, Bosch, UPS, 3M, and Autodesk, creating solutions that seamlessly integrate causal reasoning and ML. His research, published in top conferences, focuses on developing AI-powered interactive systems for domain experts. He also holds a master's degree in design computing from the University of Pennsylvania, USA.

I am deeply grateful to my parents, brother, and my wife for their consistent faith in me. Their support, along with the confidence and patience they instilled, made this gift to the community possible.

About the reviewer

Harshita Asnani is an accomplished applied scientist, specializing in data-driven decision-making across teams. She holds a master's degree in applied data science from Syracuse University and possesses expertise in ML, deep learning, and AI. Harshita excels in developing ML solutions throughout the ML/AI life cycle, from data collection to model deployment and evaluation. She has specialized knowledge in recommender systems and experience with graph ML and causal inference techniques. Harshita is passionate about leveraging advanced analytics, and she aims to solve complex challenges and enhance organizational performance through innovative data solutions.

Table of Contents

3

Initiating R with a Basic Causal Inference Example   33

Part 2: Practical Applications and Core Methods

4

Constructing Causality Models with Graphs   59

5

Navigating Causal Inference through Directed Acyclic Graphs 81

6

Employing Propensity Score Techniques 103

7

Employing Regression Approaches for Causal Inference 129

8

Executing A/B Testing and Controlled Experiments   161

9

Implementing Doubly Robust Estimation   191

Part 3: Advanced Topics and Cutting-Edge Methods

10

Analyzing Instrumental Variables   209

14

Harnessing Causal Forests and Machine Learning Methods   287

15

Implementing Causal Discovery in R   309

Preface

Hello, dear readers! I'm thrilled that you've picked up this book and are considering diving in. You might be wondering what it's all about. As the title suggests, it's about causal inference and applying it in R. But why is learning that important? Well, in today's data-driven world, understanding causality has become more critical than ever. This book is tailored for anyone with data who wants to go beyond simple correlations and discover the true causal relationships in their workstream. Whether you're an analyst, data scientist, machine learning engineer, or researcher, you'll find the tools and techniques you need to conduct rigorous causal analysis using R in this book. This knowledge will empower you to make well-informed and impactful decisions. Now, why use R? Because it's one of the best platforms for data science, offering a vast array of ready-to-use libraries, strong community support, and comprehensive tools to explore your causal ideas.

Essentially, this book guides you through the core and advanced principles of causal inference, providing practical, hands-on examples in R. You'll learn the following:

- How to handle complex datasets
- How to apply causal models
- How to interpret results to uncover the underlying causes of observed patterns

By deep-diving into scenarios modeled after real-world case studies, you'll gain a deep understanding of how to leverage causal inference to solve pressing business challenges, optimize processes, and improve outcomes across various industries.

Our goal is to equip you with the knowledge and skills to confidently apply causal inference techniques in your own work setting. The book covers the following:

- The fundamentals of causal inference and its application in R
- The basics of causal reasoning and representations using directed acyclic graphs
- Advanced topics such as propensity score methods, instrumental variables, causal forests, and causal discovery

By the end of this journey, you'll be well-prepared to conduct in-depth causal analyses, distinguish causation from correlation, and transform data into actionable insights using R.

Who this book is for

This book is for individuals who work hands-on with data and seek to glean causal insights. It is particularly valuable for analysts, data scientists, and machine learning engineers handling complex datasets who need to perform robust causal analysis using R. Researchers and academics aiming to enhance their understanding of causal inference techniques will also benefit from the comprehensive approach presented here. Through practical examples and detailed explanations, this book will deepen your ability to use R to understand data through causal inference, enabling you to identify causal relationships and make informed decisions. While a basic familiarity with the R programming language is advantageous, it is not imperative to grasp the content presented in this book.

What this book covers

Chapter 1, *Introducing Causal Inference*, lays the foundation for understanding causal inference, differentiating between association and causation. It also highlights the importance of causal questions in various scenarios. Through discussions, we will explore the historical underpinnings of causality and further dive deep into the technical aspects of foundational concepts in causality.

Chapter 2, *Unraveling Confounding and Associations*, summarizes the critical concept of confounding variables and the various challenges involved in identifying them. This chapter clarifies the distinctions between correlation, association, and causation. Further, it explores various effective strategies to manage confounding in observational data. Additionally, it addresses common biases and enlightens key assumptions in causal inference, providing a comprehensive understanding of the factors that can impact causal relationships.

Chapter 3, *Initiating R with a Basic Causal Inference Example*, introduces the fundamentals of using R for causal analysis. This chapter guides you through setting up your R environment and walks you through solving a basic causal inference problem. You'll apply causal inference techniques to a case study, leveraging various R packages to understand how these tools work in practice, laying a solid foundation for more advanced analyses.

Chapter 4, *Constructing Causality Models with Graphs*, explores the use of Directed Acyclic Graphs (DAGs) in representing causal relationships and identifying causal effects. Through a practical case study in R, you will learn how to apply these concepts to real-world data, providing a hands-on understanding of how graphical models can enhance your causal analysis.

Chapter 5, *Navigating Causal Inference through Directed Acyclic Graphs*, discusses advanced techniques involving DAGs, covering critical concepts such as chains, colliders, and immoralities. This chapter provides an in-depth exploration of graph-based causal structures, including essential methods such as back door and front door adjustments. The chapter concludes with a practical case study involving a grocery store scenario, demonstrating the application of these advanced DAG techniques using R.

Chapter 6, Employing Propensity Score Techniques, introduces a powerful method for causal analysis, focusing on the use of propensity score techniques. This chapter covers key applications such as matching, weighting, and integrating propensity scores with causal diagrams. You'll learn how these techniques help uncover the nuances of causal relationships, particularly when dealing with heterogeneity in data, providing a clearer understanding of how different factors influence outcomes.

Chapter 7, Employing Regression Approaches for Causal Inference, explores the selection and application of appropriate regression models for causal analysis, with a focus on model diagnostics and assumptions. We will cover both linear and non-linear regression-based approaches for causal inference, as well as learning about model diagnostics and assumptions.

Chapter 8, Executing A/B Testing and Controlled Experiments, shows you how to gain expertise in designing, conducting, and analyzing A/B tests and controlled experiments within the context of causal inference. This chapter is crucial in understanding how you can actually apply causal knowledge in real-world applications through experimental data collection. You will learn the specifics of controlled experiments, common pitfalls, and how to apply your knowledge in R.

Chapter 9, Implementing Doubly Robust Estimation, provides an in-depth exploration of the concept of doubly robust estimation, highlighting its strengths and comparing it with other causal inference techniques. This chapter demonstrates how to apply doubly robust methods using R, showcasing their resilience and effectiveness. For example, it shows that even if one of the models in this approach is not entirely accurate, the technique still manages to yield reliable and robust results, making it a powerful tool for causal analysis.

Chapter 10, Analyzing Instrumental Variables, introduces the concept of instrumental variables and their critical role in identifying causal effects when direct manipulation isn't possible. You will learn how to identify valid instrumental variables, understand their assumptions, and apply instrumental variable analysis using R. This chapter specifically demonstrates the power of instrumental variables in uncovering causal relationships that might otherwise remain hidden, providing you with robust tools for causal analysis in complex scenarios.

Chapter 11, Investigating Mediation Analysis, dives into a technique that allows you to understand the mechanisms through which causal effects operate. You will learn mediation analysis, which shows you how to identify and measure mediation effects, explore different types of mediation models, and apply these mediation techniques in R. This chapter is essential for those looking to disentangle the pathways and processes that link causes to their effects, offering deeper insights into causal relationships.

Chapter 12, Exploring Sensitivity Analysis, focuses on sensitivity analysis, a crucial tool for assessing the robustness of your causal inferences. This chapter covers the principles of sensitivity analysis, practical methods for its implementation in R, and how to interpret the results. By the end of this chapter, you will be equipped to evaluate the reliability of your causal conclusions, ensuring that your findings remain valid even under varying assumptions.

Chapter 13, Scrutinizing Heterogeneity in Causal Inference, explores the concept of heterogeneity, highlighting how different subgroups can experience varying causal effects. This chapter guides you through identifying and estimating heterogeneous treatment effects, providing insights into how these differences can impact your analysis. By using R, you will learn to customize interventions and strategies for specific groups, thereby enhancing the effectiveness and precision of your causal analyses.

Chapter 14, Harnessing Causal Forests and Machine Learning Methods, introduces causal forests, an advanced machine learning approach to estimate heterogeneous causal effects. You will learn how causal forests differ from traditional machine learning models and how to implement them using R. This chapter combines causal inference with modern machine learning techniques, providing powerful tools to uncover complex causal relationships and tailor interventions.

Chapter 15, Implementing Causal Discovery in R, delves into causal discovery methods, which aim to uncover causal relationships directly from data. In this final chapter, you will explore various causal discovery algorithms, learn about their strengths and limitations, and apply them in R. This chapter provides a comprehensive overview of how to use data-driven approaches to identify potential causal structures, equipping you with the skills to analyze and interpret complex datasets where the underlying causal relationships are not immediately apparent.

To get the most out of this book

To fully benefit from this book, it is recommended you have some understanding of basic statistical concepts, including distributions, hypothesis testing, and confidence intervals. Familiarity with fundamental data analysis techniques, such as linear and logistic regression, may help further (but are not necessary), as these will provide a basis for understanding the more advanced causal inference methods covered in this book.

In terms of programming, having some experience with R is highly advantageous but not essential. You should be comfortable with basic R operations, such as data manipulation using `data.frame`, handling different data types, and using common functions and packages such as `dplyr` and `ggplot2` for data analysis and visualization. If you are new to R, don't worry – the book includes introductory sections on setting up R and guiding you through essential R skills. However, having a basic programming mindset will help you grasp more complex coding tasks.

Regarding system requirements, you will need a laptop or desktop computer capable of running R and RStudio Desktop, the popular integrated development environment (IDE) for R. A computer with at least 8 GB of RAM is recommended to efficiently handle larger datasets and perform computations required for causal analysis. While R itself is not resource-intensive, some of the analyses and models discussed in the book, especially those involving large datasets or complex simulations, will benefit from a modern multi-core processor and ample storage space. Additionally, having a stable internet connection will be useful for downloading R packages and accessing online resources.

Setting up the coding platform R Studio is detailed step by step in *Chapter 3*.

Software/hardware covered in the book	Operating system requirements
RStudio Desktop	Windows, macOS, or Linux
R	

> **Important note on sensitivity and analytical focus of the practical examples**
>
> The practical examples covered in this book may explore complex societal issues, however, our focus is solely on learning about the analytical methods. The insights gained from the examples are intended to demonstrate statistical and computational techniques, not to provide definitive conclusions about real-world problems.

If you are using the digital version of this book, we advise you to type the code yourself or access the code from the book's GitHub repository (a link is available in the next section). Doing so will help you avoid any potential errors related to the copying and pasting of code.

Download the example code files

You can download the example code files for this book from GitHub at `https://github.com/PacktPublishing/Causal-Inference-in-R`. If there's an update to the code, it will be updated in the GitHub repository.

We also have other code bundles from our rich catalog of books and videos available at `https://github.com/PacktPublishing/`. Check them out!

Conventions used

There are a number of text conventions used throughout this book.

`Code in text`: Indicates code words in text, database table names, folder names, filenames, file extensions, pathnames, dummy URLs, user input, and Twitter handles. Here is an example: "We use the `lm` function for our outcome model as before but will adjust how we calculate the standard errors using the `sandwich` package."

A block of code is set as follows:

```
library(MASS)
set.seed(123)
ad_data <- data.frame(ad_spend = runif(100, 100, 1000),
                      sign_ups = rpois(100, lambda=20))
# Poisson regression model
model <- glm(sign_ups ~ ad_spend, data=ad_data, family="poisson")
summary(model)
```

When we wish to draw your attention to a particular part of a code block, the relevant lines or items are set in bold:

```
x <- c(1, 2, 3)
factor_x <- structure(x, class = "factor",
                      .Label = c("low", "medium", "high"))
```

Bold: Indicates a new term, an important word, or words that you see on screen. For instance, words in menus or dialog boxes appear in **bold**. Here is an example: "Found at the top left (accessible via **File | New File | R Script**), this is where you write and save longer code blocks or scripts for complex analyses."

> **Tips or important notes**
> Appear like this.

Get in touch

Feedback from our readers is always welcome.

General feedback: If you have questions about any aspect of this book, email us at customercare@ packtpub.com and mention the book title in the subject of your message.

Errata: Although we have taken every care to ensure the accuracy of our content, mistakes do happen. If you have found a mistake in this book, we would be grateful if you would report this to us. Please visit www.packtpub.com/support/errata and fill in the form.

Piracy: If you come across any illegal copies of our works in any form on the internet, we would be grateful if you would provide us with the location address or website name. Please contact us at copyright@packt.com with a link to the material.

If you are interested in becoming an author: If there is a topic that you have expertise in and you are interested in either writing or contributing to a book, please visit authors.packtpub.com.

Share Your Thoughts

Once you've read *Causal Inference in R*, we'd love to hear your thoughts! Scan the QR code below to go straight to the Amazon review page for this book and share your feedback.

https://packt.link/r/1-837-63902-7

Your review is important to us and the tech community and will help us make sure we're delivering excellent quality content.

Download a free PDF copy of this book

Thanks for purchasing this book!

Do you like to read on the go but are unable to carry your print books everywhere?

Is your eBook purchase not compatible with the device of your choice?

Don't worry, now with every Packt book you get a DRM-free PDF version of that book at no cost.

Read anywhere, any place, on any device. Search, copy, and paste code from your favorite technical books directly into your application.

The perks don't stop there, you can get exclusive access to discounts, newsletters, and great free content in your inbox daily

Follow these simple steps to get the benefits:

1. Scan the QR code or visit the link below

https://packt.link/free-ebook/9781837639021

2. Submit your proof of purchase
3. That's it! We'll send your free PDF and other benefits to your email directly

Part 1: Foundations of Causal Inference

This part introduces the core principles of causal inference, focusing on distinguishing causation from association and correlation. It covers fundamental concepts such as confounding variables, biases, and assumptions in causal analysis, providing a solid theoretical base. Additionally, it introduces the use of R for basic causal inference, preparing you for practical applications using R.

This part has the following chapters:

- *Chapter 1, Introducing Causal Inference*
- *Chapter 2, Unraveling Confounding and Associations*
- *Chapter 3, Initiating R with a Basic Causal Inference Example*

1
Introducing Causal Inference

In this inaugural chapter, let's explore the topic of **causal inference** a bit. For some, this may be a new topic; for others, it might be somewhat familiar. However, whether you find this topic intimidating or not depends less on your existing statistical knowledge and more on your interest in the subject and your consistent effort throughout the book.

Our exploration begins with three pivotal questions: What exactly is causal inference? Why is it indispensable? How can it be effectively utilized? To clarify these concepts, we'll use both fictitious and real-life scenarios.

Approach this chapter with unhindered curiosity and an open mind. Be prepared to encounter concepts and terminology that might initially seem abstruse. Don't worry, though—we will be with you every step of the way, ensuring you understand everything clearly and thoroughly as we explore causal inference together.

In this chapter, we will cover the following topics:

- Defining causal inference
- Historical perspectives on causal inference
- Why do we need causality?
- Is it an association or really causation?
- Deep diving into causality in real-life settings
- Exploring the technical aspects of causality

Defining causal inference

Picture yourself as a teacher contemplating a curious phenomenon among high school students: the relationship between sleeping late and catching the school bus. An initial hypothesis might be, "Sleeping late causes students to miss their school bus." This stems from a personal experience: I slept late and consequently missed the bus. However, this hypothesis might be challenged upon observing graduate students, who, despite sleeping late, consistently catch their buses.

This scenario exemplifies the complex, often misleading nature of causality. The observed association—sleeping late and missing buses—doesn't inherently imply causation. Here, various other factors could be at play. Perhaps high school students have earlier bus schedules, or graduate students, despite sleeping late, can wake up early and not miss their transportation. It's plausible that the initial observation of sleeping late and causing one to miss the bus was a mere coincidence, not a universal rule.

Causal inference in this context goes beyond the superficial observation of relationships. It goes deeper, exploring whether one factor (sleeping late) actively influences another (missing the bus). This is not just about associating events but understanding thoroughly the underlying mechanisms that link them. Is the relationship direct, or are there hidden variables that mediate this connection?

Practically, this kind of analysis is vital. Consider a school administrator who, based on initial observations, might advocate for earlier bedtimes to ensure students catch their buses. However, a more nuanced causal analysis could reveal that the issue isn't bedtime but perhaps bus scheduling or student time management. Making decisions based on superficial associations could lead to ineffective or even counterproductive policies.

In statistical terms, particularly when employing tools such as R, causal inference is the methodology that allows us to rigorously test these relationships. It helps us differentiate between mere assumptions and substantiated causality.

Reflecting on our initial hypothesis, a thorough causal analysis would involve examining all potential variables—bus schedules, student routines, and even the difference between high school and graduate lifestyles. Only a comprehensive study can reveal whether sleeping late directly causes a student to miss the bus or is just a coincidental factor in a student's life. Now that we have defined what causal inference is, let's learn where it came from.

Historical perspective on causal inference

In Hellenic philosophy, ancient Greece played a pivotal role. Pre-Socratic philosophers such as Thales and Heraclitus explored the nature of change and causality. Ancient Greek philosophers contributed significantly to systematic approaches to understanding causality by examining cause-and-effect relationships. They introduced important causal concepts, including the idea that nothing comes from nothing (attributed to Parmenides). While not originating the principle of sufficient reason, their work laid the foundations for later philosophical developments. Their ideas have influenced our understanding of causation, though modern concepts have evolved significantly, incorporating insights from various traditions, scientific advancements, and mathematical frameworks.

Aristotle, however, provided a more structured approach to causality with his *four causes* theory:

- **Material cause**: The material from which something is made (e.g., the bronze in a statue)
- **Formal cause**: The essence or shape of something (e.g., the design of a statue)
- **Efficient cause**: The initiator of change or stability (e.g., the sculptor of the statue)
- **Final cause**: The intended purpose of an object (e.g., the statue's artistic or religious function)

Aristotle's framework significantly advanced the systematic study of causality, influencing philosophical and scientific thought for centuries.

In Eastern philosophies, causality was similarly a significant concept. Hindu texts, such as the Upanishads, explore causality within material and spiritual realms, often using metaphors such as a spider spinning a web to depict inherent causation.

Buddhism introduces *Pratītyasamutpāda*, or **dependent origination**, conceptualizing the interdependence and interconnectedness of all phenomena. This principle suggests that everything arises in dependency on conditions, forming a complex causal network.

The initial forays into the concept of causality by ancient civilizations provided the cornerstone ideas that would ultimately give rise to the scientific method. While these early thinkers were not conducting causality studies as understood in contemporary terms, their philosophical examination of cause and effect laid the groundwork that profoundly influenced later generations. Their efforts exemplify the varied approaches through which human societies have endeavored to comprehend the linkages between actions and their consequences—a pursuit that persists in the intricate causal analyses of modern times.

In studying causal inference, it's important to acknowledge the many thinkers who have improved our understanding of causality. This field has many key ideas, ranging from basic philosophical concepts to complex mathematical models.

20th-century statisticians and economists, such as Ronald A. Fisher, Jerzy Neyman, Egon Pearson, and Donald Rubin, laid significant groundwork in the field of causal inference. Fisher was instrumental in experimental design, particularly with his contributions to randomization and analysis of variance. Neyman and Pearson developed a framework for hypothesis testing and the Neyman-Rubin causal model, which are fundamental to contemporary causal inference methods. Rubin, co-developing the **Rubin causal model** (**RCM**) [3], introduced pivotal concepts such as potential outcomes and propensity score matching (both covered in this book), essential in observational studies.

In thought leadership within causal science, Judea Pearl [7] and James Heckman stand out. Pearl, renowned in artificial intelligence and statistics, formulated the do-calculus and a theoretical framework for causal relationships using graphical models. Heckman, a Nobel laureate in economics, made significant strides in understanding causal relationships through his work on selection bias and the Heckman correction.

Furthermore, these pioneers teach the foundational elements of causal inference:

- Mill's methods [4] provide logical strategies for causal identification in observational studies.

- Fisher's *Design of Experiments* [5] emphasizes the importance of randomization and control in establishing causality

- The **Neyman-Pearson framework** [6] underlines statistical rigor and meticulous hypothesis testing for causal conclusions

- Rubin's **potential outcomes** [3] approach stresses understanding what might have happened under different scenarios

- Pearl's *Causal diagrams for empirical research* [7] introduce graphical models for managing various confounding factors

- Heckman's *Handbook of Econometrics* [8] addresses selection bias and endogeneity in economic and social data analysis

We have provided a comprehensive overview of causality, highlighting its distinction from simple association. A pertinent inquiry emerges: Why is the knowledge of causality essential? Let's discuss it in the next section.

Why do we need causality?

Beyond understanding the theoretical and historical underpinnings, one must ponder the practical necessity of causality first. We shall discuss examples of the ubiquitous application of causality across various industries. For instance, enterprises leverage causal inference techniques to gain deeper insights into customer behaviors, needs, and preferences. They employ these methods to elucidate both natural and anthropogenic phenomena. Mastery of causal inference equips you with an extremely powerful tool, rendering you an invaluable asset in any team or organizational context. Your proficiency in this domain can significantly contribute to the overarching objective of delivering value to stakeholders.

Let's discuss further why causality is not only an intellectually rewarding area but also practically indispensable.

In medical and public health arenas, causal inference is vital for assessing treatment efficacy. **Randomized controlled trials** (**RCTs**) stand as the pinnacle of causal inference methods, isolating drug effects from external variables. RCTs are experiments where participants are randomly assigned to an intervention or control group to measure the intervention's effects, minimizing bias for reliable results.

A pertinent example is its role during the COVID-19 pandemic, where causal inference underpinned the evaluation of vaccines' efficacy and safety, informing critical decisions on their approval and distribution strategies, thereby saving lives.

Economists utilize causal inference to decode market behaviors and policy impacts. Analyzing the effects of a minimum wage hike on employment, for instance, requires separating the causative effects from economic trends and other policy shifts. Techniques such as difference-in-differences analysis enable economists to extract causal insights from observational data, influencing policies that impact millions.

In business, causal inference informs the effectiveness of marketing efforts and strategic decisions. A/B testing, a direct application of causal principles, guides companies in optimizing profits and enhancing customer experience. By comparing conversion rates from different advertising campaigns, businesses can determine which strategies are more effective.

Statistically, causality is paramount for accurate data interpretation. Identifying associations between variables is one aspect, but establishing causation is a more complex and significant task. This distinction shapes the conclusions and recommendations derived from data analysis.

In survey methodology, understanding causality is critical. When analyzing survey data, statisticians must discern potential causal links between variables to avoid erroneous conclusions based on merely correlated associations. Causal inference, therefore, is not just a statistical tool but a fundamental approach to deciphering the dynamics of various phenomena. Now, in the next section, let's learn about more critical aspects of causality.

Is it an association or really causation?

It's tempting to attribute causality to superficial observations, mistaking mere associations for causation. Take, for instance, the observation that social media posts made later in the day receive fewer likes and comments, suggesting reduced engagement. One might hastily conclude that the timing of these posts is the causal factor. However, without rigorous statistical testing, such claims remain speculative. In this book, we will teach you how to conduct these necessary tests, distinguishing between simple association and true causation.

In statistics, we discuss association, causation, and *correlation*. While correlation is often used interchangeably with association in everyday conversations, they have distinct meanings in statistical contexts. So, what is the difference between association and correlation?

In causality, **association** encapsulates a general linkage between two variables, without explicitly characterizing the nature or magnitude of this relationship. This concept encompasses both linear and non-linear associations. Contrastingly, **correlation** denotes a specific statistical measure, exemplified by metrics such as Pearson's correlation coefficient, which quantifies the strength and direction of a linear relationship between variables. The coefficient's value spans from -1 to 1, indicating a spectrum from strong negative to strong positive linear relationships.

Now that the distinction is clear between association and correlation, let's see what the relationship between correlation and causation is.

To understand this, let's examine a real-life example. Picture the observed increase in motorcycle accidents coinciding with a rise in rainfall. This simultaneous upsurge could suggest a strong positive linear correlation, particularly if their correlation coefficient hovers near 1. However, it's crucial to recognize that *correlation does not equate to causation*. Increased rain does not cause motorcycle accidents.

To unpack this further, we need to consider the role of confounding variables, a concept we will explore more comprehensively later in this chapter. In our case, the overarching weather conditions serve as a potential confounder. Rainy days are associated with more motorcycle accidents. However, rainy days also often come with other weather factors, such as strong winds and poor visibility. These overall weather conditions affect both rainfall and driving safety. Why is this important? Because bad weather not only increases rainfall but also creates hazardous driving conditions. This dual impact can lead to a spike in motorcycle accidents, primarily due to slippery roads, reduced visibility, and challenging driving environments.

Thus, while rainfall and motorcycle accidents are correlated, the actual causative factor may be the broader weather patterns, a subtle yet significant distinction in data interpretation (see *Figure 1.1*).

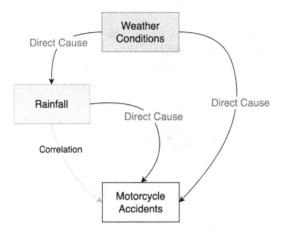

Figure 1.1 – A directed acyclic graph (DAG)

This diagram illustrates that while there's a direct path from Rainfall to Motorcycle Accidents, there's also another path from Weather Conditions to Motorcycle Accidents through Rainfall. Additionally, Weather Conditions directly influences Motorcycle Accidents. This is a case of confounding.

> **Confounding variable**
>
> It is an external variable that is not the primary focus of a study but can affect both the **independent variable** (the variable you are changing) and the **dependent variable** (the variable you are measuring).

The primary objective of causal inference is to ascertain whether observed correlations genuinely reflect causal relationships, by meticulously controlling for potential confounding factors. While both association and correlation can hint at possible causal connections, they do not inherently validate causality. Consequently, further empirical investigation, through experimental or quasi-experimental methodologies, is frequently necessitated to establish causal links definitively.

Did you know that our tendency to interpret past experiences as causality is a classic human trait? It's quite amusing, yet deeply ingrained in our psychology, to mix up association with causation.

Let's consider the age-old practice of burning the midnight oil before exams or project deadlines, often linked to better grades or outcomes. This belief stems from those all-nighters that seemed to coincide with academic triumphs. However, what really leads to higher grades? It might not just be the long hours spent studying but factors such as the precision of your work, regular class attendance, the flair in your presentations, and being punctual. This shows us that the supposed connection between night-time study marathons and academic success might just be a classic case of mistaking association for causality, without accounting for other important elements.

Globally, there's a popular notion that kids studying STEM (science, technology engineering, or mathematics) subjects are on a fast track to higher earnings. This originates from observing that folks with a STEM background often land lucrative jobs. But let's pause and think—is it just the STEM education driving higher income? Other elements, such as the quality of education, networking, where you live, economic factors, individual talents outside STEM, or even one's social and economic background, play a significant role. This example shines a light on the potential error of directly linking STEM education with higher income, without considering these extra factors, and reminds us not to confuse correlation or association with causation.

Going back to associations, you should know by now that *association* means events or variables occurring together, while *causation* implies one variable brings about a change in another. Understanding this difference is critical.

Let us go deeper into understanding how causality is fundamentally embedded in the fabric of our everyday world, encompassing a vast array of industrial settings, diverse use cases, and problem formulations.

Deep dive causality in real-life settings

Let's walk through a study titled *Inked into Crime? An Examination of the Causal Relationship between Tattoos and Life-Course Offending among Males from the Cambridge Study in Delinquent Development* [1]. It's about a study to ascertain whether or not there exists a causal connection between the presence of tattoos and the propensity for criminal behavior across one's lifespan. Analyzing data from 411 British males, the researchers utilized **propensity score matching**—a statistical technique frequently used to ascertain causal inference (which we shall learn about later in this book, in *Chapter 6*). This approach meticulously dissects the complexity between the ink on skin and the propensity for crime, offering a more refined perspective on this age-old debate.

Rooted in the shadow of 19th-century criminological thought, specifically the theories of Lombroso, tattoos have long been cast in the dim light of criminality. This study, however, peels back layers of historical bias and cultural assumptions, examining how tattoos have been portrayed across both academic and pop culture spectrums. It's crucial to note here that while there exists a tangible correlation between tattoos and a gamut of criminal behaviors and psychological markers—such as impulsivity and substance abuse—this link is more correlational than causal.

Based on the causal analysis, the study concludes that tattoos and crime are linked not by causality but through shared risk factors and personality traits. This study clearly elucidates how a strong correlation does not necessarily imply causality. Now, one may wonder how this study can be utilized across industries. Well, that's a great question. It can actually be used in many settings, such as in human resources and employment practices; the study's conclusion that tattoos do not have a causal link to criminal behavior could encourage businesses to reassess and potentially modify their employment policies. This adjustment could result in a more inclusive hiring process, expanding the talent pool by eliminating biases against tattooed individuals. Similarly, in marketing and advertising, these findings can be instrumental in dismantling stereotypes associated with tattooed individuals, aiding in the creation of advertising content that is both inclusive and diverse, reflective of a more nuanced understanding of tattoos as personal or cultural expressions rather than indicators of criminal tendencies. Furthermore, the tattoo industry can harness these findings to address and mitigate stigmas associated with tattoos, employing this information in marketing strategies to foster a normalized perception of tattoos in professional and social contexts. For industries such as insurance and risk assessment, the study's conclusions provide a critical perspective for refining risk profiling models, ensuring that tattoos are not erroneously factored as indicators of criminal propensity, leading to more equitable and accurate risk assessments.

By looking at how the study separates correlation from causation, you can develop a more critical and analytical mindset. This is important in causal inference, where understanding details and mechanisms is crucial.

The next case study helps you engage more deeply with causal analysis principles.

The paper titled *Causal or Spurious: Using Propensity Score Matching to Detangle the Relationship between Violent Video Games and Violent Behavior* [2] undertakes a rigorous investigation into the often-debated link between the playing of violent video games and the manifestation of violent behavior, non-violent deviance, and substance use. Utilizing propensity score matching, the researchers learned a more nuanced understanding of causality in this context.

The study's initial findings, based on an unmatched sample (where participants are not paired based on similar characteristics), indicate a noticeable correlation: children who engage in violent video games exhibit a higher propensity for various forms of deviant behavior. If you don't understand what an unmatched sample is, don't worry now. We'll cover it later in the book. Coming back to the study, the trend is observed across genders, with males showing a particularly heightened likelihood of engaging in non-violent deviance, violent acts, and substance use. Females, while following a similar pattern, exhibit slightly lower rates. However, when they looked deeper using propensity score matching to create a quasi-experimental framework, a striking shift emerged. For males, the apparent negative effects attributed to playing violent video games dissipate significantly in the matched sample, suggesting that the initial correlational relationships might be spurious rather than indicative of a causal link.

In terms of gender differences, the study underscores a fascinating divergence. The propensity score matching, while effectively nullifying most of the significant correlations for males, reveals persistence in certain deviant behaviors among females. This includes an increased likelihood of engaging in group fights and carrying weapons, hinting at a potential causal connection between playing violent video games and certain types of violent behavior in females. Conclusively, the study challenges the prevalent notion of a robust causal link between violent video games and violent behavior. A robust causal link would be a strong, consistent relationship where changes in one factor (such as playing violent video games) directly and reliably lead to changes in another (such as violent behavior), even when accounting for other potential influences. The study here posits that for males, personality and background factors are likely more influential in the observed correlations with deviant behavior. For females, although there is some evidence of causality, it is not as pronounced as previously believed. This research serves as a critical reminder of the importance of dissecting underlying factors in the analysis of complex social phenomena, urging a reconsideration of commonly held beliefs about the impact of violent video games.

This work also offers strategic insights that are of significant utility to the video game industry and its related sectors. In advocacy and legal defense, the findings provide a robust foundation for the industry to counter regulatory or legislative actions that might be predicated on the assumption of a direct causal link between violent video games and aggressive behavior, particularly in male demographics. This evidence, showcasing a more complex and less definitive connection, empowers the industry to resist measures that would unjustifiably censor or restrict video game content. Additionally, in the sphere of marketing and public relations, these insights afford an opportunity to reshape the public narrative surrounding video games. By leveraging the study's findings, the industry can challenge and mitigate negative stereotypes, potentially enhancing the public image of video games and broadening their appeal across a more diverse audience base.

Moreover, the nuanced understanding of gender-specific impacts of violent content highlighted in the study can inform content development strategies within the industry. This includes tailoring game design elements to cater to or responsibly address different demographic groups, thus fostering more conscientious content creation. In parallel, these insights can guide the industry in enhancing parental guidance and educational campaigns. Such initiatives could involve refining age-rating systems or creating informative material to assist parents and guardians in making more informed choices regarding video game suitability for their children. Beyond these immediate applications, there is a scope for the industry to delve deeper into research and development, building upon these findings to explore the broader behavioral impacts of video games. This proactive approach could not only contribute to the development of socially responsible gaming products but also equip the industry to pre-emptively address potential controversies or regulatory challenges. Furthermore, the potential for collaboration with the healthcare and educational sectors emerges as a significant opportunity, wherein video games could be utilized as tools for positive development in therapeutic or educational settings. Lastly, these findings can enable the video game industry to engage more effectively with a range of stakeholders, including policymakers, educators, and advocacy groups, ensuring that discussions and policies related to media consumption and youth behavior are informed by comprehensive, evidence-based insights. By strategically leveraging these findings, the video game industry stands to not only safeguard its interests but also contribute meaningfully to the broader societal discourse on the impact of media consumption on behavior.

Drawing inspiration from the previous examples, can you recall two or three instances from your own life or work experiences where you might have employed causal inference to explore the underlying causes of certain events? Taking our exploration into causality further, let's explore the technical aspects of this topic. This exploration aims to equip you with the necessary knowledge and skills to adeptly apply these principles in your specific use cases.

Exploring the technical aspects of causality

From the previous section, it is evident that causal inference involves employing observational or experimental data to establish causal links, utilizing various statistical methods and theories to measure the influence of one variable (the "treatment" or "intervention") on another (the "outcome" or the "effect").

From a statistical vantage point, it focuses on estimating the **counterfactual**, hypothesizing the outcomes in alternate scenarios where the treatment was absent. This necessitates assumptions about data and underlying mechanisms, including the exclusion of unmeasured confounders. Let's go over these concepts one by one.

Counterfactual analysis

Counterfactual analysis involves exploring "what-if" scenarios to understand the effects of actions that didn't occur. It is used to estimate the causal impact of interventions by imagining alternative outcomes. In e-commerce, for example, this method helps assess new sales strategies by using statistical techniques to estimate what would have happened to customers who didn't experience the strategy. It answers the question: what if the strategy hadn't been implemented? By combining rigorous causal inference methods with practical business insights, this approach effectively evaluates strategies.

Central to this approach is the **potential outcomes** concept, positing that each subject has a hypothetical outcome for every potential treatment level. The causal effect is the variance between these potential outcomes. However, since only one outcome per subject is observable (the one pertaining to the actual treatment), causal inference techniques aim to deduce what would have occurred under different treatment scenarios.

Causal inference methods range from randomized experiments, the benchmark for controlling confounders, to techniques used in observational studies, such as matching, stratification, instrumental variables, regression discontinuity designs, difference-in-differences methods, and causal diagrams (such as DAGs). These methods seek to emulate randomized experiment conditions and extract causal conclusions from non-experimental data.

Ultimately, causal inference empowers you to quantify the causal effect as accurately and unbiasedly as possible, grounded in the available data and justifiable assumptions.

Simpson's paradox

Next, we learn a unique phenomenon crucial to the deeper understanding of causality. It is called **Simpson's paradox**, and it represents a statistical conundrum where a trend observed in separate groups vanishes or reverses when these groups are amalgamated. This paradox highlights the challenges and potential missteps in interpreting causality from observational data.

To exemplify Simpson's paradox and its impact on causal analysis, consider the case of university admissions with these statistics (see *Figure 1.2*).

In a university comprising the literature and engineering departments, an investigation into potential gender bias in admissions is conducted. The applicant data for the past year is analyzed:

- In the literature department, of 100 male applicants, 60 are admitted (60% admission rate), while out of 200 female applicants, 150 are admitted (75% admission rate)

- In the engineering department, 450 of 900 male applicants (50% admission rate) and 20 of 100 female applicants (20% admission rate) are admitted

In examining the admissions data department-wise, we observe a preference for female applicants within the literature department, contrasted starkly by the engineering department's noticeable inclination to favor male applicants.

Combining the figures from both departments presents the following picture:

- **Total male applicants**: 1,000 (100 in literature, 900 in engineering)

- **Total female applicants**: 300 (200 in literature, 100 in engineering)

- **Total accepted males**: 510 (60 in literature, 450 in engineering)

- **Total accepted females**: 170 (150 in Literature, 20 in engineering)

This amalgamation yields these overall admission rates:

- **Male admission rate**: 51% (510/1,000)

- **Female admission rate**: 56.7% (170/300)

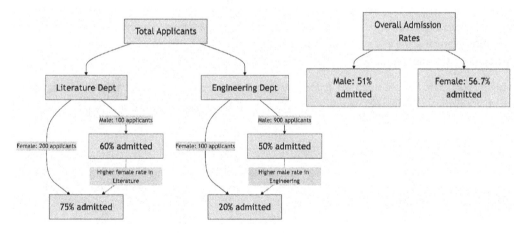

Figure 1.2 – Simpson's paradox in the context of university admissions

Despite the apparent bias against females in the engineering department and favor toward them in the literature department, the combined data misleadingly suggests a higher overall admission rate for females.

Surprisingly, when viewed collectively, the data suggests a marginally higher admission rate for females than males. This seems to contradict the specific trends observed in each department and might hint at an absence of gender bias against women or even a potential bias against men. Simpson's paradox arises when group sizes vary significantly and outcomes within each group differ, leading to misleading aggregate results. In this case, the disparity in the number of male and female applicants across departments and the varying admission rates create an illusion that contradicts the trends seen in individual departments.

However, this apparent paradox stems from the disparate sizes and admission rates of the departments involved. A greater number of men apply to the engineering program, which has a notably lower admission rate, while a larger contingent of women seeks entry into literature, where the acceptance rate is significantly higher. When merged, these figures disproportionately elevate the overall female admission rate, thanks to their dominant presence in the more lenient literature department. This scenario masks the stark gender bias evident in engineering.

This instance is a classic demonstration of Simpson's paradox, underscoring the intricacies involved in drawing conclusions from aggregated data, especially in causal analysis. To accurately discern whether gender affects admission rates, it's essential to dissect the data by relevant categories, such as departments in this case. A departmental breakdown reveals potential biases that are not apparent when data is lumped together.

Such scenarios highlight the need for careful analysis in causal studies. It's vital to consider underlying variables, such as departmental choice, that can significantly alter outcomes. Overlooking these elements can lead to false interpretations of causal relationships.

In summary, effective causal analysis requires meticulous management of confounding variables, data stratification, and an acute awareness of phenomena such as Simpson's paradox, which can skew our understanding and interpretations.

Moving forward, let's make a slight yet crucial distinction between the kinds of variables we may find as we solve complex problems of causality.

Defining variables

In statistical analysis, particularly when dissecting causal relationships, the concepts of confounding and lurking variables play a pivotal role. These elements can significantly skew how we interpret data, often leading to misleading conclusions:

- **Confounding variables**: These are the behind-the-scenes actors influencing both the independent variable (what we think is causing the change) and the dependent variable (the change we're observing). This variable can create the illusion of a causal link where none exists or can hide a real connection. They are a frequent headache in observational studies, where control over variables is limited. Take, for instance, a study probing the connection between regular exercise and heart health. Here, diet could be a confounder, impacting both a person's exercise routine and their heart health.

- **Lurking variables**: These are the stealthy variables not initially included in your study but still capable of affecting both your independent and dependent variables, thus potentially derailing your study's conclusions. Think of them as hidden influencers, similar to confounders, but often overlooked or unidentified in your analysis. For example, when examining how education level impacts income, geographic location could be a lurking variable. It might influence both educational opportunities and income levels but goes unnoticed in the study's framework.

In a nutshell, **lurking variables** are hidden factors affecting the variables of interest, while **confounding variables** are known factors influencing both dependent and independent variables.

Overlooking confounding and lurking variables can confuse us in understanding causal links. Identifying and managing these variables is crucial, especially in observational studies where the luxury of randomization isn't available. Being vigilant about these variables ensures the integrity and accuracy of your conclusions. Remember, in the world of data, what you see isn't always what you get!

Summary

This chapter introduced the concept of causality and its importance across various fields. A brief historical overview acknowledged the contributions of ancient philosophers and modern statisticians in developing causal inference methods. We started by defining causal inference and distinguishing it from association and correlation with practical examples. We also touched on complex ideas such as potential outcomes, confounding variables, and Simpson's paradox, explaining how they affect causal studies.

Finally, the chapter underscored the importance of causal inference in making informed decisions in our data-driven world. This foundation prepares you for a deeper exploration of causal inference in subsequent chapters.

References

1. *Inked into Crime*: https://www.sciencedirect.com/science/article/abs/pii/S0047235213001189.

2. *Causal or Spurious: Using Propensity Score Matching to Detangle the Relationship between Violent Video Games and Violent Behavior*: https://www.researchgate.net/publication/257252863_Causal_or_Spurious_Using_Propensity_Score_Matching_to_Detangle_the_Relationship_between_Violent_Video_Games_and_Violent_Behavior.

3. Imbens, G.W., Rubin, D.B. (2010). *Rubin Causal Model*. In: Durlauf, S.N., Blume, L.E. (eds) Microeconometrics. The New Palgrave Economics Collection. Palgrave Macmillan, London. https://doi.org/10.1057/9780230280816_28.

4. Mill's methods, `https://beisecker.faculty.unlv.edu/Courses/Phi-102/Mills_Methods.htm`.

5. Peter Armitage, *Fisher, Bradford Hill, and randomization*, International Journal of Epidemiology, Volume 32, Issue 6, December 2003, Pages 925–928, `https://doi.org/10.1093/ije/dyg286`.

6. BIOS 6611 (2021, July 31) , *Neyman-Pearson Approach to Statistics*. YouTube. `https://www.youtube.com/watch?v=4boPRRK1OGY`.

7. Judea Pearl, *Causal diagrams for empirical research*, Biometrika, Volume 82, Issue 4, December 1995, Pages 669–688, `https://doi.org/10.1093/biomet/82.4.669`.

8. *Characterizing Selection Bias Using Experimental Data*, J. Heckman, Hidehiko Ichimura, Jeffrey A. Smith, Petra E. Todd.

2

Unraveling Confounding and Associations

In this chapter, we deepen our knowledge of causal inference, exploring more complex aspects of the theory, including an overview of treatment effects. We also clarify the often-muddled concepts of confounding and associations, using real-world examples to illustrate how associations are frequently misinterpreted as causality. We introduce a mathematical framework designed to clearly distinguish between confounding, associations, and causality.

A key distinction is drawn between statistical and causal inference, particularly in the context of infinite data. In addition, we discuss two common strategies to mitigate confounding and highlight various biases inherent in causal analysis. Alright, we are all set to explore these intricate concepts in detail.

The following are the topics covered in this chapter:

- A deep dive into associations
- Causality and a fundamental issue
- The distinction between confounding and associations
- The concept of bias in causality
- Assumptions in causal inference
- Strategies to address confounding

A deep dive into associations

Traditional statistical methods primarily focus on the uncertainty in finite datasets. On the other hand, infinite data theoretically eliminates this uncertainty, allowing precise calculations of statistical measures such as means and variances. Infinite data eliminates statistical uncertainty by providing all possible observations, thus removing sampling error and variability.

However, causal inference faces unique challenges that persist regardless of data volume. Infinite data can accurately depict correlations between variables, but it does not clarify the direction or nature of these relationships. This is where causal inference becomes crucial, requiring methods and theories beyond mere pattern recognition to uncover true causal dynamics. It's essential to grasp that statistical association and/or correlation is not synonymous with causation. Distinguishing causal relationships from statistical associations requires more than abundant data; it demands an in-depth understanding of the interactions between variables.

To fully understand causality, it is essential to first grasp the different types of associations that exist. These associations can be categorized as positive, negative, or null. We will define these categories mathematically, emphasizing the relationships between variables as indicated by correlation and regression coefficients.

Mathematical concepts are fundamental to causality, where not knowing the basics can cause even experienced researchers to sometimes struggle with accurately discerning causation from mere association. Establishing a mathematical foundation enables you to grasp the essentials, laying the groundwork for a deeper understanding. As we progress through the book, the focus will shift toward more extensive use of R programming, but it is crucial that the initial chapters concentrate on the mathematical underpinnings and fundamentals.

Let's consider associations in a practical context: the relationship between traffic congestion patterns. Our variables of interest are **the day of the week** and **the time of day**. We will use a simple regression model to explain the associations. Before that, let's briefly go through what regression models look like first.

In a simple linear regression model, the following equation helps us understand the relationship between variables:

$$Y = \alpha + \beta X + \epsilon \qquad\qquad (1)$$

Here, Y is the dependent variable we want to predict, X is the independent variable we think influences Y, α is the y intercept, β is the slope of the line, and ϵ is the error term accounting for unexplained factors. These models are useful because they provide a simple, interpretable way to predict and understand how different factors affect traffic severity.

Let's now focus on various kinds of association:

- **Positive association**: Here, an increase in one variable coincides with an increase in another, manifesting as a positive correlation coefficient or a positive slope in a regression model (explained in depth later in the chapter).

  In the context of traffic, heightened congestion on certain days (such as Mondays) typifies a positive association. In the model $\left(Y = \alpha + \beta_1 X_1 + \epsilon \right)$, with (Y) representing traffic jam severity and (X_1) denoting days, a positive (β_1) for Monday indicates increased congestion.

Similarly, escalating traffic as the day advances (for instance, during morning peak hours) also demonstrates a positive association. In the model $(Y = \alpha + \beta_2 X_2 + \epsilon)$, a positive (β_2) denotes worsening traffic as the day unfolds.

- **Negative association**: This is marked by a reciprocal relationship, where an increase in the value of one variable leads to a decrease in the value of others. A negative (β_1) value might imply lighter traffic on certain days compared to others, such as Sundays. Conversely, a negative (β_2) suggests that congestion lessens as the day progresses.

- **Null (or no) association**: This indicates a lack of linear correlation between two variables. It doesn't imply a complete absence of any relationship but rather a lack of linear connection. A β_1 value close to 0 on different days suggests that the day has little impact on traffic congestion severity. Similarly, a β_2 value around 0 indicates that time has a negligible effect on traffic intensity.

Grasping these distinctions is crucial for understanding how variables interact in statistical models, especially in the causal world. Next, let's peel the complex layers of causality further.

Causality and a fundamental issue

The goal in this chapter is to clarify and simplify concepts that, though they may seem clear and straightforward in daily conversation, reveal a layer of complexity when expressed mathematically. Our approach is informed by the *Neyman-Rubin causal model* [1], often referred to as the potential outcomes framework. This framework is not just an academic exercise but also a practical tool to understand how specific actions lead to real-world outcomes.

Imagine you live in a loud/noisy neighborhood and consider moving to a quieter one so you can better focus on your studies. *The key question is: does moving to a quieter place actually cause an increase in your concentration?*

Consider this: you move and find your concentration improves. But it's important to question whether this improvement might have happened even if you hadn't moved. If the answer is yes, then the move itself might not be the main reason for your better focus, challenging its role as a direct cause.

In a simpler form, consider two scenarios. In the first, moving to a quieter area helps with better focus, highlighting the significant impact of a calm environment on focus. In the second, staying in a noisy place continues to disrupt focus. These situations exemplify the concept of potential outcomes. Here, *outcome* (Y) means your focus level: $Y = 1$ for improved focus and $Y = 0$ for no improvement or decline. Treatment (T) represents the decision to move $(T = 1)$ or stay $(T = 0)$. The potential focus outcomes, $Y(1)$ and $Y(0)$, vary depending on the scenario. The first scenario suggests equal focus regardless of moving $Y(0) = Y(1) = 1$, while the second indicates improved focus only if you move $Y(0) = 0$, $Y(1) = 1$.

It is vital to distinguish potential outcomes from observed outcomes; we cannot witness all potential outcomes simultaneously, as the realized outcome depends on the actual treatment condition.

The observed outcome is the actual result we see after a specific action is taken. In our example, if you moved to a quieter neighborhood, the observed outcome would be your actual focus level after the move. This is what we can directly measure.

Potential outcomes, however, represent all possible results under different treatment conditions. They include both what did happen and what could have happened. In our scenario, there are two potential outcomes: focus level if moved ($Y(1)$) and if stayed ($Y(0)$). We can only observe one, while the other remains a hypothetical counterfactual. This concept helps us think about causality more systematically, considering both actual and possible outcomes.

Let's deep dive into this further in the next section, where we learn about the individual treatment effect.

Individual treatment effect

Expanding our view to include a broader population, causal inference often considers a varied group. Taking our neighborhood example, let's discuss **individual treatment effects (ITEs)**.

Imagine a community where each member contemplates moving to a quieter area to enhance their focus. For any one person in this group, let's say the $k = 8th$ individual, we define their decision to move (treatment), their distinct characteristics (covariates), and their level of focus (outcome) as T_8, X_8, and Y_8, respectively. The ITE for this eighth person is expressed as follows:

$$ITE_8 = Y_{8(1)} - Y_{8(0)} \qquad (2)$$

Here, $Y_{k(1)}$ represents the outcome observed if the treatment occurs, that is, they relocate, and $Y_{k(0)}$ represents the outcome observed if the treatment does not occur, that is, they do not move. Across a population, the potential outcome under a specific treatment condition, denoted as $Y(C)$, varies among individuals due to their different reactions to the same treatment. However, $Y_{k(C)}$ is deemed non-random, mirroring the unique circumstances and attributes of the eighth individual.

Understanding ITEs is crucial to causal inference. For instance, if relocating to a quieter neighborhood significantly enhances an individual's focus, $Y(1) - Y(0)$ being positive, it suggests a positive causal impact of the relocation. Conversely, if focus levels remain unchanged regardless of the move ($Y(1) - Y(0)$ equals 0), it indicates an absence of causal effect.

This exploration into potential outcomes and ITEs forms the basis for tackling the intricate challenges in causal inference, setting it apart from areas focused solely on association or prediction.

In causality, the decision to move or stay exemplifies a fundamental dilemma: we can observe the outcome of either relocating, $Y(1)$, or remaining, $Y(0)$, but not both for the same individual. This embodies the core challenge in causal inference – *the impossibility of witnessing all potential outcomes for an individual, which hinders direct measurement of causal effects* ($Y(1) - Y(0)$). Unlike predictive fields such as machine learning, causal inference deals with potential, observed, and counterfactual outcomes, the latter being crucial yet unobservable. A counterfactual outcome is the potential result that would have occurred for an individual if they had received a different treatment than the one they actually received. It's the 'what if" scenario that can never be directly observed in reality.

Identification is another key concept in causal inference. It involves methods to infer causal quantities from observable data, acknowledging that only one potential outcome can be observed in any given scenario. Direct measurement of causal effect is not feasible due to this inherent limitation. Identification is a process of approximating causal quantities using available data.

Once a causal effect is identified, we can estimate it using observable data. In this, we remove bias and align correlation with causation under certain conditions. If treatment and control groups are fundamentally similar, the average outcome of the untreated group, had they received the treatment, should align with the treated group. Now, you may ask, what is a control group? A control group is a subset of study participants or data points that do not receive the treatment being studied. It serves as a baseline to compare the effects of the treatment.

Mathematically, this approach aims to neutralize bias, isolating the treatment's effect on the treated group. This methodology is essential in drawing reliable causal inferences from data.

Average treatment effect

While we cannot directly observe ITEs, we can explore the **average treatment effects (ATEs)** across a population. In our neighborhood scenario, this involves averaging the difference observed in focus levels between those who moved and those who didn't:

$$ATE = E[Y(1) - Y(0)] \tag{3}$$

The symbol $E[Y]$ signifies the expected or average value of these outcomes.

However, this calculation is not straightforward due to the missing counterfactual data – an issue that is inherent in causal inference. We might see some individuals' focus levels post-move, $Y(1)$, but lack data on their focus had they not moved, $Y(0)$, and vice versa. It's tempting to use the associational difference, $E[Y \mid T = 1] - E[Y \mid T = 0]$, as a stand-in for ATE, but this approach is flawed due to potential confounders, which create non-causal paths influencing both the decision to move (treatment) and the focus levels (outcome).

Ignorability

To bridge the gap and calculate the ATE accurately, we lean on the assumption of ignorability. This posits that treatment assignment is as random as flipping a coin, implying no underlying factors skewing the decision to move or not, which would also affect focus levels. This assumption is potent yet challenging to justify in real-world situations.

Mathematically, ignorability simplifies the ATE calculation:

$$ATE = E[Y(1)] - E[Y(0)] = E[Y \mid T = 1] - E[Y \mid T = 0] \tag{4}$$

In this equation, $E[Y \mid T = 1]$ is the average outcome among those who received treatment, and $E[Y \mid T = 0]$ is the average outcome among those who did not. This formula depicts the ATE as deduced from observable data, presuming the treatment groups are exchangeable and devoid of confounding variables.

This means that the potential outcomes are independent of the treatment assignment.

Through identification, we deduce a method for estimating the ATE using data that is observable. For example, under the assumption of random treatment assignment (ignorability), the average outcome among those treated (observed data) can act as a stand-in for $E[Y(1)]$, and similarly for the untreated group for $E[Y(0)]$. In summary, identification in causal inference is the art of leveraging assumptions and observable data to approximate a causal effect that cannot be directly measured due to the inherent unobservability of all potential outcomes for an individual.

Exchangeability

Another lens to view the independence of the treatment assignment is through exchangeability, which implies that the treatment and control groups (the group that did not get the treatment) are comparable, and swapping their treatment statuses wouldn't change the outcomes. Achieving this requires *controlling for* or *adjusting for* certain variables, ensuring conditional exchangeability where groups are comparable post-adjustments.

In essence, exchangeability ensures that the treatment groups (treated and untreated) are comparable in every aspect except for the treatment itself. This comparability is vital for attributing observed outcome differences directly to the treatment, rather than to pre-existing disparities between the groups.

Diving deeper into the mathematical nuances of exchangeability and its relation to the independence of treatment assignment from potential outcomes, we can represent this assumption as follows:

$$Y(0), Y(1) \perp T \tag{5}$$

This equation states that the potential outcomes, $Y(0)$ (if untreated) and $Y(1)$ (if treated), are independent, $\perp$, of whether treatment, T, is applied. Essentially, the treatment assignment should not provide any insights into the possible outcomes.

In real-world settings, achieving ignorability is a steep mountain to climb due to confounders affecting both treatment and outcomes. Randomized experiments often come to the rescue here, ensuring random treatment assignments devoid of confounder influence. In essence, ignorability and exchangeability are two sides of the same coin in causal inference. They provide different conceptual views but ultimately convey the same mathematical idea. While crucial, they demand careful consideration and often experimental designs to hold water.

Randomization to the rescue

Within the sphere of causal inference, independence is deemed a vital assumption for equating associations with causation. Causal inference pipelines typically unfold in two fundamental phases:

- **Identification**: This stage is centered around conceptualizing how to translate the causal quantity of interest into measurable terms using observable data. It's essentially laying the conceptual groundwork, discerning what needs to be measured, and devising a method to measure it with the available data.

- **Estimation**: Here, the focus shifts to the actual application of data to approximate the previously identified causal quantity. This phase bridges the gap between theoretical concepts and practical data analysis to extract meaningful insights.

In causality, the concept of treatment randomization can be exemplified through a credit card approval (outcome) test. In this test, applicants are randomly chosen for approval without regard to their job status, race, demographics, gender, or other characteristics. This method of randomization is pivotal for several reasons. First, it makes the treatment assignment (here, say, mortgage payments via a smartphone app, T_1, versus payments through a bank teller in person, T_2) very similar to a random event, such as flipping a coin. This ensures the treatment is assigned independently of other variables in our analysis. Then, the independence of the treatment from other variables, brought about by randomization, is key. It fulfills a necessary condition for directly linking observed relationships with causation. Evidently, randomization greatly diminishes, if not entirely removes, bias (we'll learn more about this later in the chapter).

Before treatment is assigned, all outcomes are just possibilities. Random selection ensures that treatment distribution isn't affected by participants' existing traits or potential outcomes. This means everyone has an equal chance, making the results more reliable. Randomization in causal inference serves as a robust mechanism. It upholds the independence assumption, minimizes bias, and enhances the accuracy of causal effect estimates. This highlights the criticality of meticulously planning and executing experiments or interventions in causal studies to maintain the integrity and reliability of the findings, particularly for those new to the field.

Given this essential role of randomization in ensuring unbiased and accurate causal estimates, let's expand on some concepts that we already know about.

The distinction between confounding and associations

We will focus here on understanding the distinction between confounding and association, as touched upon briefly in the previous chapter. Recall that confounding occurs when an extraneous variable, not accounted for in the analysis, affects both the independent (treatment) and dependent (outcome) variables in a study, thereby masking the genuine causal connection. Such a confounder can give rise to a false association, potentially misleading the understanding of the direct influence of the independent variable on the dependent variable. For example, in a study examining the relationship between sports wear sales and fishing accidents, weather could be a confounder, influencing both variables and creating a false association. On the other hand, an association denotes a noted linkage between two variables, which does not necessarily signify a cause-and-effect relationship and does not inherently account for the presence of any external influencing variables.

Let's consider a practical example involving a study on the impact of a fitness program (*treatment* (T)) on weight loss (*outcome*(Y)), where the initial fitness level (*variable*(X)) could be a confounder. Now, what makes (X) a confounder? The following criteria are pivotal for correctly identifying the confounders:

- **Association with treatment**: A confounder (X) must exhibit an association with the treatment (T). This is mathematically evident when (X) significantly predicts (T) in a model, represented as follows:

$$T = \delta_0 + \delta X + v \qquad (6)$$

  Here, T represents the treatment variable. In this example, it's the fitness program enrollment (treated versus untreated). X is the confounder variable. In this example, it's the initial fitness level of individuals. δ_0 is the intercept term, representing the expected value of T when X is 0. δ is the coefficient that quantifies the relationship between X and T. It represents how much T changes for a one-unit change in X. v is the error term, representing all other factors that affect T but are not included in the model.

  (δ) should be statistically significant, implying it is distinct from 0. In our example, the likelihood of individuals enrolling in the fitness program may be influenced by their initial fitness levels. This implies a relationship between the initial (X) *and* (T).

- **Association with the outcome**: Additionally, (X) must be linked to the outcome (Y). In the predictive model for (Y), the coefficient of (X), denoted as (γ), should carry statistical significance:

$$Y = \alpha + \beta T + \gamma X + \epsilon \qquad (7)$$

  Here, (γ) is indicative of the association between the confounder (X) and the outcome. (Y) should notably differ from 0, while (ϵ) is the error component. The term (α) is the intercept, representing the expected value of (Y) when all other variables are 0. The coefficient (β) is the causal effect of interest, often referred to as the treatment effect.

  For weight loss, the initial fitness level can affect how easily an individual loses weight, thus showing a connection with the weight loss outcome (Y).

Now, what is association?

To further grasp the distinction between confounding and association in causal inference, we'll commence with the preceding mathematical framework, focusing on the impact of the treatment (T) on the outcome (Y), while considering the potential influence of the confounder (X).

If we had not considered the confounding variable (X), then our model would be as follows:

$$Y = \alpha + \beta T + \epsilon \qquad\qquad (8)$$

Here, the link between (T) (treatment) and (Y) (outcome), represented by (β), does not automatically indicate a causal relationship. It merely points to some form of connection between these variables, also called **association**. The reasons for this relationship can be varied: (T) might cause (Y), (Y) might cause (T), both could be influenced by another variable, or the association might be purely coincidental (a case of spurious correlation).

To summarize, a confounder (X) is a variable that affects both the treatment (T) and the outcome (Y), resulting in a misleading association between them. Not accounting for a confounder such as (X) can lead to a misleading or non-causal connection between (T) and (Y). When a confounder is present, the observed relationship (β) between (T) and (Y) is skewed, not truly representing the causal impact of (T) on (Y). The aim of causal inference is to discern the direct influence of (T) on (Y). This is achieved by accounting for (X) in the statistical model, thus adjusting for its effect.

For accurate causal inference, it's imperative to pinpoint and adjust for confounders. This helps in precisely estimating the causal impact of a treatment or intervention. Identifying and counteracting confounders is a fundamental aspect of causal analysis. This ensures that the evaluated association between (T) and (Y) genuinely represents a causal link. Achieving this often involves statistical techniques such as stratification, matching, or more sophisticated methods such as propensity score analysis (covered in *Chapter 6*). These methods strive to equalize the influence of confounding factors across different treatment groups, providing a more accurate picture of the treatment's true effect. In essence, distinguishing between mere association and true causation hinges on effectively dealing with confounding variables, a critical step in the robust analysis of causal relationships. Essentially, discerning true causation from mere correlation depends on effectively managing confounding variables. For example, in healthcare studies, such methods adjust for patient age, lifestyle, or pre-existing conditions, isolating the treatment's effect (e.g., the effect of a new drug) from these confounders.

Next, we will address various forms of bias that can still be present in causality studies.

Discussing the concept of bias in causality

An estimator is a tool used to measure specific parameters, such as the average effect of a treatment. When an estimator is biased, it means it regularly deviates from the true value it's supposed to measure. In terms of ATE, a biased estimator either consistently overestimates or underestimates the actual impact of the treatment. This concept is crucial in separating mere associations in data from true causation. Bias becomes apparent when the estimates we get from the data do not match the actual causal effects we are interested in.

Estimating ATE involves a thought experiment. We need to imagine two scenarios:

- Scenario 1, what would have happened to the treated group if they hadn't received the treatment?
- Scenario 2, what would have happened to the untreated group if they had received the treatment?

In statistical terms, we represent these hypothetical outcomes as $Y(0)$ for the treated group without treatment, and $Y(1)$ for the untreated group with treatment. By comparing these imagined outcomes with the actual results, we can calculate the true impact of the treatment. Typically, we use the observed outcomes of the treated to estimate $Y(1)$ and those of the untreated for $Y(0)$. However, if inherent differences exist between these groups, this methodology results in biased ATE estimates. The estimated ATE might not truly reflect the causal effect of the treatment.

This approach requires a shift from traditional analysis methods, which only focus on observed outcomes, to considering potential outcomes – the scenarios that could have unfolded under different conditions. This paradigm shift is essential in causal inference, fostering a deeper and more nuanced understanding of the causality behind certain outcomes.

To illuminate the concept of bias in causal inference through mathematical terms, we again refer to the ATE, defined as follows:

$$ATE = E\,[Y(1)] - E\,[Y(0)] \tag{9}$$

Here, $Y(1)$ and $Y(0)$ represent the potential outcomes with and without the treatment, respectively. The symbol $E[]$ signifies the expected or average value of these outcomes.

As said, in practical settings, it's not feasible to observe both $Y(1)$ and $Y(0)$ for the same subject. As a result, we often rely on observed outcomes as proxies. This limitation is crucial to understand as it shapes our approach to studying causal relationships in real-world situations.

For example, consider a case where girls who apply more makeup might be more focused on their appearance. The treatment here is the application of more makeup, while the control group consists of those who use less or no makeup.

At first sight, it might appear that the increased focus on appearance is directly linked to the amount of makeup used. However, this interpretation could be oversimplified. It's feasible that girls who apply more makeup (the "treated" group) inherently possess a greater inclination toward emphasizing their appearance, independent of the amount of makeup used. Therefore, these "treated" individuals may not be directly comparable to those who use less or no makeup (the "untreated" group):

- $E[Y \mid T = 1]$ (the average outcome for girls applying more makeup) approximates $E[Y(1)]$
- $E[Y \mid T = 0]$ (the average outcome for girls not applying makeup) approximates $E[Y(0)]$

Should girls who apply makeup differ significantly from those who don't, such as an inclination toward self-appearance, this will lead to bias in the ATE estimation. The bias in the estimator $\widehat{ATE}$ can be expressed as follows:

$$\widehat{ATE} = E[Y \mid T = 1] - E[Y \mid T = 0] \tag{10}$$

This estimate $\widehat{ATE}$ diverges from the true ATE due to the non-comparability of treated and untreated groups. Note the following two points:

- $E[Y \mid T = 1]$ might exceed $E[Y(1)]$ due to the inherent advantages of girls inclined towards self-appearance (for example, in their attitude towards personal appearance), leading to biased estimates of ATE.

- $E[Y \mid T = 0]$ could vary from $E[Y(0)]$, reflecting differences in the untreated group

Here, $E[Y \mid T = 1]$ and $E[Y \mid T = 0]$ denote the average outcomes for the treated and untreated groups, respectively. Yet, the ideal measure of causation is: $ATE = E[Y(1)] - E[Y(0)]$.

Basically, the average outcome for makeup users might be higher than what we'd expect if makeup were the only factor, because these girls might already have advantages related to their interest in appearance. The average outcome for non-users might not accurately represent what would happen if they did use makeup, because they might have different characteristics compared to the makeup-using group.

To elucidate the gap between association and causation, envisage substituting the observed outcomes in the association equation with potential outcomes. The treated group's observed outcome is $Y(1)$, and for the untreated, it is $Y(0)$. By incorporating and then subtracting the counterfactual outcome, $Y(0)$, for the treated group, the equation morphs into the following:

$$E[Y \mid T = 1] - E[Y \mid T = 0] = (E[Y(1) \mid T = 1] - E[Y(0) \mid T = 1]) + (E[Y(0) \mid T = 1] - E[Y(0) \mid T = 0]) \tag{11}$$

This formulation captures the intricacies of causal inference. The initial term, $E[Y(1) \mid T = 1] - E[Y(0) \mid T = 1]$, reflects the treatment effect on the treated, whereas the latter term, $E[Y(0) \mid T = 1] - E[Y(0) \mid T = 0]$, signifies the bias. Such bias emerges from inherent differences between the treated and control groups, unrelated to the treatment.

Achieving an unbiased estimation hinges on ensuring that the treated and untreated groups are comparable, often realized through randomization or statistical adjustments for confounding variables. This process involves recognizing and accommodating the factors that influence both the treatment assignment and the outcome. In the next section, we'll focus on assumptions seen in causal inference.

Assumptions in causal inference

Causal inference is fundamentally built upon a foundation of carefully constructed assumptions. Assumptions represent the underlying beliefs about the origins of our data. Often, these assumptions are not directly verifiable by the data itself, which necessitates the need to pre-suppose their existence. Identifying these assumptions is a critical challenge, and this section aims to provide clear guidance on how to do so.

It's not a surprise anymore that central to causal inference is the task of identifying causal effects. This is distinctly different from the challenges of estimation found in traditional statistics and machine learning. Identification involves determining whether it's possible to learn a causal effect from the data, based on the underlying assumptions. Once these effects are identified, estimation—common to both causal inference and traditional statistics—aims to quantify the size or nature of these effects.

So, you understand that assumptions are the cornerstone of causal inference. They delineate the parameters within which causal effects can be discerned and quantified from the data. The cogent expression and substantiation of these assumptions are imperative, as described here:

- **Independence or ignorability**: This assumption posits that the allocation to treatment is not influenced by potential outcomes, given certain covariates (X). Formally, we can frame it as $(T \perp \{Y(0), Y(1)\} | X)$. This implies that post adjusting for (X), treatment assignment seems random. In **randomized controlled trials (RCTs)**, this is often inherently satisfied, whereas in observational studies, it demands meticulous consideration of confounding variables. To recall, RCTs are experimental designs where participants are randomly assigned to either a treatment or a control group to rigorously evaluate the effects of an intervention.

- **Positivity or overlap**: Here, the assumption is that every individual has a non-zero chance of receiving each treatment level, conditional on covariates, represented as $0 < P(T = t \mid X = x) < 1$ for all (t) and (x). This is crucial to ensure representation across all covariate combinations in both treatment and control groups.

- **Stable unit treatment value assumption (SUTVA)**: SUTVA is a critical concept in causal inference, consisting of two main parts:

  - Firstly, it includes the principle of non-interference between units. This suggests that the potential outcome for any unit is not affected by the treatment given to other units. Essentially, it implies that the treatment of one unit doesn't impact another's outcome. Mathematically, this means the outcome (Y_i) for a unit (i) is independent of the treatment (T_j) given to any other unit (j).

  - Secondly, SUTVA encompasses the principle of consistency. This dictates that the observed outcome of a unit must match one of its potential outcomes, depending on the treatment it receives. Put simply, the effect we observe in a unit should be the one we expect based on the treatment that the unit actually got.

  Both elements of SUTVA offer a systematic and logical basis for correctly measuring and interpreting causal effects.

It is imperative to understand and internalize the following additional assumptions, which are crucial to absorbing the concepts learned in this book:

- **Linear model presumption**: In linear modeling, we assume a direct linear relationship between variables, as in $Y = \alpha + \beta T + \gamma X + \epsilon$, where the treatment ($T$), covariates ($X$), and outcome ($Y$) are linearly related. Other assumptions in this equation include error terms having constant variance (homoscedasticity) and normal distribution for some inferences.

- **Absence of measurement error**: This assumption underscores that the treatment and outcome variables are measured accurately without error, as measurement inaccuracies can skew causal effect estimates.

- **Consistency assumption**: The consistency assumption holds that the observed outcome for an individual under a treatment aligns with their potential outcome if they were to receive that treatment. Mathematically, for an individual receiving treatment (T), the observed outcome should align with the potential outcome ($Y(T)$).

  This assumption can be breached if the treatment has multiple versions that are unaccounted for. Consider a bank's study on the impact of a new account type on customer retention. The treatment – opening a new account – varies in incentives (interest rates, fees, rewards). Treating this as a binary variable without considering incentive variations breaches the consistency assumption.

In summary, understanding and meticulously applying these assumptions are crucial to the accurate interpretation of causal relationships in studies. These foundations not only guide the analytical process but also ensure the validity and reliability of the conclusions drawn. Next, let's understand confounding a bit better.

Strategies to address confounding

As discussed, addressing confounding is essential to ensure that estimated effects genuinely reflect the true causal relationship, devoid of influence from extraneous factors. In the following subsections, we present two prevalent statistical methods that are often employed to tackle confounding: regression adjustment and propensity score methods.

Regression adjustment

This method is a staple in controlling for confounding in observational studies. Luckily, we have already applied this method in this chapter. As you have seen previously (in the *Individual treatment effect* section), the primary idea is to integrate potential confounders as covariates into a regression model, thereby separating the effect of the treatment or exposure of interest from the influences of the confounders. We saw the impact of a treatment (T) on an outcome (Y), alongside a collection of confounders (X). A typical linear regression model might be formulated as follows:

$$Y = \alpha + \beta T + \gamma X + \epsilon \qquad (12)$$

Here, (α) is the intercept, (β) denotes the treatment coefficient or the slope of the line of the linear equation, (γ) encompasses coefficients for the confounders, and (ϵ) represents the error term. Crucially, by incorporating (X) in this model, the coefficient (β) is adjusted to reflect the effect of (T) on (Y), factoring in the confounders (X). This method relies on correct model specification and the inclusion of all relevant confounders. Remember that assessing the model's functional form and potential interactions between variables is also vital.

What do we mean by interactions here? An interaction occurs when the influence of one independent variable on the dependent variable varies based on another independent variable's level or value. This contrasts with scenarios where two independent variables independently affect the dependent variable. In a regression model featuring two independent variables, X_1 and X_2, and an outcome (Y), an interaction term would be $(X_1 \cdot X_2)$ (their product). The model could be represented as $Y = \alpha + \beta_1 X_1 + \beta_2 X_2 + \gamma(X_1 \cdot X_2) + \epsilon$. Here, (γ) is the coefficient for the interaction term.

Suppose you're exploring how exercise (T) affects weight loss (Y), and you hypothesize that dietary habits (X_1) and age (X_2) might modulate this effect. You're also intrigued by how the combination of exercise and dietary habits differentially impacts weight loss:

- **Model without interaction**: A simple model might look like this:

$$Y = \alpha + \beta T + \gamma_1 X_1 + \gamma_2 X_2 + \epsilon \qquad (13)$$

Here, (α), (β), (γ_1), (γ_2), and (ϵ) are coefficients and error terms, as previously described. The model presumes a uniform effect of exercise across all dietary habits and age groups.

- **Model with interaction**: To explore the varying impact of exercise across dietary habits, include an interaction term:

$$Y = \alpha + \beta T + \gamma_1 X_1 + \gamma_2 X_2 + \delta(T \times X_1) + \epsilon \qquad (14)$$

Here, $\delta(T \times X_1)$ is the interaction term, exploring how exercise's effect on weight loss shifts with dietary habits. Including $\delta(T \times X_1)$ allows for the possibility that exercise might be more effective for weight loss with certain dietary habits than others. Recognizing these interactions can inform more effective, tailored weight loss strategies, acknowledging the complexity of how exercise and diet interplay.

Considering interactions between variables in regression models, especially when dealing with confounders, is essential. This approach deepens our understanding of how different variables are related and greatly impacts the interpretation of statistical analysis outcomes.

Propensity score methods

In observational studies, where randomization is not a viable option, propensity score methods emerge as a crucial statistical technique. On the other hand, an RCT naturally ensures that treatment and control groups are comparable in all respects, except for the treatment, effectively controlling for both known and unknown confounders. In contrast, observational studies lack random treatment allocation, which often leads to a correlation with confounding factors, complicating the isolation of treatment effects.

We'll look into this more in detail later in *Chapter 6*, but an overview is presented here for clarity:

- **Defining the propensity score**: The propensity score is the likelihood of receiving a treatment based on observed covariates, mathematically expressed as $P(T = 1 \mid X)$. Here, (T) signifies the treatment status (1 for treated, 0 for untreated) and (X) represents a range of observed covariates. The primary objective of the propensity score is to balance these covariates between treatment groups, thereby diminishing confounding effects.

- **Calculating propensity scores**: Commonly, a logistic regression model is utilized:

$$P(T = 1 \mid X) = \frac{\{e^{((X\beta))}\}}{(1 + e^{((X\beta))})} \qquad (15)$$

Here, $X\beta$ indicates the linear combination of covariates. These estimated scores represent the probability of each individual receiving treatment, considering their covariates.

How do we employ this score for confounder adjustment? Here are a few ways:

- **Matching**: Treatment group individuals are paired with similar individuals from the control group based on propensity scores, creating matched sets with similar covariate distributions

- **Stratification**: The sample is segmented into categories based on propensity scores, allowing for treatment effect analysis within each category

- **Weighting**: Subjects are assigned weights according to their propensity scores, forming a synthetic sample where covariate distribution is balanced between treatment and control groups

By employing matching, stratification, or weighting based on propensity scores, covariates (X) are equilibrated between treatment and control groups, echoing the conditions of an RCT. These methods ensure similar covariate distribution across groups, thereby reducing the bias stemming from confounding factors, and rendering the causal inference regarding the treatment effect more credible. Without randomization in observational studies, propensity score techniques offer a statistical pathway to manage confounding. By ensuring covariate balance across treatment and control groups, these methods approximate randomized experimental conditions, enhancing the accuracy of causal conclusions. This approach presumes the absence of unmeasured confounders and posits that every subject has the probability of receiving either treatment.

In addressing confounding, both regression adjustment and propensity score techniques are highly effective. The choice between these two methods largely depends on the unique aspects of the study, such as the nature of the data and the known confounders. Each method comes with its own set of assumptions and limitations. Therefore, it is wise to use a combination of these approaches to ensure the credibility of the causal conclusions drawn from the study.

Summary

In this chapter, we investigated the complexities of causal inference, focusing on confounding and associations. We explored this by understanding various types of associations – positive, negative, and null – and their implications in different contexts, such as traffic patterns. We also emphasized the fundamental challenges in causal inference, notably in scenarios with infinite data where statistical associations still fail to reveal causality's direction or nature.

We dealt with ITEs in a population, using a neighborhood-moving scenario to illustrate how different people react uniquely to the same treatment. This led to a discussion on ATEs and the difficulties in their calculation, especially due to missing data in causal inference. We differentiated between confounding and association, using a fitness program's impact on weight loss as an example. We elaborated that a link between treatment and outcome doesn't necessarily imply causation.

We discussed key assumptions in causal inference, such as ignorability, positivity, and SUTVA. These assumptions are crucial in ensuring the validity of causal interpretations. The chapter also highlighted the importance of addressing biases, explaining how biased estimators can lead to incorrect conclusions about causal relationships. Finally, we reviewed strategies to address confounding, particularly through regression adjustment and propensity score methods. These techniques are vital in observational studies to emulate the conditions of RCTs and achieve credible causal inferences.

In summary, this chapter laid a foundational understanding of confounding, associations, and the nuances of causal inference. It highlighted the importance of distinguishing true causation from mere statistical associations and offered practical strategies for addressing confounding in research and application.

In the next chapter, we'll learn how to use the programming language R in the world of causal analysis.

References

1.   Imbens, G.W., Rubin, D.B. (2010). Rubin Causal Model. In: Durlauf, S.N., Blume, L.E. (eds) *Microeconometrics. The New Palgrave Economics Collection*. Palgrave Macmillan, London. https://doi.org/10.1057/9780230280816_28

3

Initiating R with a Basic Causal Inference Example

As we journey from the theoretical foundations to the empirical applications of causal inference, this chapter marks a significant transition. Herein, we shall dive deep into the utilization of the R programming language as a tool for applying the concepts previously discussed. This chapter caters to both those newly acquainted with R and those seeking to refine their existing knowledge. Our objective is to explain the core principles of R programming within the context of causal analysis.

In this chapter, we start with the basics of R, including setting up your workspace and writing simple scripts. You'll learn about data types, basic operations, and essential functions in R. We'll then apply basic causal inference methods using real-world data examples. This includes data preparation, exploratory analysis, implementing simple causal models, and interpreting results. This chapter features carefully selected code snippets to highlight key concepts and enhance your understanding.

In this chapter, we will cover the following topics:

- Setting up the R environment
- Basic R programming concepts
- Implementing causal inference in R
- Case study – a basic causal analysis in R

Technical requirements

You can find the code examples for this chapter in this book's GitHub repository: `https://github.com/PacktPublishing/Causal-Inference-in-R/tree/main/chap_03`.

What is R? Why use R for causal inference?

R is a powerful and versatile programming language and environment designed for statistical computing and graphics. It is widely used among statisticians, data analysts, and researchers for data manipulation, calculation, and graphical display.

Now the question is, why use R for causal inference? Well, firstly, we can use any programming language that is equipped with the right tools for statistical analysis, especially causal analysis. However, R's primary strength lies in its vast range of statistical functions. It is equipped to handle various causal inference methods, from basic t-tests to more complex techniques such as propensity score matching and instrumental variable analysis. R has numerous packages specifically designed for causal analysis, such as `mediation`, `MatchIt`, `CausalImpact`, and many others. These packages simplify the implementation of complex causal inference techniques. Furthermore, R excels in handling and manipulating data, which is crucial in causal analysis. This includes capabilities for data cleaning, transformation, and preparation for analysis.

R is open source, meaning it is free to use and benefits from a large community of users who contribute to its vast array of packages and libraries. These packages extend R's capabilities, making it an impactful tool for a wide range of statistical methods and data analysis techniques. R's advanced graphical capabilities facilitate sophisticated data visualization, crucial for exploring data and effectively communicating results in causal inference studies. Additionally, R's robust community offers extensive resources, including forums, tutorials, and user-contributed documentation, providing valuable support for learning and troubleshooting.

Getting started with R

In this venture, the first step is setting up the R environment. This involves installing two key components: R itself and RStudio, a popular **integrated development environment** (**IDE**) that makes using R easier and more efficient.

Setting up the R environment

R can be downloaded from the **Comprehensive R Archive Network** (**CRAN**) [1] by Windows and macOS users. The installation process is straightforward: run the downloaded file and follow the on-screen instructions, accepting the default settings that are suitable for most users. For Linux and Unix systems, you can install R packages using their package management tool. Please follow the tutorial here for supportive guidance [2].

Once R is installed, the next step is to install RStudio, which provides a user-friendly interface for working with R. Download RStudio from its official website [3, 4], selecting the free version, the RStudio Desktop Open Source license.

Navigating the RStudio interface

Figure 3.1 shows the interface of RStudio, color-coded with different panels: (**1**) **Script Editor**, (**2**) **Console**, (**3**) **Environment**, **History**, and **Connections**, and (**4**) **Files**, **Plots**, **Packages**, **Help**, and **Viewer**.

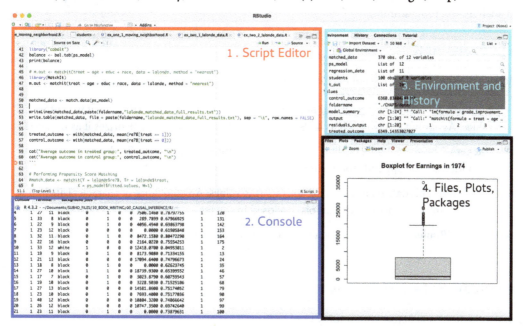

Figure 3.1 – An overview of the RStudio interface

After installing R and RStudio, you'll find RStudio's interface divided into several key panels:

1. **Script Editor**: Found at the top left (accessible via **File | New File | R Script**), this is where you write and save longer code blocks or scripts for complex analyses. To run code, press *Ctrl + Enter* (*Cmd + Return* on a Mac), either for the whole script or a selected part.

2. **Console**: Usually located at the bottom left, this is where you execute R code. You can type commands here and see the output immediately. It's ideal for quick tests and analyses.

3. **Environment** and **History**: In the top-right panel, the **Environment** tab displays your created variables, data frames, and functions. The **History** tab logs all executed commands.

4. **Files**, **Plots**, **Packages**, **Help**, and **Viewer**: The bottom-right panel serves multiple functions. **Files** allows file navigation, **Plots** shows graphs from your R code, **Packages** manages R extensions, **Help** provides R documentation, and **Viewer** displays local web content.

Understanding these panels is essential for efficient work in RStudio (we'll use it in the next sections), with each playing a vital role in your data analysis process.

Now that we know the tools we will use, let's learn R in detail.

Basic R programming concepts

For novices starting with R, it is imperative to initially grasp the fundamental aspects of various data types and data structures you can use. Such an understanding is critical for conducting proficient data analysis and programming.

Data types in R

Following are the commonly-used data types in R:

- **Numeric**: Representing decimal or floating-point numbers, numeric is R's default for numerical values. Assigning a value such as 4.5 or 10.2 in R automatically categorizes it as numeric.

- **Integers**: These are whole numbers, devoid of fractional components. To designate a number as an integer, one appends the L suffix, as in 4L.

- **Characters**: This data type is used for text or string values. Any sequence of characters, including letters, numbers, spaces, and symbols, enclosed in quotes (" ") is treated as character data.

- **Logical (Boolean)**: The logical data type, often referred to as Boolean, represents binary values: TRUE or FALSE.

Advanced data structures

Here are a few data structures commonly used in R:

- **Vectors**: These are one-dimensional arrays limited to homogenous elements, thereby facilitating efficient vectorized operations. We can create a vector containing the numeric data type with the c(1, 2, 3) command.

- **Matrices**: These are two-dimensional arrays, akin to vectors but with a dimension attribute, requiring uniformity in element type. Created using the matrix function, such as matrix(1:4, nrow=2, ncol=2) for a 2x2 matrix, matrices are instrumental in linear algebra and organizing data in two-dimensional structures.

- **Data frames**: In R, data frames are similar to relational database tables or Excel spreadsheets. A data frame is capable of holding diverse data types across its columns, similar to a database table. Consider the data.frame(name=c("Alice", "Bob"), age=c(25, 30)) instance, which constructs a data frame with two distinct columns: one for names (of the character type) and another for ages (of the numeric type).

- **Lists**: Lists in R are an enhanced version of vectors, capable of encapsulating elements of varied types and structures. Take, for example, `list(name="Alice", age=25, grades=c(80, 90))`, which generates a list comprising a character string, a numeric value, and a numeric vector.

- **Factors**: In R, factors are crucial for encapsulating categorical data. They adeptly represent nominal data, which lacks an intrinsic order (e.g., `"apple"`, `"banana"`, and `"cherry"`), as well as ordinal data that possesses a natural sequence (such as `"low"`, `"medium"`, and `"high"`).

- **Structure**: The `structure()` function in R serves as a dual-purpose instrument, enabling users to inspect and alter the internal structure of R objects. This functionality is crucial for comprehending the organization of data and tailoring data objects for specific analytical needs. For inspection purposes, `structure()` reveals comprehensive details of an object, encompassing its type, length, attributes (such as factor levels or matrix dimensions), and its contents, which are vital for understanding intricate objects and for debugging processes. Moreover, `structure()` can be employed to modify object attributes, exemplified by transforming a numeric vector into a factor with designated levels:

```
x <- c(1, 2, 3)
factor_x <- structure(x, class = "factor",
                      .Label = c("low", "medium", "high"))
```

Here, `factor_x` becomes a factor with three levels: `low`, `medium`, and `high`, corresponding to the numeric values of 1, 2, and 3, respectively.

Packages in R

Packages in R are collections of R functions, compiled code, and sample data that extend R's functionality, and are stored in a standardized format under the `library` directory. To work with packages in R, follow these steps:

- **Installation**: We install packages with the `install.packages('package_name')` command. Replace `package_name` with a valid package name, such as `ggplot2`.

- **Finding packages**: Check CRAN's online directory [5].

- **Package documentation**: We access package documentation with the `help(package = "'package_name'")` command.

Mastering these concepts is key to leveraging R's capabilities for data analysis. Practice with the provided examples to enhance your understanding and skills.

Preparing for causal inference in R

Next, we'll implement an example causal inference problem in R. Take our previous problem, as discussed in *Chapter 2*, of a group of students changing their home location from a noisy neighborhood to a quieter place. The university observed the change might have brought improvement in their grades and performance in class. They assign you as a researcher to learn whether there is any causal link between moving neighborhoods and improved grades.

Preparing and loading data

We use a toy dataset representing students who moved from a noisy to a quieter neighborhood. The data is located in the Git repository. The dataset includes grades before and after the move, noise levels in the neighborhood, and other factors such as study hours, part-time job status, and family income.

Let's transform the data for our analysis.

Specifically, we begin by reading a CSV file named `student_data.csv` (provided in the Git repository) into a data frame called `students`. We then check for missing values in the dataset using the `sum(is.na())` function. Next, we convert several variables—`noise_level_pre`, `noise_level_post`, and `part_time_job`—into factors to properly represent categorical data. Finally, we calculate a new variable, `grade_improvement`, by subtracting the pre-move grade from the post-move grade. These steps are essential for preparing the data for analysis, ensuring proper data types, and creating a key variable for assessing student performance changes:

```
#set work directory
setwd(dirname(rstudioapi::getSourceEditorContext()$path))
# Simulating a dataset
set.seed(123)  # Setting a seed for reproducibility
foldername = "./data/"
students <- read.table(file = paste(
  foldername, "student_data.csv"), sep = ",", header = TRUE)


# Data cleaning and transformation
# Check for missing values
sum(is.na(students))
students$noise_level_pre <- as.factor(
  students$noise_level_pre)
students$noise_level_post <- as.factor(
  students$noise_level_post)
students$part_time_job <- as.factor(
  students$part_time_job)
students$grade_improvement <- students$post_move_grade -
  students$pre_move_grade  # Calculating grade improvement
```

Exploratory data analysis (EDA)

We will now explore the data to understand its distribution and identify patterns or anomalies. Let's visualize the data using this code :

```
plot(students$noise_level_pre, students$pre_move_grade,
    main="Pre Noise Level vs. Pre Move Grades",
    xlab="Pre Noise Level ", ylab="Pre Move Grades")
# Scatter plot for study hours vs grade improvement
ggplot(students, aes(
  x = study_hours, y = grade_improvement)
  ) + geom_point() + geom_smooth(method = "lm")
```

We get the following graph:

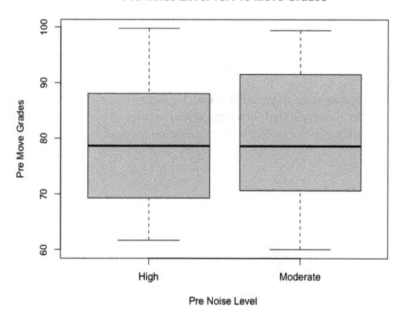

Figure 3.2 – The relationship between pre-noise levels (high and moderate) and pre-move grades

As seen in the preceding chart (*Figure 3.2*), we see higher grades in students living in the **Moderate** noise level compared to the **High** noise level neighborhoods. As seen in the following scatterplot (*Figure 3.3*), the observed data also suggests a slight negative correlation between the number of study hours and grades, indicating that increased study time may marginally lower grades.

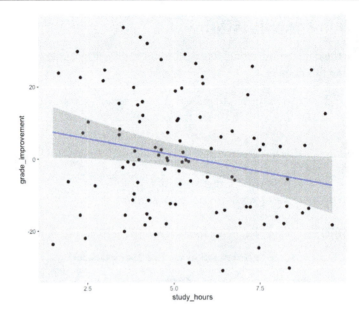

Figure 3.3 – The relationship between grade improvement and study hours

This counterintuitive finding suggests potential underlying factors that merit further investigation. It raises the possibility that prolonged study hours might lead to fatigue and decreased cognitive agility, adversely affecting exam performance. Further data collection and analysis are warranted to explore these hypotheses.

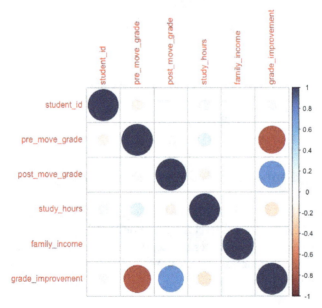

Figure 3.4 – The correlation between the numeric variables in the students dataset

Next, let's understand the correlation between numerical variables. The correlogram displayed here (*Figure 3.4*) assesses relationships between variables, utilizing circle sizes for correlation strength and color for direction—blue for positive and red for negative. Correlation coefficients range from -1 to 1, with values near the extremes indicating stronger relationships. There is a pronounced positive correlation between `pre_move_grade` and `post_move_grade`, implying consistency in grades relative to relocation. A slight negative correlation is observed between `study_hours` and `grade_improvement`, hinting at potential adverse effects of increased studying. `family_income` correlates positively with both `post_move_grade` and `grade_improvement`, suggesting economic status may influence academic outcomes. For more complex relationships, consider using pair plots or heatmaps. Check the code repository provided to see all R commands you can run for EDA and visualization.

Simple causal inference techniques

Having covered EDA in R, which provided us with a solid understanding of our data's structure and characteristics, we now transition to implementing causal inference models in R. This shift marks a crucial progression from data exploration to the rigorous analysis of causal relationships.

Comparing means (t-tests)

Causal inference determines whether a change or intervention, such as moving houses, causes an effect, such as a change in grades. It's important to ensure the effect isn't due to other factors or chance. A t-test is commonly used for causal inference. It compares the average (mean) of an outcome, such as grades, before and after an intervention. This test checks whether the intervention, such as moving, significantly impacts the outcome. For instance, comparing grades before and after moving can reveal whether the move significantly affected academic performance.

Building the model

In a paired t-test such as in the following code snippet, each subject is their own control. This design is effective for causal inference as it minimizes variability due to individual differences. By comparing the same subject's performance before and after an intervention, the test effectively isolates the impact of the treatment. The t-test generates a **p-value**, which measures the strength of the evidence against the null hypothesis (that there is no effect):

```
# T-test to compare means
t.test(students$pre_move_grade, students$post_move_grade, paired=TRUE)
```

In causal studies, a low p-value (usually below 0.05, sometimes 0.01) indicates the observed effect is likely not due to chance, suggesting the intervention had a causal effect. In that, a significant p-value (< 0.05) implies a noticeable difference in outcomes, such as grades, before and after the intervention.

There are some limitations: t-tests can't adjust for other confounding variables that may influence the outcome unless the study is carefully designed. For this, we may use regression, explained later in this chapter.

Coming back to t-tests, they indicate statistical significance but do not prove causation on their own; they only suggest a significant association. Understanding these aspects is crucial in interpreting t-test results in research, especially for beginners in statistical analysis.

Interpreting results

The test compares students' grades before and after moving (from `students$pre_move_grade` to `students$post_move_grade`).

Here's how to interpret the results:

- **t-value** (`-0.38694`): The t-value measures the size of the difference relative to the variation in the sample data. A negative t-value indicates that the mean of the first group (pre-move grades) is lower than the mean of the second group (post-move grades), though this alone doesn't tell us about the significance of the difference.

- **Degrees of freedom** (`df = 99`): This represents the number of independent pieces of information the data provides. It's calculated based on the sample size. Here, it suggests a sample size of `100` (since degrees of freedom for a paired t-test are typically one less than the sample size).

- **p-value** (`0.6996`): The p-value tells us about the probability of observing a result as extreme as the one obtained if the null hypothesis (no difference between the groups) were true. A p-value of `0.6996` is quite high, well above the common significance level thresholds such as `0.05` or `0.01`. This means there's a 69.96% chance of observing such a difference (or more extreme) if there really is no difference between the pre-move and post-move grades. Therefore, we fail to reject the null hypothesis and conclude that the difference is not statistically significant.

- **Confidence interval** (`-3.840042` to `2.586771`): This interval estimates the range within which the true mean difference between the two sets of grades likely falls. Since this 95% confidence interval includes 0, it suggests that the true mean difference could be 0 (no effect), further supporting the lack of statistical significance. A 95% confidence interval provides a range within which we can be 95% confident the true population parameter lies. It is used here to assess the statistical significance of the mean difference, with an interval including 0 indicating no significant effect.

- **Mean difference** (`-0.6266354`): This is the average difference in grades before and after moving. The negative sign indicates that the post-move grades are, on average, lower than the pre-move grades, but given the lack of statistical significance, this difference is not considered to be statistically meaningful.

To conclude, the paired t-test results suggest that there is no statistically significant difference in the student's grades before and after moving.

Regression analysis

Regression analysis is especially valuable in studies such as examining the impact of changing neighborhoods on grade improvement while considering other influential factors. Regression is a statistical method that examines the relationship between a dependent variable and one or more independent variables, allowing researchers to adjust for multiple factors simultaneously. Unlike t-tests, regression can account for confounding variables that may influence the outcome, making it a more versatile tool for analyzing complex relationships in data [10].

Confounding variables (those that affect the outcome besides the primary variable of interest) are effectively controlled in regression analysis, enabling more accurate isolation and comprehension of the primary independent variable's effect. Regression not only quantifies the strength and direction (positive or negative) of each variable's influence on the outcome but also provides a nuanced understanding of causality through its coefficients.

Building the model

In essence, while the t-test is limited to mean comparisons between two groups, regression analysis delves into the relationship between a dependent variable and multiple independent variables. Unlike the t-test, regression can handle a variety of dependent variable types and model complex interactions and non-linear relationships, offering a broader range of analytical possibilities and adjusting for potential confounders in ways that the more basic t-test cannot. The following code writes the model construction and output summarization task online:

```
lm_model <- lm(grade_improvement ~ noise_level_pre +
                 noise_level_post + study_hours +
                 part_time_job + family_income,
              data=students)
summary(lm_model)  # Summarizing the linear model
```

In the preceding, the coefficients for `noise_level_post` can indicate how much moving to a quieter neighborhood is associated with grade improvement, controlling other factors.

Interpreting results

Interpreting the results of this regression model involves examining the coefficients, their significance levels, the model's overall fit, and the implications for causal inference. Let's break down the output. The following numbers are calculated by the `lm` function and can be found from the `summary()` statement in the code:

- **Formula**: The model predicts `grade_improvement` based on `noise_level_pre`, `noise_level_post`, `study_hours`, `part_time_job`, and `family_income`.

- **Residuals**: The residuals' summary indicates how the predicted values differ from the actual values. The range from `-33.527` to `32.074` suggests some variability in the model's predictions.

- **Intercept** (`1.123e+01`): The expected `grade_improvement` when all other variables are zero. However, interpreting the intercept in isolation is not always meaningful, especially if zero values for predictors are not realistic.

- `noise_level_pre - Moderate` (`1.347e+00`): Being in a moderately noisy environment before moving is associated with a 1.347 unit increase in grade improvement compared to being in a high-noise environment (the reference level), though this is not statistically significant (p = 0.6829).

- `noise_level_post - Moderate` (`-2.722e+00`): Moving to a moderately noisy environment (as opposed to a low-noise environment) is associated with a 2.722 unit decrease in grade improvement, but again, this is not statistically significant (p = 0.4086).

- `study_hours` (`-1.750e+00`): Each additional hour spent studying is associated with a 1.750 unit decrease in grade improvement, which is statistically significant (p = 0.0449). The negative coefficient for `study_hours` is somewhat counterintuitive and significant, suggesting a need to explore this relationship further, possibly considering interaction effects or non-linear relationships.

- `part_time_job - Yes` (`4.687e+00`): Having a part-time job is associated with a 4.687 unit increase in grade improvement compared to not having a part-time job, but this is not statistically significant (p = 0.1485).

- `family_income` (`-6.338e-05`): An increase in family income is associated with a slight decrease in grade improvement, but this is not statistically significant (p = 0.6913).

- **Residual standard error**: Approximately `15.97`, indicating the average distance of the data points from the fitted line.

- **Multiple R-squared** (`0.07643`): About 7.64% of the variance in `grade_improvement` is explained by the model. This is relatively low, suggesting that other factors not included in the model might be influencing grade improvements.

- **Adjusted R-squared** (`0.0273`): Adjusted for the number of predictors, it's lower, indicating that some of the predictors might not be contributing much to the model.

- **F-statistic** (`1.556`): Tests the overall significance of the model. The corresponding p-value (`0.1803`) suggests the model is not statistically significant at conventional levels.

- **Noise level and grades**: The coefficients for `noise_level_pre` and `noise_level_post` are not statistically significant, suggesting that within this model, there's no clear evidence that noise level changes are causally related to grade improvements.

- **Other variables**: The coefficients for `part_time_job` and `family_income` are not significant, indicating they don't have a statistically significant impact on grade improvement in this model.

This regression model suggests that the noise level changes are not significantly associated with grade improvements, based on the provided data and model specification. The significant negative relationship with study_hours is intriguing and warrants further investigation. The model's overall low R-squared measure indicates that there are other important factors not captured by the model that might be influencing grade improvements.

For causal inference, it's crucial to remember that regression analysis can indicate associations but does not definitively prove causality. The results should be interpreted in the context of the study design, potential confounders, and the plausibility of the causal mechanisms.

In summary, the t-test is principally employed for mean comparisons across two distinct groups, whereas regression analysis investigates the nexus between a dependent variable and one or several independent variables. T-tests, constrained by their design, assess the influence of a binary independent variable without accounting for extraneous variables, in contrast to regression's capability to manage multivariate contexts and adjust for potential confounders. Regression analysis offers more flexibility and complexity, allowing for different types of dependent variables and the modeling of interactions and non-linear effects, unlike the t-test, which is more simplistic and limited in scope.

Propensity score matching

When we're looking at how changing neighborhoods affects students, **propensity score matching (PSM)** is really the way to go, better than regular regression. Why? PSM is great at making sure the students who moved and those who didn't are similar in key ways, such as how much they study or their family's income, which helps us avoid biased conclusions. If you don't understand PSM fully just yet, don't worry as we'll cover it in detail in *Chapter 6*. For now, follow along to understand the application in R.

Building the model

Here's how it works: we use a special calculation (propensity scores) to figure out each student's odds of moving. Then, we pair up students with similar scores, kind of like setting up students on a study date! This way, we make sure the groups we're comparing are really alike, and our findings on how moving affects their grades are as reliable as they can be. This helps us understand whether the observed effect is due to the neighborhood change or other factors. In R, we build the PSM model using the MatchIt library. The following code describes how to put it into action:

```
library(MatchIt)
m.out <- matchit(noise_level_post ~ study_hours +
                    part_time_job + family_income,
                method = "nearest", data = students)
matched_data <- match.data(m.out)  # Getting the matched dataset
```

Interpreting results

Interpreting the results from the matched data (see code) obtained through propensity score analysis involves understanding the individual variables and their relationship to the outcome of interest, in this case, grade_improvement. Let's first dive into the relevant metrics to look for in the PSM analysis:

- distance reflects the propensity score, representing the probability of a student moving to a quieter neighborhood based on their characteristics.

- weights shows the weighting applied to each case to balance the comparison groups.

- subclass is an identifier for matched groups or strata in the analysis.

Are students who moved to quieter neighborhoods consistently showing more positive grade improvement than their matched counterparts? This would suggest a causal effect. Is the change in grades substantial or marginal? Large differences in grade improvement post-move would strengthen causal claims. For example, if students with high study hours show more improvement regardless of noise change, the effect might be more attributable to study habits than noise levels. If certain subclasses (groups of matched pairs) show different trends, it might indicate other variables at play influencing the grades. Let's go over the important points one by one:

- **Individual cases**: You can examine how individual students' grades changed after moving. For example, a student with a significant increase in grade_improvement after moving to a Low noise level post-move might suggest a beneficial effect of a quieter environment. For instance, Student 1, who remained in a Moderate noise level, showed a grade improvement of 12.496 points. Student 3, moving from a Moderate to Low noise level, shows a smaller improvement of 3.185 points. This suggests that the change in noise level might not be the only influencing factor, as improvements vary even within similar noise level changes.

- **Overall trend**: Look for general patterns in the grade_improvement column. Are there many students with significant positive changes? Are these changes more common in certain noise_level_post categories? You can try it on your own by visualizing the data in a suitable chart.

- **Match quality**: The distance (propensity score) and subclass columns help assess the quality of the matches. Lower distances indicate better matches. Check whether the matched students are indeed similar in their baseline characteristics.

- **Trend analysis**: It's a bit of a mixed bag when we look at how things changed. Take Student 2, for instance: their grades dropped quite a bit (-18.219), and that happened without any change in the noise around them. Then there's Student 6 who moved from a Moderate to Low noise area and saw their grades jump up by 33.792. These kinds of ups and downs suggest that there's more at play than just noise levels changing. If we line up the students who went from High or Moderate noise areas to quieter Low ones against those who stayed put, we might spot some patterns. But right now, it's not clear-cut; some students are doing better, others not so much.

- **Additional factors**: The influence of `study_hours`, `part_time_job`, and `family_income` cannot be ignored. These could confound the relationship between noise level changes and grade improvements. For instance, `Student 3` and `Student 6`, both moving to `Low` noise levels, show different outcomes possibly influenced by other factors such as study hours or family income.

The data does not conclusively show that changing neighborhoods (specifically to quieter noise levels) causes grade improvement. While some students show improvements, others do not, or even show a decrease in grades, despite similar changes in their environment. The presence of mixed outcomes within similar noise level change categories suggests that other unaccounted variables might be influencing grade improvements. Based on the provided data, we cannot conclusively establish a causal relationship between moving to quieter neighborhoods and improved grades. The results are mixed and suggest the influence of multiple factors on students' academic performance. For a robust causal inference, further statistical analysis and consideration of additional variables would be necessary.

In this section, we learned various causal inference techniques. Next, let's hone our skills more with a case study in R.

Case study – a basic causal analysis in R

Let's dive into a practical example from causal inference using R. Our objective is to evaluate the effect of attending Catholic schools on students' standardized math scores. The challenge here is that our dataset is observational, meaning there's a risk of selection bias. This bias arises when the differences in outcomes (such as educational achievement) between students from Catholic and public schools are due to pre-existing characteristics, not just the school type.

The dataset [7] includes information on student demographics (mother's age, family income, and number of residences) and educational background (mother's education level and whether the student attended a Catholic or public school). The key variables we're interested in are whether a student attended a Catholic school (`catholic`) and their standardized math scores (`c5r2mtsc_std`). This example highlights how to tackle potential biases to uncover the true impact of Catholic schooling on math achievement.

We use PSM to investigate the causal effect of Catholic education on standardized math scores. This method is particularly suitable for our observational data, which is prone to selection bias and confounding due to differences in school type. PSM helps to emulate a randomized experiment, reducing these biases by matching students based on relevant covariates. Our goal is to provide a clearer picture of Catholic education's impact, offering insights into its advantages or disadvantages compared to public education. This study not only deepens our understanding of educational practices but also demonstrates the value of PSM in making causal conclusions from observational studies.

Data preparation and inspection

The dataset contains a variety of columns with different data types and some missing values. Here's an overview:

- `childid` (`object`): Identifier for the child. No missing values.

- `catholic` (`int64`): Indicates whether the child is Catholic (1) or not (0). No missing values.

- `race` (`object`): Descriptive category of the child's race. No missing values.

- `race_white` (`int64`): Indicates whether the child is white (1) or not (0). No missing values.

- `race_black` (`int64`): Indicates whether the child is black (1) or not (0). No missing values.

- `race_hispanic` (`int64`): Indicates whether the child is Hispanic (1) or not (0). No missing values.

- `race_asian` (`int64`): Indicates whether the child is Asian (1) or not (0). No missing values.

- `p5numpla` (`float64`): Unknown. 14.22% missing values.

- `p5hmage` (`float64`): Age of the child's mother. 14.58% missing values.

- `p5hdage` (`float64`): Age of the child's father. 29.17% missing values.

- `w3daded` and `w3momed` (`object`): Unknown. Both have 12.40% missing values.

- `w3daded_hsb` and `w3momed_hsb` (`float64`): Unknown. 28.98% and 14.37% missing values, respectively.

- `w3momscr` and `w3dadscr` (`float64`): Unknown. 33.90% and 32.24% missing values, respectively.

- `w3inccat` (`object`): Category of household income. 12.40% missing values.

- `w3income` (`float64`): Household income. 12.40% missing values.

- `w3povrty` (`float64`): Unknown. 12.40% missing values.

- `p5fstamp` (`float64`): Indicates whether the family receives food stamps (1) or not (0). 13.76% missing values.

- `c5r2mtsc` and `c5r2mtsc_std` (`float64`): Child's standardized math scores and its standardized value. No missing values.

We will overcome the missing value issue by ignoring the cells that have missing values for this example. You may wonder why. We've chosen to ignore cells with missing values as it's a straightforward way to handle incomplete data. This approach ensures we're only working with complete cases, which simplifies our analysis and helps us avoid potential biases or inaccuracies that could come from more complex methods such as imputation. It's worth noting, though, that this method might reduce our sample size.

Understanding the data

Let's start with some EDA using R. First, load the data into R as a data frame named `data`. The code snippet creates a new data frame, `data_anonymized`, that excludes the `childid` column to not let personal identity information come in the way. You can now use `data_anonymized` for further analysis without concern for exposing personal identity information:

```
setwd(dirname(rstudioapi::getSourceEditorContext()$path))
data <- read.csv(file = "./data/ecls.csv")
data
# Remove the columns to anonymize the data
data_anonymized <- data[, !(names(data) %in% c("childid"))]# View the
first few rows of the anonymized dataframe
head(data_anonymized)
```

Next, we load the `dplyr` and `ggplot2` libraries for data manipulation and visualization. We then calculate summary statistics (mean, standard deviation, minimum, and maximum) for all numeric columns in our dataset, ignoring any missing values. Finally, we count the occurrences of different categories in the `catholic` column using the `table` function. This gives us a clear overview of our data's structure and distribution:

```
library(dplyr)
library(ggplot2)
Library(MatchIt)
# Summary statistics for numeric columns
summary_statistics <- data_anonymized %>%
  reframe(across(where(is.numeric), ~ list(
    mean = mean(.x, na.rm = TRUE),
    sd = sd(.x, na.rm = TRUE),
    min = min(.x, na.rm = TRUE),
    max = max(.x, na.rm = TRUE))))

# Counts for 'catholic' columns
catholic_counts <- table(data_anonymized$catholic)
```

Next, let's write some code to visualize the data.

We create histograms for four numeric variables: mother's age, father's age, household income, and child's math score. We set up a 2x2 grid layout for our plots using `par(mfrow = c(2, 2))`. Then, we use `ggplot2` to create a histogram for the mother's age distribution, with customized aesthetics and labels for clarity:

```
# Histograms for numeric variables: Age of mother and father,
household income, and child's math score
par(mfrow = c(2, 2))
```

```
# Mother's age distribution
ggplot(data_anonymized, aes(x = p5hmage)) +
  geom_histogram(binwidth = 1, fill = "skyblue",
                 color = "black") +
  labs(title = "Distribution of Mother's Age",
       x = "Mother's Age", y = "Frequency") +
  theme_minimal()

# Father's age distribution
ggplot(data_anonymized, aes(x = p5hdage)) +
  geom_histogram(binwidth = 1, fill = "gold", color = "black") +
  labs(title = "Distribution of Father's Age",
       x = "Father's Age", y = "Frequency") +
  theme_minimal()
```

Let's now make two more histograms using ggplot2: one for household income and another for the child's math score. For the household income distribution, we use a light green fill color, while for the math score distribution, we use salmon. Both histograms are styled with minimal themes and clear labels to make the data easy to interpret:

```
# Household income distribution
ggplot(data_anonymized, aes(x = w3income)) +
  geom_histogram(fill = "lightgreen", color = "black") +
  labs(title = "Distribution of Household Income",
       x = "Household Income", y = "Frequency") +
  theme_minimal()

# Child's math score distribution
ggplot(data_anonymized, aes(x = c5r2mtsc)) +
  geom_histogram(fill = "salmon", color = "black") +
  labs(title = "Distribution of Child's Math Score",
       x = "Math Score", y = "Frequency") +
  theme_minimal()
```

We then create a bar plot using ggplot2 for Catholic affiliation. The Catholic distribution plot uses light coral bars and clearly labels the categories (0 for No, 1 for Yes).

```
# Catholic distribution
ggplot(data_anonymized, aes(x = factor(catholic))) +
  geom_bar(fill = "lightcoral") +
  labs(title = "Distribution of Catholic",
       x = "Catholic (0 = No, 1 = Yes)", y = "Count") +
  theme_minimal()
```

These generate summary statistics and visualize key aspects of this dataset. Adjust the column names if you want to explore some other columns.

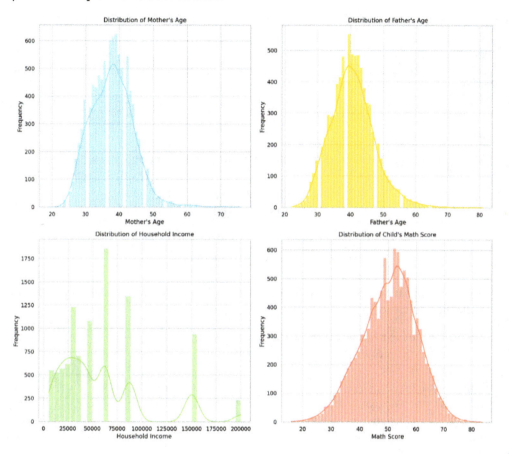

Figure 3.5 – The distribution of various columns from the dataset

Performing causal analysis

To provide a comprehensive solution for PSM, we will include pre-analysis using non-matched data, estimating the propensity score, executing a matching algorithm, examining covariate balance in the matched sample, and estimating treatment effects. We'll use the MatchIt package for PSM, and visualize and evaluate our findings. The code is inspired by the works on PSM as seen here [6, 7].

First, we calculate the difference in means for the outcome variable (`c5r2mtsc_std`) between students who went to a Catholic school and those who did not:

```
data_anonymized %>%
  group_by(catholic) %>%
  summarise(n_students = n(),
            mean_math = mean(c5r2mtsc_std, na.rm = TRUE),
            std_error = sd(c5r2mtsc_std, na.rm = TRUE) /
            sqrt(n_students))
```

We get the following results:

catholic	n_students	mean_math	std_error
<int>	<int>	<dbl>	<dbl>
0	9568	-0.0306	0.0104
1	1510	0.194	0.0224

Figure 3.6 – The summary statistics

Once you see the summary statistics as seen here, we have to choose relevant pre-treatment covariates and calculate their difference in means between the two groups:

```
pre_treatment_covariates <- c(
  'w3income', 'p5numpla', 'w3momed_hsb')
data_anonymized %>%
  group_by(catholic) %>%
  summarise_at(vars(one_of(pre_treatment_covariates)),
               ~mean(.x, na.rm = TRUE))
```

We get the following result:

catholic	w3income	p5numpla	w3momed_hsb
<int>	<dbl>	<dbl>	<dbl>
0	54889.	1.13	0.464
1	82074.	1.09	0.227

Figure 3.7 – The summary statistics from the provided code

The next step is to estimate the propensity score by fitting a **logistic regression model**. Logistic regression is the most common method used to estimate propensity scores, where the treatment assignment is regressed on observed baseline characteristics. Logistic regression is widely used for binary outcomes, modeling the log odds of an event as a linear combination of predictor variables [8].

We then start by constructing a logistic regression model. This model is tailored to predict the likelihood of belonging to a specific group (in this case, `catholic`) based on several predictors, such as income (`w3income`), the number of places lived (`p5numpla`), and the mother's education level (`w3momed_hsb`). The logistic regression is performed in R using the `glm` function, specifying a binomial family to denote the logistic nature of the model:

```
model_ps <- glm(catholic ~ + w3income + p5numpla + w3momed_hsb,
                family = binomial(), data = data_anonymized)
```

You might be wondering what these propensity scores are. What is their distribution? Propensity scores, the outcome of this model, represent the probability of treatment assignment conditional on observed covariates. They are pivotal in balancing the distribution of covariates across treatment groups, allowing for a more unbiased estimation of treatment effects.

To explore and understand the distribution of these propensity scores, we visualize them using a histogram, distinguishing between the groups:

```
data_anonymized$pscore <- predict(
  model_ps, newdata = data_anonymized, type = "response")
ggplot(data_anonymized, aes(
  x = pscore, fill = as.factor(catholic))) +
  geom_histogram(alpha = 0.6, position = "identity", bins = 30)+
  labs(x = "Propensity Score", y = "Frequency", fill = "Group")+
  theme_minimal()
```

After estimating the propensity scores, we proceed to match individuals from the treatment and control groups based on their propensity scores using nearest-neighbor matching. This method pairs individuals from the treatment group with similar individuals from the control group based on their propensity scores:

```
# Remove rows with any NA values in the specified covariates
data_cleaned <- data_anonymized %>%
  filter(complete.cases(catholic, race_white, w3income,
                        p5numpla, w3momed_hsb))

# Further remove rows with non-finite values in numeric covariates
numeric_covariates <- c( "w3income", "p5numpla")
data_cleaned <- data_cleaned %>%
  filter(sapply(data_cleaned[numeric_covariates], is.finite)
         %>% rowSums() == length(numeric_covariates))

# run matchit() on data_cleaned
mod_match <- matchit(
  catholic ~ w3income + p5numpla + w3momed_hsb,
  method = "nearest", data = data_cleaned) # or data_imputed
```

Visual inspection always helps in examining the results, as you may know very well by now. For that, plot the covariate balance before and after matching. Evaluating the balance of covariates before and after matching is crucial to ensure that the matching process has effectively balanced the groups, allowing for a fair comparison. We visually inspect this balance using jitter (*Figure 3.8*) and histogram plots.

Histogram plots are particularly useful here because they provide a clear visual representation of the distribution of covariates across groups, making it easy to spot any imbalances or shifts that may have occurred during the matching process:

```
# Using the MatchIt package for visualization
plot(mod_match, type = "jitter")
plot(mod_match, type = "hist")
```

We get the graph as follows:

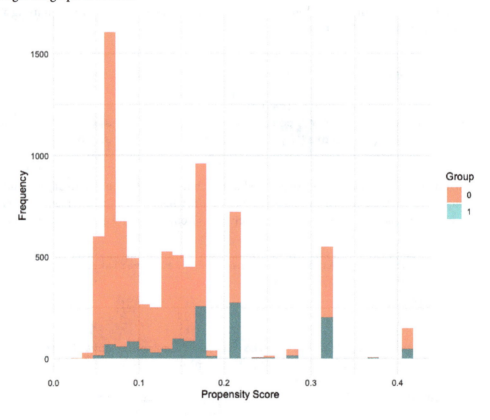

Figure 3.8 – The distribution of propensity scores

To quantitatively assess the balance of covariates, we compute the mean of each pre-treatment covariate for both groups in the matched sample:

```
matched_data <- match.data(mod_match)
matched_data %>%
  group_by(catholic) %>%
  summarise_at(vars(one_of(pre_treatment_covariates)),
               ~mean(.x, na.rm = TRUE))
```

Next, we examine the standardized differences between groups for each covariate, assessing how well the matching process has aligned the covariates across treatment and control groups.

To determine the treatment effect (here, the impact of being Catholic), we use a t-test and a linear regression on the matched data. These analyses measure the outcome differences between the two groups:

```
with(matched_data, t.test(c5r2mtsc_std ~ catholic))

# You can also use a linear model for a more detailed analysis
lm_effect <- lm(c5r2mtsc_std ~ catholic, data = matched_data)
summary(lm_effect)
```

Through this process, we carefully assess the effect of being Catholic on our outcome of interest, leveraging a matched sample to minimize bias from confounders in our treatment effect estimates.

Visual tools such as **histograms** and **jitter plots** are instrumental in assessing the distribution of propensity scores and ensuring covariate balance across treatment and control groups. Post-matching, an evident balance enhancement signals a successful match. The final treatment effect estimation reveals the influence of Catholic school attendance on standardized math scores, factoring in the selected covariates.

Crucially, the effectiveness of a propensity score analysis hinges on the judicious choice of covariates that impact both the assignment to treatment and the outcomes, coupled with a rigorous evaluation of covariate balance after matching.

Summary

This chapter offered you a comprehensive introduction to R programming within the context of causal inference. It began with a foundational understanding of R, including setting up the working environment, basic programming concepts, data types, structures, and functions. The chapter then transitioned into the practical application of causal inference techniques, such as t-tests, regression analysis, and PSM, using various datasets for illustration.

Specifically, the chapter addressed the implementation of PSM analysis in R, with an evaluation of the effect of attending Catholic schools on students' standardized math scores. The chapter emphasized the importance of statistical techniques in establishing causality and provided guidance on interpreting PSM results. Further depth on PSM and causal analysis in R will be covered in subsequent chapters.

References

1. *The Comprehensive R Archive Network*: `https://cran.rstudio.com/`

2. *Install R from Source* (for Linux): `https://docs.posit.co/resources/install-r-source/`

3. *Getting Started* (setting up RStudio): `https://rc2e.com/gettingstarted#intro-GettingStarted`

4. *Download RStudio Desktop*: `https://posit.co/download/rstudio-desktop/`

5. *Available Cran Packages by Name*: `https://cran.r-project.org/web/packages/available_packages_by_name.html`

6. *PSM Tutorial*: `https://simonejdemyr.com/r-tutorials/statistics/tutorial8.html`

7. *Git for PSM Tutorial*: `https://github.com/sejdemyr/ecls`

8. Hosmer, D. W., & Lemeshow, S. (2000). *Applied Logistic Regression* (2nd ed.). New York: Wiley

Part 2: Practical Applications and Core Methods

This part focuses on applying key causal inference methodologies to real-world scenarios. It covers the use of Directed Acyclic Graphs (DAGs), propensity score techniques, regression models, and doubly robust estimation to analyze causal relationships. Practical guidance on conducting A/B testing and controlled experiments is also provided, with an emphasis on implementation using R.

This part has the following chapters:

- *Chapter 4, Constructing Causality Models with Graphs*
- *Chapter 5, Navigating Causal Inference through Directed Acyclic Graphs*
- *Chapter 6, Employing Propensity Score Techniques*
- *Chapter 7, Employing Regression Approaches for Causal Inference*
- *Chapter 8, Executing A/B Testing and Controlled Experiments*
- *Chapter 9, Implementing Doubly Robust Estimation*

4

Constructing Causality Models with Graphs

In the previous chapters, we really got our hands dirty with some complex stuff – causality, confounding, association, and how they play out in statistics. We didn't just talk theory; we got practical, diving into sample regression and propensity score matching, mostly using R, to see how these concepts come alive in real research. Now, we'll head into the world of graphical models that explain causal relationships in a different way.

This chapter mixes graph theory – which is all about dots and lines and how they connect – with our focus on figuring out cause and effect. Our goal? To give you a toolkit that makes sense of causal connections in various situations. We'll start with graph theory basics and gradually get into the nitty-gritty of dynamic causal models.

The topics covered in this chapter include the following:

- Basics of graph theory
- Graph representations of variables
- Graphical causal models
- Case study example of a graph model in R

Technical requirements

You can find the code examples for this chapter in this book's GitHub repository: https://github.com/PacktPublishing/Causal-Inference-in-R/tree/main/chap_04.

Basics of graph theory

When it comes to making sense of complex networks, there is graph theory, which contains nodes and edges as its bare bones. You might have seen a web of interconnected points – these are nodes or vertices, and they represent all sorts of things, depending on what you're looking at. Now, you may ask, what are the lines or edges connecting them? They represent the relationships or interactions between nodes. This is the core foundational concept behind graph theory, and it is radically utilized in areas as varied as computer science, biology, and social sciences.

Let's paint a picture with some examples. In social network analysis, think of each node as a person. The edges are the ties that bind them – friendships, work relationships; you name it. It's like mapping out your own social universe. Now, switch gears to biology. Here, nodes could be proteins, and the edges show us how they interact, kind of like a molecular dance. This isn't just academic stuff – it's a real-deal look into how intricate networks, whether among people or proteins in biological sciences, weave the fabric of these systems. Next, let's learn about types of graphs.

Types of graphs – directed versus undirected

In the study of graphical structures, we primarily distinguish between two types of graphs: undirected and directed, each serving a unique purpose in representing relationships:

- **Undirected graphs**: These graphs (*Figure 4.1b*) are analogous to bi-directional pathways. The links (or edges) between various points (termed nodes) lack a designated direction. Such graphs are aptly employed in scenarios where relationships are reciprocal or where directionality is not the primary focus. They are basically a two-way communication channel where information flows in both directions:

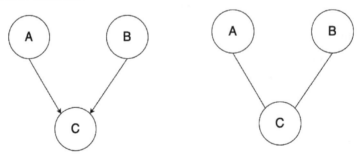

Figure 4.1 – (a) Directed graph and (b) undirected graph

- **Directed graphs (or digraphs)**: Contrasting with undirected graphs, directed graphs (*Figure 4.1a*) embody a one-way directional flow. Each connection in these graphs is like an arrow, pointing from one node to another, thereby clearly illustrating the flow of influence or causation. This type of graph is particularly instrumental in delineating scenarios where one event or factor directly influences another.

To illustrate directed graphs, one might consider a citation network. Here, articles are represented as nodes, and a directed edge from one article to another signifies a citation from the former to the latter. In contrast, an undirected graph can exemplify a co-authorship network where edges simply indicate collaborative relationships without suggesting any directional influence:

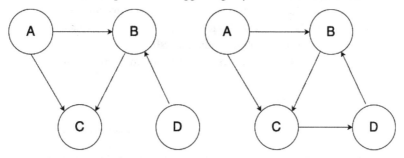

Figure 4.2 – (a) Directed acyclic graph (b) directed cyclic graph

Picture a scene with four points: *A*, *B*, *C*, and *D*. So, if *A* impacts *B* and *C*, *B* impacts *C*, *D* impacts *B* (*see Figure 4.2a*), then we can say it's a **directed acyclic graph** (**DAG**). But imagine if there's another edge between *C* to *D*, saying *C* impacts *D*; then, clearly there is a cycle(*Figure 4.2b*). We cannot classify that as a DAG. This setup is super important, especially in causal inference, where figuring out who influences whom – without any confusing feedback loops – is key. In the world of directed graphs, these kinds of graphs are a special type called a DAG. *Acyclic* means no going in circles.

Visualize that a path in this graph is a journey from one point to another, hopping from node to node along edges. In a regular path, you can zigzag any which way. But in a DAG, it's like following a series of one-way streets; you've got to move in just one set direction. So, if there's a one-way route from *A* to *B*, *A* becomes the origin – the ancestor – and *B* is the endpoint – the descendant. This way, there's a clear line of who's influencing whom. Now that we know about DAGs, let's learn about graph typologies.

Other graph typologies

We will now talk about weighted graphs. Think of them like a social network where the edges have values, showing how strong or frequent the interactions are – these values give more context to each connection. Unweighted graphs are simpler; they just show who's connected to whom without any extra frills. As for how these graphs are constructed, there are connected ones, kind of like a big web of roads connecting every city, and then there are unconnected graphs, where some points are like isolated islands, not linked by the same network. In such graphs, not every node can be accessed from every node, like an island in the real world.

In some analyses, the direction of connections in a graph isn't always clear-cut. That's where **complete partially DAGs (CPDAGs)** come in handy. They're part of a larger group of graphs with partly directed edges, used in fields such as network science and with certain **graph neural networks**. These CPDAGs let researchers map out possible causal pathways without pinning down a specific direction for each connection. Take, for example, studying a new virus's spread; factors such as individual behavior, environmental conditions, and genetic factors all intertwine. CPDAGs help visualize these relationships, even when the exact cause-and-effect order isn't fully understood, such as how personal hygiene and population density might influence virus transmission without knowing which causes which.

Why we need DAGs in causal science

DAGs hold a pivotal role in causal analysis for a multitude of compelling reasons:

- **Causal structure depiction**: DAGs can depict causal relationships in an unambiguous and intuitive manner. The directed edges represent potential causal influences from one variable to another, a feature indispensable for grasping the implications of variable alterations. The non-circular nature of DAGs is essential, as it precludes the existence of feedback loops. This characteristic assures that a cause does not become an indirect effect of itself, avoiding illogical or contradictory outcomes. It resonates with our inherent perception of causality, where the effect cannot precede its cause.

- **Tool for unraveling confounders and mediators**: DAGs are instrumental in pinpointing confounders and mediators (covered more in *Chapters 5* and *11*) within causal pathways. Confounders, as we know, are variables influencing both cause and effect that can lead to misleading associations if overlooked. Mediators, conversely, are vehicles for the causal effect. DAGs facilitate the construction of statistical models that accurately estimate these causal effects. Through the application of d-separation rules (discussed in *Chapter 5*), DAGs enable the assessment of conditional independence among variables. This is crucial for differentiating causation from mere correlation and understanding the conditions that foster independence between variables.

- **Elucidating intricate causal networks**: The complexity of real-world causal systems is often daunting. DAGs serve as a roadmap, illustrating cause-and-effect relationships in diverse areas, from biology to economics. By presenting these connections visually and mathematically, DAGs provide a clearer understanding of intricate causal webs, which is invaluable in scientific research where challenging assumptions is crucial.

- **Guiding causal inference and discovery**: Moreover, DAGs are practical tools for identifying and analyzing cause and effect. They guide researchers in selecting and managing variables in their studies, playing a pivotal role in uncovering new causal links. Essentially, DAGs act as a foundational framework in various fields, helping to distinguish genuine causal relationships from mere correlations.

As we have learned why we need DAGs in causal science, naturally, you may want to know how to represent them in both theory and practice. Let's dig into it next.

Graph representations of variables

Using graphs is like drawing a map of how different factors influence each other. Each circle (or node) in our doodle is a clue (a variable), and each line (or edge) is a "this leads to that" arrow. These doodles can be as simple as a child's stick figure or as complex as a subway map during rush hour.

For a bit of fun, let's say we're investigating if staying in school makes you richer. Our graph would have circles labeled *Time Spent in School*, *Smarty Pants*, *Kind of Job*, and *Size of Paycheck*. Arrows would point from *Time Spent in School* to *Smarty Pants*, and from there to *Size of Paycheck* (*see Figure 4.3*), suggesting that more school might lead to brainier brains and fatter wallets. But remember – in the twisty world of cause and effect, it's not always so straightforward; sometimes, the plot thickens!

For our learning purposes, we can also visualize the preceding causality using a graph, as seen here. The edges and directions represent the source node that causes the target node:

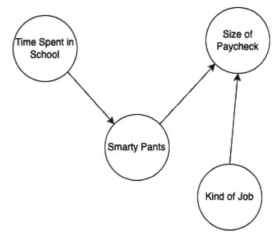

Figure 4.3 – The relationship between variables in terms of causality

Mathematical interpretation

Imagine that each person in the school – students, teachers, even the janitor – is like a dot on this web. And the lines connecting these dots? They're the "hellos," high-fives, and "do your homework" of daily school life.

Let's break it down, shall we?

- **Dots (or nodes)**: Think of each student as a dot or a node. So, in a school with 100 students, you've got 100 nodes, like a giant connect the dots – oops, nodes. In graph terms, we call these nodes $(v_1, v_2, v_3, ... , v_{100})$. Nodes represent entities or objects in a network—such as students in this case.

- **Lines (or edges)**: These lines are the secret handshakes and borrowed pencils of the school world. In graph theory, an **edge** connects two nodes and signifies a connection between the entities, like a friendship or an order of authority. They come in two flavors:

    - **Two-way streets (undirected edges)**: This is the "I'll share my lunch if you share yours" kind of friendship. If students *A* and *B* are buddies, we draw a line between them that goes both ways because friendship is a two-way street. In graph terms, this is an **undirected edge**, meaning both *A* and *B* are connected equally because friendship is reciprocal. Mathematically, we denote this as an undirected edge (A, B).

    - **One-way roads (directed edges)**: This is more like an "I'm the boss of you" relationship. If student *A* is the class president over *B*, we draw an arrow from *A* to *B*. It's a one-way road where *A* gets to do the bossing. This is called a **directed edge**, and it's represented by an arrow pointing from one node to another, such as $A \rightarrow B$.

- **Paths (or the who-tells-who chain)**: Imagine a rumor starting from one dot and zigzagging across to another. In our math world, this is a path. If the rumor goes from *A* to *B* to *C*, it's like a game of telephone where the message hops from one person to the next. In mathematical terms, a path is an ordered set of edges that allow traversal from one node to another, such as $A \rightarrow B \rightarrow C$.

Now, when we talk about figuring out who really caused the latest cafeteria food fight, that's where causal inference comes in. In terms of causality, our initiative aims to trace the connections to determine who is influencing whom within the school, and this is where DAGs come in handy.

Representing graphs in R

In R's versatile toolkit, graph representation is achieved through several methods, each tailored to specific needs.

Edge-list representation

This method simplifies graph representation into a list format, where each edge, a fundamental graph component, is defined by a node pair. This approach is both elementary and efficient for certain graph-based analyses:

```
# Example of an edge list
edge_list <- matrix(c("A", "B", "B", "C", "C", "A"),
                    ncol = 2, byrow = TRUE)
# Printing the edge list
print(edge_list)
```

In the provided code snippet, the edge list elucidates connections between nodes: *A* linked to *B*, *B* to *C*, and *C* back to *A*. Edge lists are space-efficient and simple for sparse graphs but can be slow for complex operations and lack information on unconnected nodes.

To represent a weighted graph—where each edge has a weight (for example, a distance, cost, or strength of relationship)—we simply extend the edge list by adding a third column that specifies the weight for each edge. This third column would contain numerical values representing the weight or strength of each connection. See the R example here:

```
# Example of a weighted edge list
weighted_edge_list <- matrix(c(
  "A", "B", 2, "B", "C", 3, "C", "A", 1),
  ncol = 3, byrow = TRUE)
# Printing the weighted edge list
print(weighted_edge_list)
```

Adjacency matrix

An adjacency matrix employs a square matrix format for graph representation. In this matrix, the presence or absence of adjacency between vertex pairs within the graph is indicated by the matrix's elements. This method provides a structured and comprehensive view of graph connectivity.

In such a matrix, the digit "1" signifies the existence of an edge connecting the respective nodes, whereas "0" denotes the absence thereof. You can assign weights to an edge by replacing "1" with a number representing the weight.

Adjacency matrices enable fast edge lookups and provide a comprehensive graph view but are space-intensive for sparse graphs and can face scalability issues with large node numbers. Adjacency matrices and edge list representation both have distinct strengths and weaknesses, making them suitable for different types of graph representations and operations.

Here's an example of an adjacency matrix:

```
# Example of an adjacency matrix
adj_matrix <- matrix(c(0, 1, 1,
                       1, 0, 1,
                       1, 1, 0), nrow = 3, byrow = TRUE)

# Setting row and column names for clarity
rownames(adj_matrix) <- colnames(adj_matrix) <- c("A", "B", "C")

# Printing the adjacency matrix
print(adj_matrix)
```

Utilization of graph packages

R is equipped with advanced packages, such as `igraph`, designed for intricate graph operations. The tools in this package not only facilitate the creation and modification of graphs but also provide an array of techniques for their representation and analytical examination. Let's see how in the code:

```
# Install and load igraph
if (!require(igraph)) {
  install.packages("igraph")
  library(igraph)
}
# Create a graph using an edge list
graph <- graph_from_edgelist(edge_list, directed = FALSE)
# Plot the graph
plot(graph)
```

Utilizing the `igraph` package, this code elegantly constructs a graph from an edge list. The `igraph` package is adept at managing both directed and undirected graphs, offering an extensive suite of functions for intricate graph analysis.

Each graph representation technique brings its unique strengths. Edge lists excel in simplicity and their capacity to seamlessly integrate supplementary data about nodes or edges. Conversely, adjacency matrices are instrumental for specific computational tasks such as path discovery or connectivity assessment. The `igraph` package elevates this landscape, furnishing a robust framework for advanced graph operations, encompassing sophisticated network analysis algorithms, visualization capabilities, and beyond.

Bayesian networks

Let's explore Bayesian networks [2] and causality with a statistical lens.

Say you are at a family gathering, examining interactions to discover hidden relationships. Some connections are obvious, such as parent-child links, while others are subtle, such as distant cousins. This setting helps us translate social complexities into mathematical models, crucial for analyzing cause and effect in complex systems.

In data analysis, Bayesian networks are like family trees for variables. They map out how variables are connected, with nodes representing the variables and arrows showing how one variable influences another, much like relationships at a family reunion.

Consider a Bayesian network as a DAG. In this structure, nodes marked as $(X_1, X_2, X_3, X_4, \ldots, X_n)$ represent the family's individuals, while edges (the lines interlinking the nodes) signify the subtle, and sometimes not so subtle, influences exerted within the family hierarchy. Let's build a solid foundation and learn some theory next, as we go deeper into the world of causality.

Basics of probability theory

Before getting into the theory, let's learn the basics of probability theory Before getting into the theory, let's learn the basics of probability theory, as it is fundamental to what we will cover next in the chapter. Probability quantifies the likelihood of an event, ranging from 0 (impossible) to 1 (certain). For an event A, its probability $P(A)$ is calculated as the ratio of favorable outcomes to total possible outcomes. For example, the probability of rolling a 4 on a fair six-sided die is $P(4) = 1/6$ (only one 4 in the six possible outcomes of numbers).

There are a few important rules in probability. The addition rule quantifies the probability of either of two events occurring. In this case, for mutually exclusive events, we represent it as $P(A$ or $B) = P(A) + P(B)$. For non-mutually exclusive events, it is defined as $P(A$ or $B) = P(A) + P(B) - P(A$ and $B)$. Then, there is the multiplication rule that applies to independent events $P(A$ and $B) = P(A) \times P(B)$.

Finally, there is another concept you need to know called conditional probability (further explained later in this chapter), denoted as $P(A|B)$. It measures the likelihood of event A given that event B has already occurred. For dependent events, $P(A|B) = P(A$ and $B) / P(B)$, where $P(B) > 0$. This concept is crucial for understanding complex probabilistic relationships and advanced statistical analyses.

Chain rule of probability

The chain rule of probability is used for breaking down a complex family story into smaller parts. Instead of tackling the whole story at once, we examine each relationship step by step. Mathematically, this means evaluating the probability of each variable in sequence, based on those already assessed.

In other words, this method deconstructs the probability of an entire familial tableau into a series of individual probabilities $P(X_1, X_2, X_3, X_4, ... X_n)$.

It commences with the likelihood of one member $P(X_1)$, then progressively incorporates the conditional probabilities of subsequent members $P(X_1 \mid X_2, ... X_{(i-1)})$ for $(i = 2$ to $n)$.

Exponential growth of parameters

Now, the problem is documenting every interaction at a large family gathering — it would quickly use up all your stationery. Similarly, in Bayesian networks, the number of details grows rapidly with each added variable. Efficient representation methods are needed to manage this complexity and avoid being overwhelmed.

Consider a scenario where each family member exhibits binary moods. Documenting every mood combination becomes a laborious task. Mathematically, it would require $2^{(n-1)}$ parameters to capture all possible states for n individuals. For a mere quartet, this amounts to $2^{(4-1)} = 8$ parameters, illustrating the rapidly escalating complexity with each additional member.

Furthermore, in the Bayesian world, the local Markov assumption is a critical viewpoint. This assumption is the mathematical equivalent of the adage: "To understand a person, look no further than their parents." It posits that a node x_i in any Bayesian network is conditionally independent of non-descendants, given its progenitors (that is, the parent node of a given node), $pa(X_i)$. This is expressed as $P(X_i \mid pa(X_i))$ standing apart from $P(X_i \mid non\_descendents|)$.

This assumption implies that to predict a person's behavior, one need only consider their immediate family, not their entire lineage. In Bayesian terms, it means a variable's behavior is dictated solely by its direct ancestors (or parents in the network), not by the broader network.

Complex techniques

In the context of Bayesian networks, we simplify a complex joint probability (that is, the probability of multiple events occurring simultaneously at a family reunion) into more manageable conditional probabilities (similar to calculating the likelihood of one specific event, given another has already happened). This decomposition is mathematically represented by the following formula:

$$P(X_1 \dots X_n) = \prod_{i=1}^{n} P(X_i \mid pa_i) \qquad (1)$$

Here, pa_i signifies the "parent" nodes of X_i in the graph. This method divides the overwhelming task of comprehending the entire network into digestible segments.

There is one more principle that can and should be applied to this concept: the principle of minimality. This principle posits that the narratives we derive from our family interactions are as straightforward as possible. If the relationship between two family members can be adequately explained through their immediate connections, there's no need to dig deeper into ancestral links. In Bayesian terms, this translates to not assuming extra independence among variables beyond what the network structure indicates.

Adopting the Markov and minimality assumptions in our "family" dynamics, represented by the Bayesian network, allows us to not only decipher interactions but also to understand the potential cascading effects of one member's actions on others. This shifts our focus from mere associations to deducing causality—determining if one person's actions are genuinely influencing another's response.

Consider a scenario where Aunt Alice's speech leads to Cousin Bob's tears. If we observe that Bob consistently cries following Alice's speeches, we might infer a causal link: Alice's speeches lead to Bob's emotional reaction. In Bayesian terminology, if there is a direct link (arrow) from Alice (variable X) to Bob (variable Y), this connection can be used to corroborate the influence Alice has on Bob.

In essence, Bayesian networks facilitate our transition from mere observation of data to a profound understanding of underlying cause-and-effect relationships.

Like figuring out who caused a commotion at a family reunion by examining the family tree and interactions, Bayesian networks help us understand complex data by revealing the causal relationships within it. Next, we should discuss conditional independence.

Conditional independence

Let's understand conditional independence, a concept that is crucial in causal graphs, such as Bayesian networks and other graphical models, mapping cause and effect.

Think of three variables, X, Y, and Z, as players in a game. In our causal graph, these players are nodes, and their relationships are the edges between them.

Here's the twist: conditional independence helps in finding out that once you know what Z is up to, X and Y become lone wolves, minding their own business. Mathematically, this is like saying the following:

X and Y don't spill any secrets about each other if you've already got the scoop from Z. Or, in more formal terms, this could be represented as follows:

$$P(X, Y|Z) = P(X \mid Z) P(Y \mid Z) \qquad (2)$$

This fancy equation is our way of saying that once Z enters the chat, X and Y have nothing new to add about each other.

In our causal map, if knowing Z acts like a roadblock on all routes from X to Y, then voilà – X and Y are conditionally independent given Z.

In causal graphs, especially in DAGs, there's a tool called d-separation: a method to determine if variables are independent or linked. D-separation helps us understand when a group of nodes, Z, can block or cut off all paths from node X to node Y.

Now, a path gets "blocked" by Z in a couple of interesting ways:

- First, there's the chain scenario, such as $(X \rightarrow Y \rightarrow Z)$, or the fork setup, $(X \leftarrow Y \rightarrow Z)$. Here, the node Y in the middle is either passing on the info in a straight line (that's the chain) or spreading it out from a central point (the fork).

- Then, there's the curious case of the collider, such as $(X \rightarrow Y \leftarrow Z)$. This is where two separate paths crash into each other at Y.

- X and Z are independent unless we condition on Y or its descendants.

If Y successfully blocks every single path from X to Z, then in the world of causal graphs, we say X and Z are d-separated by Y. This means they're conditionally independent – basically, they're not spilling any beans about each other because Y has cut off all the gossip lines.

In causal inference, conditional independence helps simplify complex probabilities. It makes computations easier and clarifies causal relationships, guiding us to focus on important variables and avoid biases.

We'll dive into the details of chains, forks, and colliders with some compelling examples in *Chapter 5*. Next, let's learn about **graphical causal models (GCMs)**.

Exploring Graphical Causal Models

Graphical models epitomize the lexicon of causality as a tool for elucidating one's conceptual understanding of causal relationships.

Consider, for instance, the principle of conditional independence regarding potential outcomes; a concept that enables the isolation of a treatment's effect on an outcome, distinct from the influence of extraneous variables. An illustrative case is the administration of medication to patients: if only the most severely ill receive the treatment, it might erroneously appear that the medication exacerbates health conditions. This misconception arises from the conflation of illness severity with the medication's impact. Stratifying the patient population into categories based on severity and subsequently examining the medication's effect within these subgroups yields a more accurate assessment of its true impact. This process, termed "controlling for" or "conditioning on a variable," ensures that within each severity category, medication administration approximates a random assignment. Consequently, within these stratified groups, the treatment is rendered conditionally independent of the potential outcomes.

Understanding the nuances of independence and conditional independence is where the utility of GCMs becomes evident [1, 4]. These models provide a framework to visually articulate the causal dynamics at play, delineating the relationships of causation between different variables.

In causality, determining the causal impact of one set of variables upon another often necessitates discerning which variables ought to be controlled for. As previously discussed extensively, effective control of confounding variables extends beyond the purview of pure statistical analysis, requiring insights into the data-generating process. Now, what can precisely represent this process? You guessed it right; it can be done with a causal graph. Consequently, the proficiency to navigate between graphical representations and statistical properties is crucial here. Equally important is the methodology to invert this translation, transforming statistical properties into graphical representations. This skill is fundamental to the practice of causal discovery (discussed in *Chapter 15*), an endeavor focused on reconstructing the causal graph using both observational and interventional data.

Comparison with Bayesian networks

Both GCMs and Bayesian networks utilize a graph-based representation where the following applies:

- Nodes represent random variables $(X_1, X_2, \dots X_n)$
- Edges signify some form of dependency between these variables

Mathematically, they both can express the joint probability distribution of a set of variables as the product of conditional probabilities:

$$P(X_1, X_2, \dots X_n) = \prod_{(i=1)}^{n} P(X_i \mid Parents(X_i)\mid) \qquad (3)$$

Here, $(Parents(X_i))$ denotes the set of nodes with edges directed toward (X_i).

Both GCMs and Bayesian networks are acyclic, meaning they do not allow for feedback loops where a variable can be a cause of itself, either directly or indirectly. This property ensures that the models can be topologically ordered to reflect a flow from cause to effect.

However, there are some distinctions:

- **Causal directionality**: In GCMs, the edges are explicitly interpreted as causal relationships. If there is an edge from X to Y (that is, $X \rightarrow Y$), it is understood that X has a causal influence on Y. This is grounded in the theory of causality, where interventions (explained in this section) are meaningfully defined, and the effect of such interventions can be studied through counterfactual reasoning [5, 6].

- **Conditional dependencies**: Bayesian networks, on the other hand, are often used to represent conditional dependencies rather than explicit causal relationships. An edge in a Bayesian network indicates that once we know the parent nodes, the child node is conditionally independent of all other nodes, not necessarily implying a direct cause.

- **Interventions**: GCMs are typically used in conjunction with interventions to identify causal effects. In causal inference, an intervention is a deliberate action or manipulation applied to a system to change the value of a variable, allowing us to observe its causal effects on other variables. Unlike passive observation, which merely measures natural relationships between variables, an intervention actively alters a variable's value to examine the system's response. This is typically represented using the do-operator, $do(X = x)$, which signifies "intervene to set X to the value x," regardless of X's natural causes.

  For instance, in a causal model examining urban planning, we might investigate the effect of public walking and green corridors (X) on residents' well-being (Y). The intervention $do(X = 1)$ would represent the active implementation of these corridors, after which we observe the impact on Y (residents' well-being). The resulting probability, $P(Y \mid do(X = x))$, represents the probability of Y when X is set to x through intervention rather than by its natural causes. Interventions are crucial for distinguishing between correlation and causation as they disrupt the system's normal dependencies, revealing true causal relationships.

As we just learned, the do-operator (also referred to as do-calculus) symbolizes a deliberate intervention within a population. Furthermore, it is distinct from conventional probabilistic conditioning. Let's learn from this example, where conditioning on a treatment ($T = t$) implies focusing on a subset of the population who received treatment (t), and intervening with $do(T = t)$ equates to administering treatment t across the entire population.

The do-operator, symbolized as $do(T = t)$, is a key concept in GCMs, also reflected in potential outcomes notation. It's used to express interventions, such as $P(Y(t) = y)$, which can be equivalently written as $P(Y = y \mid do(T = t)|)$ or, more concisely, as $P(y \mid do(t)|)$. This notation is consistently used for denoting interventions throughout related discussions.

Evidently, in binary treatment scenarios using **Average Treatment Effect** (ATE), we utilize the do-operator to compare the means of outcomes under different treatment conditions:

$$E(Y \mid do(T = 1)\mid) - E(Y \mid do(T = 0)\mid) \tag{4}$$

In contrast to focusing solely on the means of distributions, we often analyze the full distributions, such as $P(Y \mid do(t)\mid)$. These distributions, termed interventional distributions, are conceptually distinct from observational distributions (for example, $P(Y)$ or $P(Y, T, X)$), which do not incorporate the do-operator. Observational distributions can be derived from data without conducting experiments, hence the term "observational data." For example, in marketing, suppose a company launches a new ad campaign, representing an intervention in terms of the do-operator $do(Ad\ Campaign)$. The interventional distribution, $P\big(Sales \mid do(Ad\ Campaign)\mid\big)$, shows the probability of different sales outcomes directly resulting from this campaign. Alternatively, an observational distribution such as $P(Sales, Ad\ Campaign)$ merely reflects the correlation between sales and the presence of ad campaigns in historical data without any active intervention. *The key difference is that the interventional distribution gives causal insights into the effect of the ad campaign on sales, unlike the observational distribution, which only shows correlations.*

Identifiability is a crucial concept in causal inference, where an interventional expression (with the do-operator) is considered identifiable if it can be converted into an observational expression (without the do-operator). This highlights the significant difference between interventional and observational expressions, despite their similar appearance. The do-operator indicates a shift to a post-intervention scenario; for instance, $E[Y, do(t), Z = z]$ represents the expected outcome after an intervention in a specific subpopulation. Understanding the do-operator's role is essential for advanced topics in causal inference, such as counterfactuals.

Assumptions in GCMs

Understanding how one variable, X, affects another, Y, is crucial in GCMs. In these models, edges represent causality, but statistical independence can exist where edges don't imply causality, as in Bayesian networks. This integrates counterfactuals and interventions: counterfactuals explore "what-if" scenarios, while interventions involve actual changes. Together, they allow researchers to compare hypothetical and actual outcomes, revealing causal relationships. A comprehensive exploration of GCMs requires a discussion of foundational assumptions to understand both empirical reality and theoretical models:

- The **strict causal edges assumption** in directed graphs states that each parent node is a direct cause of its child nodes, implying that a child node responds only to its immediate parent nodes. This assumption, which mirrors the principles of minimality in DAGs, asserts that if the direct causes of a node remain constant, other connected variables will not influence it, emphasizing a dependency between parent nodes and their children.

- The **non-strict causal edges assumption** in causal graphs permits the existence of parent nodes that are not direct causes of their children, only ensuring that children are not causes of their parents. This approach, similar to Bayesian networks, allows for graphs with more edges, implying fewer independence assumptions and offering a more flexible representation. Generally, causal graphs imply that parent nodes are causes of their children, and this text will typically refer to a causal graph as a DAG adhering to the strict causal edges assumption, often without explicitly using the term "strict."

- The **modularity assumption** posits that the mechanism linking causes and effects is stable and can be understood in isolation. It suggests that changing one part of the model (such as an intervention in a particular variable) doesn't alter the fundamental relationships in other parts. Consider a model predicting crop yield based on various factors such as rainfall, soil quality, and fertilizer use. The modularity assumption implies that if we change the fertilizer type, it doesn't affect the relationship between rainfall and crop yield.

  Mathematically, we can say if we intervene on a set of nodes $S \subseteq [n], 1$, setting them to constants, then for all i, we have the following:

  - If $i \notin S$, then $P(x_i \mid pa_i)$ remains unchanged.

  - If $i \in S$ then $P(x_i \mid pa_i) = 1$ if x_i is the value that x_i was set to by the intervention; otherwise, $P(x_i \mid pa_i) = 0$.

- The **Markov assumption** is fundamental in probabilistic graphical models. It states that a variable is independent of its non-descendants in a graph, given its parents. This simplifies the understanding of dependencies within the model.

  - For example, in a simple graph where smoking causes lung disease, which in turn causes increased healthcare costs, the Markov assumption implies that, given knowledge about lung disease, healthcare costs are independent of smoking.

  - In other words, once the effect (lung disease) is known, the cause (smoking) provides no additional information about the healthcare costs.

- **Causal sufficiency** refers to the assumption that all common causes of the variables under study are included in the model. In other words, there are no hidden or unobserved confounders that could influence the relationships being analyzed. In real-world scenarios, ensuring causal sufficiency can be challenging. Often, it's difficult to identify and measure all potential confounding variables.

In the next section, let's build a graph model in R.

Case study example of a graph model in R

Let's learn to build a GCM in R using a toy `social_media_data` dataset, which you can download from the Git repository. It contains simulated social media user interactions and demographics. Here are the columns in the data:

1. `userA`: A unique identifier for a social media user (presumably the influencer or content creator).

2. `userB`: A unique identifier for another user (a follower or another influencer).

3. `num_post_likes`: The number of likes a post has received.

4. `num_post_commented`: The number of comments a post has received.

5. `interests`: The categories of shared interests that `userB` and `userA` might have, such as art, design, technology, and so on.

6. `follows`: A Boolean indicating whether `userB` follows `userA`.

7. `hours_active_perday`: The number of hours per day `userB` is active on social media.

8. `probability_to_like`: The likelihood that `userB` will like a post, expressed as a probability.

9. `probability_to_comment`: The likelihood that `userB` will comment on a post, expressed as a probability.

10. `userA_gender`: The gender of `userA`.

11. `userB_gender`: The gender of `userB`.

12. `geolocation_same`: A Boolean indicating whether `userA` and `userB` are in the same geographical location.

13. `userA_numfollowers`: The number of followers `userA` has.

14. `userB_numfollowers`: The number of followers `userB` has.

15. `userA_numposts`: The number of posts `userA` has made.

16. `userB_numposts`: The number of posts `userB` has made.

Problem to solve using graphs

We're tackling the question of how social media dynamics shape user engagement – that includes likes and comments. We're curious about how user attributes, such as follower count and activity, as well as their interests and location, play into this. Causal graphs let us visually plot and ponder over these potential cause-and-effect scenarios. They're great for exploring ideas such as whether a user's popularity directly boosts engagement and if this link is influenced by their activity level or the shared interests of their followers.

What's encouraging about using causal graphs is that they go beyond guesswork. We can actually run statistical tests to see if our theories hold water. This is extremely useful for social media strategists trying to figure out what really drives engagement, helping them to fine-tune their content and interactions for a stronger online impact.

Implementing in R

To build a compelling causal graph with the given dataset columns, let's define our model considering the potential causal relationships among these variables. Given the nature of social media interactions, we can hypothesize how certain attributes might influence others. For instance, the number of posts a user has (`userA_numposts`), their number of followers (`userA_numfollowers`), and how similar their interests are to their followers (`interests`) could affect the engagement on their posts (measured by `num_post_likes` and `num_post_commented`).

Here's an example of a more refined causal model:

- **User's popularity** (`userA_numfollowers`): Influences the number of likes and comments.
- **User's activity** (`userA_numposts`): More posts might lead to more visibility and engagement.
- **Similar interests** (`interests`): If a user shares similar interests with their followers, it might increase engagement.
- **Hours active per day** (`hours_active_perday`): More active users might engage more with posts.

Let's define this causal graph and perform local tests using the `dagitty` [3] and `lavaan` packages in R. Here is the code:

```
library(dagitty)
library(lavaan)
# load the data set
setwd(dirname(rstudioapi::getSourceEditorContext()$path))
social_media_data = read.csv("./data/social_media_data.csv")

# Define the causal graph
graph <- dagitty("dag {
  userA_numfollowers -> num_post_likes
  userA_numfollowers -> num_post_commented
  userA_numposts -> num_post_likes
  userA_numposts -> num_post_commented
  interests -> num_post_likes
  interests -> num_post_commented
  hours_active_perday -> num_post_likes
  hours_active_perday -> num_post_commented
}")

plot(graph, cex = 4.5)

# Prepare the data for analysis
vars_in_model <- c(
```

```
   "userA_numfollowers", "userA_numposts", "interests",
   "hours_active_perday", "num_post_likes", "num_post_commented")
corr <- lavCor(social_media_data[vars_in_model])

# Perform local tests using the corrected covariance matrix
local_test_results <- localTests(
  graph, sample.cov = corr,
  sample.nobs = nrow(social_media_data))
plotLocalTestResults(local_test_results)
```

Here is a plot of the graph:

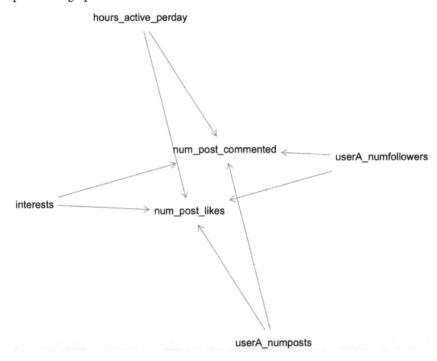

Figure 4.4 – The relationship between various variables as modeled in the daggity package

In this model, we're analyzing how a user's popularity, activity, shared interests, and active hours might causally affect the engagement on their posts. The localTests function will help us understand the validity of these hypothesized relationships by testing if the observed data fits the model. The plotLocalTestResults function will visualize the results of these tests, highlighting potential areas where the model may not fit well with the observed data.

Remember – the actual analysis and interpretation require careful consideration of the data, the model, and the statistical results. This example provides a basic framework for such an analysis.

Interpreting results

The lavaan package is designed to specify, estimate, and evaluate models that fall into the family of **structural equation models (SEMs)** [7]. These are advanced statistical techniques that enable the analysis of complex relationships among observed and latent variables. Once you have specified an SEM and estimated it using lavaan, the localTests function can be used to examine specific parts of the model to check if the observed data fits the model's predictions at a local level. It provides detailed information about each parameter in the model, such as factor loadings, regression coefficients, and variances.

The localTests function gives you a breakdown of chi-square tests for each parameter, which can help identify parts of the model that may not be fitting the data well. If the p-value for a specific test is less than the conventional threshold (for example, 0.05), it suggests that the observed data is significantly different from what the model would predict, indicating a potential issue with that part of the model:

```
                   usrA_nmf usrA_nmp intrst hrs_c_ nm_pst_l nm_pst_c
userA_numfollowers    1.000
userA_numposts       -0.160    1.000
interests             0.015   -0.026  1.000
hours_active_perday   0.036    0.035 -0.070  1.000
num_post_likes        0.073   -0.028  0.036  0.105    1.000
num_post_commented    0.002    0.019 -0.019  0.107   -0.049    1.000
```

Figure 4.5 – Correlation matrix

The preceding correlation matrix shows the output from the lavaan package, indicating the coefficients of each variable with each other:

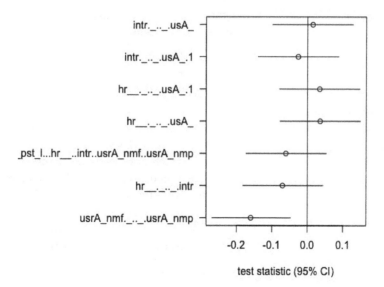

Figure 4.6 – A plot of the output (local_test_results)

Let's describe the preceding plot, showing the intercepts and coefficients. In the following list and also in *Figure 4.6*, the titles/*y*-axis labels show the intercept and coefficients between various variables in the data. Here's a detailed interpretation:

- `hr__.__._.intr`: This suggests a negative relationship (-0.06987057) between the `hours_active` variable and the intercept. The p-value is 0.227781701, which is greater than the common significance level of 0.05, indicating that the relationship is not statistically significant.

- `nm_pst_c.__.__.nm_pst_l...hr__..intr..usrA_nmf..usrA_nmp`: The negative estimate (-0.06042515) suggests a negative relationship between a complex interaction of variables involving the number of posts commented, `userA`'s attributes, and the number of followers. As we can see, the p-value, 0.300401127, is not significant.

- `usrA_nmf.__.__.usrA_nmp`: In this analysis, a correlation coefficient of -0.15986833 suggests that as the number of posts by a user (`usrA`) increases, their number of followers tends to decrease. This is an inverse relationship. The statistical significance of this finding is supported by a p-value of 0.005453336, which is less than the commonly used threshold of 0.05.

This result is somewhat unexpected; we might initially think that posting more would attract more followers. However, this data implies that the quality or relevance of posts is likely more important than quantity. It's possible that increasing the number of posts could lead to a decline in their overall quality or divergence from topics that interest the followers, resulting in a loss of followers.

The precise interpretation of these variables and their relationships would require additional context regarding the problem at hand and the specific model being used.

Summary

In causality, graphical models make it way easier to get your head around complex causal relationships, turning them into diagrams that are much simpler to understand. This is super helpful when you're trying to spot things that could skew your results, such as confounding variables, or when you're figuring out how to measure cause and effect.

These models aren't just about making things look neat. They give us a robust approach to understanding the assumptions behind our causal models and to discern when we can actually say one thing causes another. Judea Pearl's work, especially his focus on DAGs in his book *Causality: Models, Reasoning, and Inference*, [1] has really set the stage here. By applying the basics of graph theory to causality, graphical models have totally transformed how we approach and analyze what causes what. Overall, the key takeaways from this chapter are the introduction of graphical models to understand and analyze causal relationships through DAGs, essential for distinguishing causation from correlation. The chapter covered practical techniques for representing and analyzing these models in R, employing tools such as `igraph`. Through examples, it emphasized graphical models' role in exploring and validating causal hypotheses, particularly in complex data scenarios such as social media analytics.

In the next chapter, we will delve deeper into DAGs.

References

1. Pearl, J. (2000). *Causality: Models, Reasoning, and Inference.* Cambridge University Press.

2. Scutari, M., & Denis, J.B. (2014). *Bayesian Networks: With Examples in R.* Chapman and Hall/CRC.

3. Textor, J., van der Zander, B., Gilthorpe, M. S., Liśkiewicz, M., & Ellison, G. T. H. (2016). *Robust causal inference using directed acyclic graphs: the R package 'dagitty'.* International Journal of Epidemiology, 45(6), 1887–1894.

4. Pearl, J. (2009). *Causality: Models, Reasoning, and Inference.* Cambridge University Press.

5. Morgan, S. L., & Winship, C. (2015). *Counterfactuals and Causal Inference.* Cambridge University Press.

6. Imbens, G. W., & Rubin, D. B. (2015). *Causal Inference for Statistics, Social, and Biomedical Sciences.* Cambridge University Press.

7. lavaan package in R example: https://stats.oarc.ucla.edu/r/seminars/rsem/

Navigating Causal Inference through Directed Acyclic Graphs

In this chapter, we will comprehensively explore graphical representations and their pivotal role in elucidating causal relationships. Let's begin with an in-depth analysis of graph structures, where we will meticulously examine how various graph typologies imbue distinct interpretations of association and causation. Initially, our focus will be to discover the causal flow within these graphs, thereby establishing a robust foundation in the fundamental principles of this domain. This includes, but is not limited to, a thorough understanding of key concepts such as forks, chains, colliders, and immoralities.

Subsequently, we will investigate other critical mechanisms integral to better understanding causality – notably, the concepts of back door and front door adjustments. We will discuss concepts through a practical application scenario, wherein we shall leverage graph representation to model causality effectively. Utilizing the programming language R, we will actively apply the core concepts acquired in this chapter to address and resolve a specific problem.

Topics that we will cover include the following:

- Flow in graphs (i.e., chains, forks, and colliders)
- Adjusting confounding in graphs (through D-separation, the do-operator, back door, and front door)
- Practical R example (back door versus front door concepts)

Technical requirements

You can find the code examples for this chapter in this book's GitHub repository: `https://github.com/PacktPublishing/Causal-Inference-in-R/tree/main/chap_05`.

Understanding the flow in Graphs

In this section, let's dig into **Directed Acyclic Graphs (DAGs)**, which are meant to represent pathways of association and causation. The crux of this exploration is the "flow of association" within DAGs, essential for understanding node relationships, particularly their statistical dependence or independence. In this flow of association, an analysis of basic DAG structures – chains, forks, and colliders – sheds light on the (conditional) independence or dependence of node pairs, with conditional independence exemplified by the factorization of joint probabilities into conditional probabilities.

Let's contrast simple graph structures, where disconnected nodes (see node **A** and node **B** in *Figure 5.1a*) indicate statistical independence, we also examine connected nodes (node **B** and node **C** in *Figure 5.1a*), where an edge signifies an association based on the causal edge assumption (e.g., **C** is associated with **B**). This foundational principle, that suggesting a causal link implies an association, sets the stage for further exploration into more complex structures, such as three-node chain graphs, deepening our understanding of association flows and causal interpretations in DAGs.

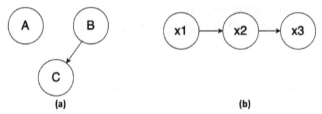

Figure 5.1 – (a) shows connected and disconnected nodes, while (b) shows a chain structure in DAGs

Chains and forks

To explain the concepts of chains and forks in DAGs, let's consider an illustrative example in a grocery store. To be precise, let's consider two scenarios that demonstrate these structures:

Chain structure

In a grocery store, the chain structure can be observed through the sequence of customer demand influencing stock levels, which in turn affects the store's decisions regarding sales promotions, something like this:

Customer demand (x_1) → stock level (x_2) → sales promotion (x_3)

In this chain (*Figure 5.1b*), the customer demand (the first node) is associated with the stock level (the second node) because higher demand typically leads to lower stock. The stock level is associated with sales promotions (the third node); this is relatable, as stores often use promotions to move products that are overstocked or understocked. The demand directly affects the stock level, and the stock level directly affects the likelihood of sales promotions. However, customer demand does not directly affect sales promotions without the mediating variable (we will discuss mediators later in this book, in *Chapter 11*), which is stock level. *Understanding this chain helps us to make decisions. For example, predicting customer demand can guide stock management, which in turn can inform promotion strategies.*

Mathematically, let's say we have nodes x_1, x_2, and x_3. Here, association flows along the chain – for example, x_1 is associated with x_3 through x_2. We can say that x_1 is conditionally independent of x_3, given x_2, represented as $x_1 \perp x_3 \mid x_2$. This is shown by factorizing the joint probability $P(x_1, x_2, x_3)$ as $P(x_1)P(x_2 \mid x_1)P(x_3 \mid x_2)$. Then, using Bayes' rule, $P(x_1, x_3 \mid x_2)$ can be broken down into $P(x_1 \mid x_2)P(x_3 \mid x_2)$.

This factorization represents the chain structure:

- $P(x_1)$ is the probability of customer demand.

- $P(x_2 \mid x_1)$ is the probability of stock level given customer demand.

- $P(x_3 \mid x_2)$ is the probability of sales promotions given the stock level.

Fork structure

Consider that a marketing campaign (x_1) is devised, which is the common cause that independently affects both the increase in the customer footfall (x_2) and the increase in sales (x_3) at a grocery store. The connections look like Marketing campaign (x_1) → increase in customer footfall (x_2), and increase in sales (x_3) (also refer *Figure 5.2a*).:

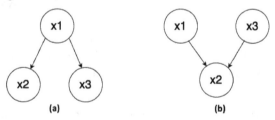

Figure 5.2 – (a) shows a fork structure in DAGs, while (b) shows a collider structure in DAGs

- *Marketing campaign (common cause – x_1)*: This could be a new advertisement or a special promotion. It is designed to attract more customers to the store.

- *Increase in customer footfall (effect 1 – x_2)*: One of the direct effects of a successful marketing campaign is an increase in the number of customers visiting the store.

- *Increase in sales (effect 2 – x_3)*: Alongside the increase in footfall, the campaign is also expected to boost sales – well, at least in an ideal world, as more customers are likely to purchase goods.

Now, how do association and conditional independence play into this scenario? The marketing campaign (x_1) is directly associated with both the increase in customer footfall (x_2) and the increase in sales (x_3). According to the rule of forks in DAGs, x_2 (the customer footfall) and x_3 (sales) are conditionally independent, given x_1 (the marketing campaign). This means if we know the nature and extent of the marketing campaign, then knowing the increase in footfall does not give us additional information about the increase in sales, and vice versa.

Understanding this fork structure is crucial for evaluating the effectiveness of marketing strategies in a grocery store. It illustrates that while both customer footfall and sales are influenced by the marketing campaign, they do not directly influence each other. This separation is vital for store managers to consider when analyzing the impact of their marketing efforts on different aspects of their business. *In a causal graph, this fork structure helps to identify how a single variable can independently cause changes in multiple variables, and it emphasizes the importance of the common cause in understanding the relationship between the affected variables.*

To summarize, a fork consists of a common cause, leading to multiple separate effects. In a fork, x_1 causes both x_2 and x_3, represented as $x_1 \rightarrow x_2$ and $x_1 \rightarrow x_3$.

Association in a fork flows from the common cause to each effect. x_2 is associated with x_3 through x_1. Mathematically, the conditional independence of x_2 and x_3, given x_1, is expressed as $x_2 \perp x_3 | x_1$. This is demonstrated by showing that the joint probability $P(x_2, x_3 | x_1)$ factorizes into $P(x_2 | x_1)P(x_3 | x_1)$.

Advantages of learning about chains and forks

Now, you may wonder why we need to learn chains and forks. There are several cogent arguments that we can make:

- *Understanding causal relationships*: These structures help distinguish between direct and indirect causal relationships. Chains illustrate direct, sequential causality (x_1 causes x_2 and x_2 causes x_3), while forks show how a single factor (x1) can independently cause multiple outcomes (x_2 and x_3). Chains reveal mediators (i.e., intermediate variables), and forks can help identify potential confounders, variables that influence both the cause and the effect, which is crucial in understanding the true causal effect.

- *Informing statistical analysis*: In forks, knowing the common cause can render the effects conditionally independent. This principle is essential for deciding which variables to control in statistical models. For a statistics beginner, "controlling" a variable in statistical analysis means accounting for its influence to isolate the effect of the main variable you're interested in studying. You can do this in, say, regression models by including the confounding variables as additional independent variables. Understanding these structures guides the selection of variables to include in regression models, ensuring that essential mediators are included (in chains) and confounders are controlled for (in forks). Furthermore, they provide a framework to test causal hypotheses and formulate new hypotheses. For instance, disrupting a link in a chain can test whether the subsequent outcomes are affected, validating the causal pathway.

- *Guiding policy and decision-making*: In chains, interventions can be targeted at various points to influence the final outcome. In forks, understanding the common cause can lead to more effective strategies that address multiple outcomes simultaneously. Understanding causal structures helps predict the ripple effects of policy changes or interventions, minimizing unintended consequences.

If that is clear, then we are perfectly poised to learn more concepts on graph structure that will help us to adjust to confounding in DAGs. Let's continue learning.

Colliders

In a collider structure, two variables independently cause a third variable – for example, $x_1 \rightarrow x_2 \leftarrow x_3$ (*Figure 5.2b*). Initially, the x_1 and x_3 variables are unconditionally independent, meaning they have no direct association with each other. This can be represented as $x_1 \perp x_3$. However, when we control for (or condition on) x_2, x_1, and x_3 become dependent. This is represented as $x_1 \not\perp x_3 \mid x_2$. The "slashed/barred $\perp$" symbol indicates dependence. Not clear? Let's consider an example involving customer satisfaction (x_2), influenced independently by store cleanliness (x_1) and product quality (x_3). The structure is store cleanliness (x_1) → customer satisfaction (x_2) ← product quality (x_3).

Store cleanliness and product quality are initially unconditionally independent – that is, knowing the level of one does not inform us about the other. However, when we consider (or control for) customer satisfaction, store cleanliness and product quality become conditionally dependent. For example, if we know that customer satisfaction is low, and we also know that the store is clean (x_1), we might infer that there's likely an issue with product quality (x_3), and vice versa.

Colliders allow for clear edge orientation in a graph. When we identify a collider structure, it unambiguously indicates the direction of causality toward the collider node. Colliders can also help orient ambiguous edges in chains and forks, providing more clarity in understanding causal relationships. Recognizing colliders in practical settings such as grocery stores reveals how independent factors influence outcomes and inform decision-making.

The mathematical properties of a collider are distinct from those of chains and forks.

Consider three variables, x_1, x_2, and x_3, where $x_1 \rightarrow x_2 \leftarrow x_3$ forms a collider at x_2. x_1 and x_3 are the parents of x_2 but have no edge connecting them. Initially, x_1 and x_3 are unconditionally independent, denoted as $x_1 \perp x_3$. This indicates that x_1 and x_3 do not influence each other directly. Using the Bayesian network factorization and marginalization of x_2, we have the following:

$$P(x_1, x_3) = \sum_{\{x_2\}} P(x_1) P(x_3) P(x_2 \mid x_1, x_3) \qquad (1)$$

Here, we can see that the joint probability of x_1 and x_3 can be decomposed into their own probabilities, multiplied by the conditional probability of x_2, given x_1 and x_3. The path $x_1 \rightarrow x_2 \leftarrow x_3$ is blocked at x_2 (it's a collider), which means that without conditioning on x_2 (or its descendants), there is no association flowing between x_1 and x_3.

Interestingly, when we condition on the collider x_2 (or its descendants), x_1 and x_3 become conditionally dependent. The conditioning on x_2 opens up a path between x_1 and x_3 because it indicates some information about the collider, which in turn creates an association between its causes (x_1 and x_3).

Deep dive into colliders and association

To explain the concept of a collider and its implications on perceived associations using a realistic scenario of holiday purchases within a family context, we will create a synthetic dataset and analyze it with R. The scenario involves three variables – the income level (x_1), the amount spent on holiday gifts (x_2), and frugality (x_3), with the amount spent on holiday gifts acting as a collider influenced by both income level and frugality.

Let's generate synthetic data in R that reflects this scenario. We'll create data where the income level and frugality are initially independent, and both influence the amount spent on holiday gifts. The synthetic data generation strategically manipulates these variables to mimic real-world dynamics, where higher income does not necessarily correlate with higher frugality but can lead to similar spending patterns, due to the collider effect:

```
set.seed(100) # For reproducibility
n <- 3000 # Number of families
# Generating independent variables
Income_Level <- rnorm(n, mean=70, sd=15) # X1
Frugality <- rnorm(n, mean=50, sd=10) # X3
# Generating the collider variable influenced by both X1 and X3
Amount_Spent <- 0.5*Income_Level - 0.3*Frugality +
  rnorm(n, mean=50, sd=20) # X2
# Combine into a data frame
data <- data.frame(Income_Level, Frugality, Amount_Spent)
# Without conditioning on the collider
cor.test(data$Income_Level,
         data$Frugality) # will give a number between 0 and 1
```

Initially, we expect no significant correlation between `Income_Level` and `Frugality`, since they were generated independently. The correlation coefficient (from the `cor.test` function), ranging from -1 to +1, quantifies the strength and direction of a linear relationship between two variables – +1 denotes a perfect positive relationship, -1 signifies a perfect negative one, and 0 signifies no linear relationship. Coefficients near +1 or -1 indicate a strong linear relationship, while those close to 0 suggest a weaker one. By focusing on families that spend a moderate amount on holiday gifts, we introduce a selection bias, conditioning on the collider:

```
# Conditioning on the collider (Amount_Spent)
library(dplyr)
moderate_spenders <- data %>%
  filter(Amount_Spent > quantile(Amount_Spent, 0.2) &
```

```
                Amount_Spent < quantile(Amount_Spent, 0.6))
cor.test(moderate_spenders$Income_Level,
         moderate_spenders$Frugality)
```

This is expected to reveal a negative association between `Income_Level` and `Frugality` within this subgroup, illustrating **Berkson's paradox**, where conditioning on a collider leads to observing an association that does not exist in the general population. By focusing only on families with a certain level of holiday spending, we've introduced a selection bias, which skews our perception of the relationship between income and frugality.

The `cor.test` *function calculates the Pearson correlation coefficient between* `Income_Level` *and* `Frugality`, *first for the entire dataset (where we expect no correlation) and then for the subgroup of moderate spenders (where we expect to find a negative correlation due to conditioning on the collider).*

The first `cor.test` should show no significant association between income and frugality, confirming their independence in the initial setup. The second test, however, is expected to reveal a negative correlation, indicating that among families with moderate spending, higher income is associated with lower frugality and vice versa. This outcome exemplifies how conditioning on a collider (the amount spent) can induce an apparent association between two otherwise independent variables, underscoring the importance of understanding causal structures in data analysis to avoid misleading inferences. Granted, as this is synthetic data, we may not see a strong correlation, but nevertheless, it proves our point of how to condition on variables by first identifying colliders. Feel free to use the same logic in your own use cases to see the difference.

This example serves as a potent demonstration of the **collider concept** and Berkson's paradox, highlighting the critical need for careful consideration of the underlying causal framework when interpreting associations in observational data. The scenario underscores the potential for selection bias that may distort perceived relationships. Next, let's discuss immorality in this context.

Immorality

The concepts of colliders and immorality in causal graphs are closely related but distinct. An immorality in a causal graph is a particular configuration involving a collider. It occurs when a child node (the collider) has two parents that are not connected by an edge (*Figure 5.2b*). In a DAG, immorality is found where two nodes, x_1 and x_3, both point to a third node, x_2 (forming a V-shape), but there is no direct edge between x_1 and x_3. *This structure is essentially a collider, with the additional specification that the parents of the collider are not connected.* This means even if the parents had an edge connecting them, we could call it a collider configuration, but not immorality.

The term **immorality** highlights the specific configuration where two causes independently contribute to a common effect without any direct causal link between the causes themselves. It's significant in graphical models because conditioning on the collider (or its descendants) in an immorality structure induces an association between the otherwise independent causes.

Every immorality involves a collider, but not all colliders constitute an immorality. A collider becomes an immorality only when its parent nodes are not connected. In other words, immorality is a specific type of collider with an additional structural constraint. Recognizing immoralities in a graph helps us to understand complex causal structures, especially when deciphering the flow of causal influence and determining appropriate conditioning sets for causal estimation.

Let's clarify these concepts and, first, revisit the holiday purchases scenario. To understand a structure of immorality, let's introduce a new variable, gift purchase satisfaction (x_4), as a common descendant influenced by both income level (x_1) and Frugality (x_3) but without a direct link between x_1 and x_3. The structure now looks like this:

Income Level $(x_1) \rightarrow$ *Gift Purchase Satisfaction* (x_4)

Frugality $(x_3) \rightarrow$ *Gift Purchase Satisfaction* (x_4)

```
# Generating a new variable x4 influenced by both x1 and x3
Gift_Purchase_Satisfaction <- 0.4*Income_Level + 0.2*Frugality +
  rnorm(n, mean=60, sd=15)
data$Gift_Purchase_Satisfaction <- Gift_Purchase_Satisfaction
# Checking correlation between X1 and X3 without conditioning on X4
cor.test(data$Income_Level, data$Frugality)
```

Initially, we would not expect a significant correlation between `Income_Level` and `Frugality`.

Let's check out the code:

```
# Checking correlation between x1 and x3 with conditioning on x4
satisfied_customers <- data %>%
  filter(Gift_Purchase_Satisfaction > quantile(
    Gift_Purchase_Satisfaction, 0.5))
cor.test(satisfied_customers$Income_Level,
         satisfied_customers$Frugality)
```

Conditioning on `Gift_Purchase_Satisfaction` (x_4) might not induce the same association as seen in the collider scenario because x_4 is a common descendant (not a collider in the traditional sense), depending on the specific dynamics of causation and association in the model. However, if x_4 acted as a proxy, revealing information about the relationship between x_1 and x_3, it might still influence our perception of their association.

The primary distinction lies in the configuration and its implications:

- **Collider** (x_2): Conditioning on the collider introduces an association between its causes due to selection bias, exemplified by Berkson's paradox.

- **Structure of Immorality** (x_4): The structure might not inherently induce an association between its causes upon conditioning. However, if x_4 serves to reveal or mask underlying causal pathways, it could still affect perceived associations, requiring careful consideration in causal analysis.

This concludes our learning on flow in graphs. With this understanding, let's proceed to discuss how we can adjust for confounding in graphs.

Adjusting for confounding in graphs

This section focuses on the sophisticated mechanisms of association and causation within DAGs, employing advanced tools such as **d-separation**, **do-operator**, and **front and back door adjustments**. These approaches are vital for recognizing and mitigating confounding factors. They enable the extraction of actionable insights, and supporting informed decision-making in areas where understanding causal structures is essential. In contexts where controlled experiments are impractical or unethical, these principles provide a solid foundation to infer causal relationships from observational data. We will progressively explore these concepts, focusing on their theoretical foundations and real-world utility.

D-separation

D-separation stands as a cornerstone concept in causal inference and graphical models. It is a structured method to ascertain conditional independencies within a DAG. The essence of understanding d-separation lies in the identification of **blocked paths** within a graph. *A path between any two nodes, X and Y, is considered blocked by a set of nodes, Z, if it encounters either a chain or a fork where an intermediary node is conditioned upon (included in Z), or if it traverses a collider that, along with its descendants, is not conditioned on.* This mechanism of blocking or unblocking paths allows us to determine whether associations between variables can be attributed to causal pathways or are merely spurious correlations, obscured by other variables.

Expanding on this foundation, d-separation provides a rigorous framework to delineate when sets of variables are conditionally independent, given a third set. Formally, if all paths between sets X and Y are blocked by Z, X, and Y are said to be d-separated by Z, denoted as $X \perp_{G} Y \mid Z$ in graph G, implying their independence in the associated probability distribution, P, given Z $(X \perp_{P} Y)$. This conceptual framework is based on the Global Markov assumption, which connects the structure of DAGs with probabilistic independence. It asserts that d-separation in a graph implies conditional independence in the empirical distribution it represents.

Let's address a real-world causal inference problem using the concept of d-separation through an R example, focusing on the scenario of buying toys for kids. In this example, we will consider three variables – the parents' income (X), the child's academic performance (Y), and the amount spent on toys (Z). The causal assumption here is that both the parent's income (X) and the child's academic performance (Y) influence the amount spent on toys (Z), but there's no direct causal link between the parent's income and the child's academic performance. We aim to understand whether the parent's income influences the amount spent on toys, controlling for the effect on the child's academic performance:

- **Parent's income** (X): Affects the amount spent on toys.
- **Child's academic performance** (Y): It is assumed that higher academic performance might lead to parents buying more toys as rewards.
- **Amount spent on toys** (Z): Influenced by both X and Y.

First, we'll generate synthetic data that follows our assumptions:

```
set.seed(42) # Ensure reproducibility
n <- 3000 # Number of observations
# Generate synthetic data
Parents_Income <- rnorm(
  n, mean=50000, sd=10000) # Parent's Income (X)
Childs_Academic_Performance <- 0.5 * Parents_Income + rnorm(
  n, mean=0, sd=5000) # Child's Performance (Y)
# Amount Spent on Toys (Z) depends only on Child's Academic
Performance
Amount_Spent_on_Toys <- 200 + 0.1 *
  Childs_Academic_Performance + rnorm(n, mean=0, sd=500)
data <- data.frame(Parents_Income,
                   Childs_Academic_Performance,
                   Amount_Spent_on_Toys)
```

We aim to analyze whether Parents_Income directly influences Amount_Spent_on_Toys when we control for Childs_Academic_Performance.

Through the following code in R, let's check for d-separation:

```
# Install necessary packages if not already installed
if (!requireNamespace("ggm", quietly = TRUE)) install.packages("ggm")
if (!requireNamespace("dplyr", quietly = TRUE)
  ) install.packages("dplyr")

# Load the required packages
library(ggm)
library(dplyr)

# Specify the DAG structure
dag <- matrix(c(
  0, 1, 0, # Row 1:Parent's Income to Child's Performance
  0, 0, 1, # Row 2:Child's Performance to Amount Spent
  0, 0, 0),# Row 3:Amount Spent does not cause any of the others
  nrow = 3, byrow = TRUE,
  dimnames = list(c("X", "Y", "Z"), c("X", "Y", "Z")))

# Check d-separation between X and Z, conditioned on Y
dsep_result <- dSep(dag, first = "X", second = "Z", cond = "Y")
print(paste("D-separation result (X and Z given Y):", dsep_result))
```

This R code snippet does not directly implement d-separation but simulates the scenario, preparing the setup for a d-separation analysis. The dSep function from the ggm package is intended to be illustrative. In practice, you'd use the structure of the DAG and your knowledge of d-separation rules to determine whether conditioning on Y (the child's academic performance) blocks the path from X (the parent's income) to Z (the amount spent on toys).

If the dSep function returns TRUE (and it does), it means that X and Z are d-separated, given Y, indicating that any association between the parent's income and the amount spent on toys can be explained by the child's academic performance. *This result would suggest that, under the assumptions of our DAG, controlling for the child's academic performance is sufficient to block any spurious paths and isolate the direct effect of the parent's income on the amount spent on toys.*

```
# Calculate correlation between X and Z (unconditional)
cor_xz <- cor(data$Parents_Income, data$Amount_Spent_on_Toys)
print(paste("Unconditional correlation between X and Z:",
            round(cor_xz, 4)))

# Function to calculate partial correlation
partial_cor <- function(x, y, z) {
  res_xz <- residuals(lm(x ~ z))
  res_yz <- residuals(lm(y ~ z))
  return(cor(res_xz, res_yz))
}

# Calculate partial correlation between X and Z, given Y
partial_cor_xz_y <- partial_cor(
  data$Parents_Income, data$Amount_Spent_on_Toys,
  data$Childs_Academic_Performance)
print(paste("Partial correlation between X and Z, given Y:",
            round(partial_cor_xz_y, 4)))

# Demonstrate d-separation by stratifying on Y
data_stratified <- data %>%
  mutate(Y_strata = cut(Childs_Academic_Performance, breaks = 3)) %>%
  group_by(Y_strata) %>%
  summarise(cor_XZ = cor(Parents_Income, Amount_Spent_on_Toys))

print("Correlations between X and Z within strata of Y:")
print(data_stratified)

# Test for conditional independence
lm_model <- lm(Amount_Spent_on_Toys ~ Parents_Income +
                 Childs_Academic_Performance, data = data)
summary_model <- summary(lm_model)
print("Regression results:")
print(summary_model$coefficients)
```

The code demonstrates the concept of d-separation through various methods, including calculating unconditional and partial correlations, stratification, and regression analysis. These methods show that `Parents_Income` (X) and `Amount_Spent_on_Toys` (Z) are conditionally independent given `Childs_Academic_Performance` (Y), meaning X does not need to be controlled for when estimating the causal effect of Y on Z. The regression results align with this, as the coefficient for `Parents_Income` is not statistically significant, indicating no meaningful impact on the amount spent on toys when controlling for academic performance.

In contrast, the coefficient for `Childs_Academic_Performance` is positive and highly significant, showing a strong positive effect on spending. The analysis confirms that `Parents_Income` (X) and `Amount_Spent_on_Toys` (Z) are conditionally independent given `Childs_Academic_Performance` (Y), as indicated by the non-significant regression coefficient for X. This aligns with the d-separation concept, showing X does not directly affect Z when Y is accounted for. Overall, `Childs_Academic_Performance` is the key predictor of spending on toys, which is consistent with the causal structure.

D-separation is fundamental in understanding the underlying structure of causal relationships within a model. It helps identify which variables need to be controlled for (i.e., adjusted for) to estimate causal effects accurately. Once the structure is understood (including which variables are independent or dependent, given others), the concept of the do-operator can be used to simulate interventions and estimate their effects. The independence and dependencies identified by D-separation guide the formulation of causal queries using the do-operator. Without a clear understanding of the causal relationships and dependencies in a model, it's challenging to correctly apply the do-operator.

Now, what is a do-operator? We narrowly touched upon it in the last chapter. Let's dig deeper!

Do-operator

The do-operator embodies a pivotal concept in causal inference. It provides the mathematical framework to represent interventions within a population. This distinction between observation and intervention is fundamental to understanding the causal mechanisms that underlie empirical observations. Traditional probability conditioning, exemplified by $P(Y \mid T = t)$, merely isolates a subset of the population already subjected to treatment t, offering insight into observational associations rather than causal relationships. *In contrast, the do-operator, denoted as $do(T = t)$, transcends this observational stance, signifying an active manipulation where treatment t is uniformly applied across the population.* This shift from passive observation to deliberate intervention underscores the essence of causal inference, allowing researchers to model the outcomes of hypothetical interventions. It depicts the causal impact of variables on outcomes beyond mere associations derived from observational data.

Further, the introduction of the do-operator facilitates the distinction between observational and interventional distributions, the latter being central to the concept of identifiability and the definition of **estimands** in causal analysis. Estimands in causal analysis are the target causal quantities we aim to estimate from data. They are often expressed using the do-operator to represent interventional effects rather than mere observational associations.

Observational distributions, such as $P(Y)$, emerge from data gathered without intervention, reflecting the natural occurrence of events. Interventional distributions, however, represented by $P(Y \mid do(t))$, envision outcomes under hypothetical scenarios where interventions are applied. This conceptual framework enables the translation of interventional queries into estimable quantities within observational data, provided the conditions for identifiability are met. Hence, the do-operator not only conceptualizes a post-intervention world but also lays the groundwork for quantifying causal effects, distinguishing causal estimands (which incorporate the do-operator) from mere statistical estimands. Through this mechanism, causal inference overcomes the limitations of observational studies, offering a rigorous methodology to infer causality from data.

D-separation and the do-operator each have distinct applications and objectives that complement each other. D-separation is crucial for mapping out relationships and conditional independencies within a model, enabling you to discern whether paths between variables are open or closed. This process essential for identifying valid sets of confounders for causal adjustment. Conversely, the do-operator is practical and operational, designed for simulating interventions and quantifying their effects, thus allowing for the empirical estimation of causal relationships by manipulating exposure variables. *While d-separation is instrumental in the initial stages of model comprehension and setup, assisting in determining the suitability of variables for back door adjustments, the do-operator comes into play during the estimation phase, applying to methodologies such as Pearl's adjustment formula, front door criterion, and instrumental variable analysis to compute causal effects.* Together, these tools encapsulate the dual aspects of causal analysis – understanding the theoretical underpinnings of causal mechanisms and applying this understanding to estimate causal effects in real-world scenarios. With that, let's move on to the topic of back door adjustment to model causality using graphs.

The back door adjustment

This method is a vital technique in causal inference. *It estimates the causal effect of one variable X, on another, Y, by accounting for confounders.* The essence of back door adjustment lies in its ability to distinguish genuine causal effects from those correlations marred by confounding variables. The process begins with the identification of a set of variables, Z, that fulfills the back door criterion for the variable pair, (X, Y). This means Z must block all back door paths from X to Y – paths emanating with an arrow into X – without introducing new pathways for confounding. As such, there are two critical assumptions – (1) the absence of unmeasured confounders affecting both X and Y, and (2) the precise specification of the statistical model governing the relationships among the confounders, the treatment, and the outcome.

We will start with the construction of a DAG to visually map out the causal relationships among all pertinent variables. This visual representation aids in the meticulous identification of all back door paths from X to Y. The next step is to pinpoint a set of confounders, Z, that effectively satisfies the back door criterion by obstructing these identified paths, thereby neutralizing the influence of confounding on the causal estimation. The final step employs statistical methods, such as regression analysis, to control for Z in the causal estimation process. This adjustment enables the isolation of the causal effect of X on Y from the confounding influences. This process facilitates a more accurate and meaningful interpretation of the causal relationship between the treatment and the outcome, within the framework of observational data.

Next, we will learn about the front door adjustment.

The front door adjustment

The front door adjustment criterion emerged as a nuanced method within Judea Pearl's causal inference framework. It is designed to elucidate causal relationships through graphical models, especially when direct control over all confounders is impractical. *This technique distinguishes itself by utilizing a mediator variable (M) that bridges the exposure (X) and the outcome (Y), thereby facilitating the estimation of causal effects in the presence of unmeasured confounders between X and Y.*

For the front door criterion to be applicable, the exposure or independent variable must directly affect the mediator, which in turn directly influences the outcome or the dependent variable. This configuration requires that the path from X to Y through M is the only causal path, and there are no unmeasured confounders between X and M or between M and Y.

This setup is particularly advantageous in scenarios where traditional back door adjustments cannot be applied, due to the presence of unmeasured confounding variables that complicate the direct causal pathway from the exposure to the outcome.

The operationalization of the front door adjustment unfolds through a series of methodical steps, beginning with modeling the direct influence of X on M. Subsequently, the effect of M on Y is modeled, this time accounting for X, to ensure that the mediation effect is properly isolated. The overarching causal effect of X on Y is then derived by integrating the mediator's influence across its entire spectrum, effectively summing over the mediator variable, M, for discrete cases or integrating for continuous variables. Mathematically, this process is encapsulated in the formula

$$P(Y \mid do(X)) = \sum_{m} P(Y \mid X, M) \, P(M \mid X) \qquad (2)$$

which quantifies the probability of Y following an intervention on X, summed over all potential states of M. This formulation not only underscores the mediating role of M but also highlights the front door criterion's capacity to render causal estimates from complex interplays of variables. It offers a powerful alternative to traditional adjustment methods in the face of unmeasured confounding.

The front door adjustment method accurately captures the relationship from X to the mediator (M) and from M to Y. The crux of this method's application hinges on the mediator's ability to encapsulate the essential variation driving the causal effect from X to Y. Thus, it necessitates a mediator that is comprehensive enough to embody the relevant causal dynamics. Despite its strengths, the method's efficacy is contingent upon the precise modeling of the relationships at both stages of mediation, underscoring the importance of a mediator that is not influenced by unmeasured variables in its association with both X and Y.

Contrastingly, the back door adjustment serves as a foundational method for addressing confounding when such confounders are known and measurable. However, the front door adjustment emerges as a pivotal alternative under conditions where back door criteria are unattainable, particularly due to the presence of unmeasured confounders. By leveraging a mediator that singularly channels the causal effect from X to Y, the front door adjustment bypasses the limitations posed by unmeasured confounders, offering a pathway to causal estimation even when traditional methods falter.

Next, let's dive deeper into these techniques by an example in R code.

Practical R example – back door versus front door

Let's consider a corporate scenario where we are interested in investigating the causal effect of gender on remuneration. We will consider both back door and front door adjustments using R. In real-world corporate settings, this analysis can be complex due to various factors, such as job role, seniority, education, work experience, and potential unmeasured confounders, such as negotiation skills or networking.

Synthetic data

Here is the R code to create a synthetic dataset; let's name it `corp_data` and add relevant columns such as `gender`, `remuneration`, `promotion_opportunities`, `performance_rating`, `department`, `work_experience`, `education`, and `job_role`:

```
# Load necessary library
library(dplyr)
# Set seed for reproducibility
set.seed(123)
# Generate synthetic data
n <- 5000 # Number of observations

corp_data <- data.frame(
  gender = sample(c("Male", "Female"), n, replace = TRUE),
  # Assuming normal distribution for salary:
  remuneration = rnorm(n, mean = 70000, sd = 15000),
  # Assuming Poisson distribution for number of promotions:
  promotion_opportunities = rpois(n, lambda = 2),
  # Ratings from 1 to 5:
  performance_rating = sample(1:5, n, replace = TRUE),
  department = sample(c("Sales", "IT", "HR", "Finance",
                        "Marketing"), n, replace = TRUE),
  # Assuming normal distribution for years of experience:
  work_experience = rnorm(n, mean = 5, sd = 2),
  education = sample(c("Bachelor", "Master", "PhD"),
```

```
                                n, replace = TRUE),
  job_role = sample(c("Analyst", "Manager", "Senior Manager",
                      "Director"), n, replace = TRUE)
)
# View the first few rows of the dataset
head(corp_data)
```

This code creates a dataset of 5,000 observations, with the associated columns.

Back door adjustment in R

In this corporate scenario, we'll dig into several aspects. First, we will analyze the role of each confounder (job role, seniority, education, work experience, department, and performance rating) in the relationship between gender and remuneration. Then, we will investigate potential interaction effects. For instance, the impact of gender on remuneration might differ across departments or job roles.

As we model this causal problem, we will also include interaction terms in the regression model to capture the nuanced effects of gender across different departments and job roles. We can also consider non-linear effects, such as the potential non-linear relationship between work experience and remuneration. However, let's consider only linear relations in the following example. Let's implement all these aspects in R:

```
# Load necessary library
library(dplyr)
library(ggplot2)
library(lmtest)
library(sandwich)
# Prepare data with interaction terms
corp_data <- corp_data %>%
  mutate(gender_factor = as.factor(gender),
         department_factor = as.factor(department),
         job_role_factor = as.factor(job_role),
         gender_dept_interaction = interaction(
           gender_factor, department_factor),
         gender_jobrole_interaction = interaction(
           gender_factor, job_role_factor))

# Expanded linear model with interaction terms
model <- lm(remuneration ~ gender_factor + job_role_factor +
              department_factor + work_experience +
              performance_rating + education +
              gender_dept_interaction +
              gender_jobrole_interaction,
            data = corp_data)
```

Finally, we want to evaluate the model's fit using diagnostic plots and statistics. Specifically, we will check for potential issues such as multicollinearity, especially given the added complexity of interaction terms:

```
# Model summary
summary(model)
# Diagnostic plots
par(mfrow=c(2,2))
plot(model)
# Check for multicollinearity
vif(model)
```

Let's move on to perform sensitivity analysis (also discussed in detail in *Chapter 12*) to assess how robust the findings are to different model specifications. In this, we specifically examine the impact of excluding certain confounders or including additional variables:

```
# Sensitivity analysis: Exclude certain confounders
model_sensitivity <- lm(remuneration ~ gender_factor +
                        department_factor + work_experience +
                        education, data = corp_data)
summary(model_sensitivity)
# Compare models
anova(model, model_sensitivity)
```

In the preceding code, if you observe high values (typically above 5 or 10) from the `vif(model)` function, it indicates potential multicollinearity in the model. Extremely high VIF values or computational issues may suggest the presence of perfect multicollinearity or aliased coefficients. To address aliased coefficients in your model, scrutinize your model specification for redundancies and linear dependencies, including checking for highly correlated variables and interaction terms that cause perfect multicollinearity. Additionally, ensure adequate data for each level of categorical variables, simplify the model by removing non-essential or highly correlated variables, and consider using model selection techniques such as stepwise regression to optimize the balance between model complexity and multicollinearity.

The **analysis of variance (ANOVA)** results compare a full model with interaction terms to a simplified model without them. The increase in **Residual Sum of Squares (RSS)** is minimal, and the F-statistic of 1.111 with a p-value of 0.3476 indicates no significant difference between the models. This suggests that the additional complexity of the full model does not substantially improve the explanation of remuneration variation. Therefore, the simpler model may be more appropriate for this analysis.

Thus, the interaction terms in the model allow us to understand how the effect of gender on remuneration varies across different departments and job roles. The diagnostic plots and multicollinearity checks help assess the quality of the model. Sensitivity analysis provides insights into the robustness of our findings. By comparing models with different sets of confounders, we can understand the influence of these variables on our primary variable of interest (gender). The use of robust standard errors (using packages such as `sandwich`, discussed later in the book) can further strengthen the analysis, especially in the presence of heteroskedasticity.

In the back door case, we're primarily interested in the effect of gender on remuneration, controlling for other variables. We need to include these libraries:

```
# Install necessary libraries
library(ggplot2)
library(broom)
```

We can visualize this in several ways:

- **Coefficient plot**: Display the estimated coefficients of the gender variable from the regression model, highlighting the adjusted effect of gender on remuneration:

  ```
  # Coefficient plot for gender
  coef_df <- broom::tidy(model) %>% filter(
    term == "gender_factorMale")
  ggplot(coef_df, aes(x = term, y = estimate)) +
    geom_point() +
    geom_errorbar(aes(
      ymin = estimate - std.error, ymax = estimate + std.error),
      width = 0.1) +
    ylab("Adjusted Effect of Gender on Remuneration") +
    xlab("Gender")
  ```

- **Scatter plot with a regression line**: Plot remuneration against gender, including a regression line that adjusts for other covariates. This gives a visual representation of the adjusted relationship.

- **Residuals analysis:** Plot residuals from the model to check for any patterns or anomalies, indicating potential issues in the model:

  ```
  # Assuming the back door adjustment model is already fitted
  # Residuals plot
  ggplot(data = model, aes(x = .fitted, y = .resid)) +
    geom_point() +
    geom_hline(yintercept = 0, linetype = "dashed") +
    xlab("Fitted Values") +
    ylab("Residuals") +
    ggtitle("Residuals vs Fitted")
  ```

The coefficient plot will illustrate the magnitude and direction of gender's effect on remuneration, adjusted for confounders. The residuals plot (see *Figure 5.3*) will help assess the model's fit and identify potential issues.

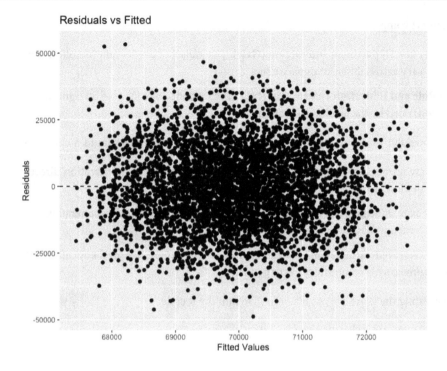

Figure 5.3 – A plot of residuals versus fitted values for assessing the model's fit

Front door adjustment in R

Now, let's solve the preceding problem from a Front door adjustment standpoint. Let's consider the following variables:

- **Exposure (X)**: The gender of the employee (male or female).

- **Outcome (Y)**: Remuneration (salary and bonuses).

- **Mediator (M)**: Promotion opportunities. This includes variables such as promotion frequency, the level of promotions, and the departments in which promotions occur.

- **Performance ratings (P)**: Annual performance scores, which could influence both promotion opportunities and remuneration.

- **Department (D)**: Different departments might have different remuneration scales and promotion practices.

- **Work experience (W)**: The total years of relevant work experience.

- **Education level (E)**: The highest educational qualification obtained.

- **Job role (J)**: Specific roles might have different baseline remuneration and promotion paths.

We can support complex interactions if we include a few other variables together.

- **Gender and Department Interaction (GxD)**: The impact of gender on promotion opportunities might vary across different departments.

- **Job Role and Education Interaction (JxE)**: The influence of a job role on remuneration might be moderated by the level of education.

While we model this problem, there are a bunch of assumptions that we should mention:

- There are unmeasured confounders affecting both gender and remuneration directly, such as inherent biases or negotiation skills.

- Performance ratings, while influencing promotion opportunities and remuneration, are assumed not to have unmeasured confounders with gender.

- Other observed variables such as department, work experience, and education level do not have unmeasured confounders with gender.

Okay, after setting the stage, we are ready to dive into the R code:

```
# Install necessary packages
install.packages("dplyr")
library(dplyr)
# Create interaction terms
corp_data <- corp_data %>%
  mutate(GxD = interaction(gender, department),
         JxE = interaction(job_role, education))


# Model 1: Effect of Gender on Promotion Opportunities
model1 <- lm(promotion_opportunities ~ gender +
                performance_rating + department + GxD +
                job_role + work_experience,
             data = corp_data)
summary(model1)
# Model 2: Effect of Promotion Opportunities on Remuneration
model2 <- lm(remuneration ~ promotion_opportunities + gender +
                performance_rating + department + GxD +
                job_role + JxE + work_experience,
             data = corp_data)
summary(model2)
# Calculate the Average Causal Effect (ACE)
# This involves predictions using both models and integrating over the
mediator (promotion_opportunities)
# Note: The ACE calculation in this complex scenario would typically
involve simulation or numerical integration techniques due to the
complexity of the model
```

In this complex scenario, the relationship between **gender** and **remuneration** is mediated by **promotion opportunities**, but the pathway is influenced by multiple other factors, including performance ratings, departmental differences, and the interaction between job roles and education levels. This complexity reflects the multifaceted nature of corporate structures and promotion dynamics.

The calculation of the average causal effect in such a complex model would likely require advanced statistical techniques, such as simulation or numerical integration, especially when dealing with interaction terms and non-linear relationships.

This example illustrates how front door adjustment can be used in a sophisticated analysis, accounting for multiple layers of a corporate structure and interactions between variables. The process underscores the importance of a comprehensive understanding of the context and the relationships between variables in causal analysis.

For the **front door case,** we focus on the mediated effect of gender on remuneration through promotion opportunities.

Key visualizations could include the following:

- **Mediation effect plot**: Let's see the path from gender to promotion opportunities and then to remuneration. This could be a series of plots showing the effect of gender on the mediator and the effect of the mediator on remuneration.

- **Path diagram**: Let's check a path diagram that shows the estimated coefficients along the paths from gender to remuneration, including the mediation effect:

```
# Install necessary library for path diagrams
library(semPlot)
# Mediation effect plot
mediation_effect <- broom::tidy(model1) %>% filter(
  term == "genderFemale" | term == "genderMale")
ggplot(mediation_effect, aes(x = term, y = estimate)) +
  geom_point() +
  geom_errorbar(aes(
    ymin = estimate - std.error, ymax = estimate + std.error),
    width = 0.1) +
  ylab("Effect of Gender on Promotion Opportunities") +
  xlab("Gender") +
  ggtitle("Mediation Effect of Gender on Promotion Opportunities")

# Path diagram visualization
semPaths(model1, whatLabels = "est",
         layout = "tree", rotation = 2)
semPaths(model2, whatLabels = "est",
         layout = "tree", rotation = 2)
```

The mediation effect plot shows the direct effect of gender on promotion opportunities. The path diagrams provide a clear visual representation of the mediation pathway, including the indirect effects of gender on remuneration via promotion opportunities.

Summary

This chapter looked at DAGs deeply and their crucial role in understanding causal relationships. We began by examining different graph structures, such as forks, chains, colliders, and their influence on interpreting associations and causations. This foundational exploration is aimed at dissecting the causal flow within these graphs, thereby establishing a solid understanding of these key concepts. The chapter further explored specific structures within DAGs, such as chains and forks, using practical examples, such as a grocery store scenario, to explain how these structures manifest in real-life situations and their implications for causal inference. We also explained colliders and the concept of immorality within DAGs, explaining how conditioning on certain variables can induce dependencies between otherwise independent variables.

The chapter then discussed more advanced topics, including back door and front door adjustments, further showcasing their applications, particularly in scenarios where certain confounders cannot be directly measured. We presented a practical application using R to demonstrate how these concepts can be applied to real-world problems, emphasizing the relevance and applicability of these theories in practical scenarios.

In the next chapter, let's continue our learning process, switching gears to focus on propensity score methods.

6

Employing Propensity Score Techniques

In this chapter, we'll familiarize ourselves with the concept of propensity scores, something we touched on lightly in *Chapters 2* and *3*. It's a vital tool for identifying confounding variables in causal inference. To be specific, propensity scores help to clear the mist in knowing which variables need conditioning by balancing confounding variables between groups of data. This methodology has power in transforming observational studies so that they resemble randomized trials, a kind of statistical practice that's crucial for solid causal conclusions.

In addition to learning new theory, we will also be getting our hands dirty with R code. We'll walk through techniques such as matching, stratification, and weighting – each of which has a unique flair. We'll also practice our theory with real-life examples, turning abstract concepts into concrete skills. By the end of this chapter, you won't just understand the "what" and "how" of propensity scores in R; you'll have the savviness to apply them confidently in your research adventures.

In this chapter, we will cover the following topics:

- Introduction to propensity scores
- Stratification and subsampling
- Propensity score matching
- Weighting in propensity score matching using R

Technical requirements

You can find the code examples for this chapter in this book's GitHub repository: https://github.com/PacktPublishing/Causal-Inference-in-R/tree/main/chap_06.

Introduction to propensity scores

A propensity score is crucial in observational studies that are designed to estimate the effect of treatments, interventions, or exposures. They function by calculating the probability of a unit, such as a person, receiving a specific treatment based on their observable characteristics. This methodology serves the vital purpose of creating a balanced comparison group, essential for causal inference in observational studies. *By doing so, propensity scores help to emulate the conditions of a* **randomized controlled trial** (**RCT**). These trials are the gold standard in experimental design but are often not feasible or ethical in every research scenario.

Imagine a large-scale multiplayer video game where players can choose to join one of two factions: the Rebels or the Empire. Each player has various attributes, such as skill level, experience points, preferred weapons, and playtime. In an ideal world (like an RCT), players would be randomly assigned to factions. However, in this game (like in observational studies), players choose their faction based on their preferences and characteristics. Now, researchers want to study the effect of being in the Rebel faction on a player's win rate. But there's a problem: players who chose the Rebel faction might be inherently different from those who chose the Empire. For example, more experienced players might prefer the Rebels. This is where propensity scores come in. They're comparable to a sophisticated matchmaking algorithm that does the following:

- For each player, it calculates the probability (propensity) of joining the Rebel faction based on their attributes.

- It then finds pairs or groups of players who have similar propensity scores but chose different factions. For instance, it might match a Rebel player with an Empire player who both had a 70% chance of joining the Rebels based on their attributes.

- By comparing these matched players, researchers can more accurately estimate the effect being in the Rebel faction has on win rates since they're comparing players who were equally likely to join the Rebels but made different choices.

This method helps to balance the "treatment" (being in the Rebel faction) and "control" (being in the Empire faction) groups by finding observations that are similar in all aspects except for the treatment received. The aim is to create a "virtual" randomized experiment from observational data. In real-world research, you would replace "factions" with treatments or interventions, and "player attributes" with observed covariates or characteristics of study participants. The goal is to create comparable groups to estimate treatment effects more accurately in situations where random assignment isn't possible. This approach allows researchers to make more reliable causal inferences from observational data, although it's important to note that it can only account for observed characteristics and may still be subject to hidden biases.

The whole point of this game is to make a fair comparison in studies where you can't just randomly assign treatments (quests) to people. Using propensity scores, the study mimics the gold standard of research; as explained previously, RCTs are the ultimate, ideal way of assigning quests in our game.

But since we can't always just randomly assign treatments in real life (sometimes, it's not safe or just wouldn't work), propensity scores are our best bet at keeping the game fair.

Taking another example, consider a scenario where a company wants to evaluate the effectiveness of a new marketing campaign. Some customers interacted with the campaign content, while others didn't. Simply comparing these two groups directly could lead to misleading conclusions because the customers who engaged with the content might inherently differ from those who didn't. This is where PSM becomes invaluable. By using PSM, the company can do the following:

- Calculate each customer's propensity to interact with the campaign based on observable characteristics (for example, past purchase behavior, demographics, and online activity).
- Match customers who interacted with the content to similar customers who didn't, based on these propensity scores.
- Compare the outcomes (for example, purchase rates and customer lifetime value) between these matched groups.

This approach allows businesses to more accurately estimate the true effect of their marketing efforts, controlling for pre-existing differences between customer groups. By addressing selection bias, propensity scores enable more reliable causal inferences from observational data, helping businesses make data-driven decisions with greater confidence.

By adjusting for these propensity scores, researchers aim to address the selection bias that occurs when the characteristics of the treatment and control groups differ at the baseline. We're going to dive deeper into the intricacies of these scores next.

A deep dive into these scores

The propensity score – let's say, $e(H)$ – is defined as the probability of treatment assignment conditional on observed covariates, H. Rosenbaum and Rubin (1983) [1] showed that conditioning on the propensity score is equivalent to conditioning on the covariates themselves in terms of removing confounding.

The calculation of the propensity score is based on the assumption that $Y(t) \perp T \mid H$, which means that the potential outcome $Y(t)$ is independent of the treatment, T, given the covariates, H. This is a critical assumption. It states that potential outcomes, $Y(t)$ (that is, what would happen under each treatment condition), are independent of the treatment assignment given the covariates, H. As we already know, this is known as the conditional independence assumption or no unmeasured confounders assumption. It means that once you control for H, the treatment assignment is as good as random, and the covariates are balanced. The idea of achieving "balance" means that the distribution of covariates is similar between treatment groups. This balance is crucial for making valid causal inferences.

Rosenbaum and Rubin's result, $Y(t) \perp T \mid e(H)$, mathematically translates into achieving balance on covariates through the propensity score, making $e(H)$ a univariate confounder that represents the multivariate confounder, H. They showed that conditioning on the propensity score is as effective as conditioning on the covariates themselves to remove confounding. This is a powerful result because it simplifies the process of adjusting for many covariates, which can be complex and computationally demanding. So, what do we do when we compute these scores?

In practice, you can use propensity scores for matching, stratification (concepts that we will cover later in this chapter), or as covariates in regression models. The goal is to compare outcomes between treatment groups that are similar in terms of their propensity scores, thus approximating a randomized experiment.

That said, it's important to note that the effectiveness of propensity scores hinges on the assumption that all confounders are measured and included. Unmeasured confounding can still bias the results.

Analyzing distribution differences in subgroups

A key step in propensity scores is to check for overlap in the propensity score distributions across treatment groups. This involves comparing probability densities of $e(H)$ between groups. Overlap suggests that for each individual in one treatment group, there is a comparable individual in the other group.

Don't get it? No problem. Consider a study examining the effect of an educational program on voting behavior. If we're comparing voters with and without the program (for example, Republican versus Democratic in an election in USA), checking for overlap ensures that for every Republican voter with a certain propensity score, there's a Democrat with a similar score, ensuring comparability. If the overlap isn't sufficient (indicating a violation of the positivity assumption, as explained later in this chapter), one approach is to exclude individuals from the analysis who don't have comparable counterparts in the other group. This method restricts the analysis to the subgroup where overlap exists, but it may limit the generalizability of the findings.

Let's add more drama to the voting scene. Think of a situation between Democrat and Republican voters concerning a pro-life policy. Assume that the treatment, T, is voting for a pro-life policy. Let $T = 1$ represent Republicans and $T = 0$ represent Democrats. Confounders ($H1$, $H2$) could represent demographic or socio-economic factors that might influence both party affiliation and voting behavior on the pro-life policy. For example, it could be age, a continuous variable that's generated from a normal distribution. Similarly, $H2$ could be a location, such as urban or rural, and thus is a binary variable that's generated from a binomial distribution.

The propensity score (e) is estimated using logistic regression, which models the probability of voting for the policy based on the confounders (logistic regression model is explained later in this chapter). We'll visualize the propensity score distributions for both groups to check for overlap, which is crucial for propensity score methods. The treatment variable, T, is generated based on the propensity score. This simulates a scenario where the likelihood of receiving the treatment is influenced by the confounders.

Here's how the code could be adapted:

```
# Function to model and visualize propensity score distributions
# Simulate confounders - demographic/socio-economic factors
# H1 might represent age,
# H2 could be a binary variable like urban/rural
H1 <- rnorm(1000)
H2 <- rbinom(n = 1000, size = 1, prob = 0.3)

# Assuming these confounders influence the propensity to
# vote for a pro-life policy
e <- exp(H1 + 3 * H2 - 1.5) / (1 + exp(H1 + 3 * H2 - 1.5))
# Simulating voting behavior based on the propensity
# Republicans (T=1) more likely to vote for pro-life
# Democrats (T=0) less likely
T <- rbinom(n = 1000, size = 1, prob = e)
# Fit a propensity score model - logistic regression
e <- fitted(glm(T ~ H1 + H2, family = binomial))
```

The preceding code snippet generates two confounding variables ($H1$ and $H2$) to represent demographic factors such as age ($H1$, continuous) and urban/rural residence ($H2$, binary). Then, it computes the propensity to vote for a pro-life policy using a logistic regression formula, based on the confounders. This score indicates the likelihood of voting for the policy for each individual.

In the next step, it uses the propensity scores to simulate whether individuals vote for the policy (Republican, $T = 1$) or against it (Democrat, $T = 0$) by employing a binomial distribution ($T < -rbinom(n = 1000, size = 1, prob = e)$).

Finally, it fits a logistic regression model to estimate the propensity scores more formally, relating the voting behavior to the confounders:

$$(e < -fitted(glm(T{\sim}H1 + H2, family = binomial))$$

Next let's plot the results:

```
# Plotting the density of propensity scores for both groups
plot(c(0, 1),
     range(density(e[T == 1], bw = .05)$y,
           density(e[T == 0], bw = .05)$y),
     type = "n",
     xlab = "Propensity Score",
     ylab = "Density",
     main = "Propensity Score Distribution: Republicans vs Democrats")

# Add density lines for each group
```

```
lines(density(e[T == 1], bw = .05),
     lty = 1, col = "blue") # Republicans
lines(density(e[T == 0], bw = .05),
     lty = 2, col = "red") # Democrats
# Add a legend for clarity
legend("topright", c("Republicans", "Democrats"),
       lty = c(1, 2), col = c("blue", "red"))
```

Takeaways

The preceding code plots the density of propensity scores for both the treatment and control groups. Before proceeding with matching, check for sufficient overlap in the propensity score distributions between the treatment and control groups. To assess the overlap in propensity scores between Republicans and Democrats in this example, examine where the two curves intersect or come close. This overlapping area reflects the similarity in propensity scores between the groups. To quantify the overlap, compute the integral (area under the curve) for each group and calculate the overlapping area. More overlap indicates greater similarity. Alternatively, the overlap coefficient can measure the overlap between the two distributions. Lack of overlap indicates that there are regions where treated and untreated subjects don't have comparable counterparts (refer to *Figure 6.1*):

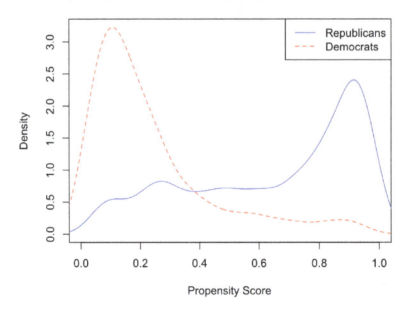

Figure 6.1 – The propensity score distribution between Republican and Democrat voters

In practical scenarios, similar to the logistic regression model used here, you would fit a model to your real dataset to estimate the propensity score. Key factors include selecting relevant confounders and choosing an appropriate model. Various models beyond logistic regression can estimate propensity scores, each with distinct advantages and limitations. Generalized linear models are flexible and familiar but assume linearity, while machine learning methods such as random forests and gradient boosting capture complex relationships but lack interpretability and demand higher computational resources. Regularized regression models manage multicollinearity and feature selection effectively but introduce bias, and neural networks offer scalability and flexibility but are prone to overfitting and are less interpretable.

Once you've estimated propensity scores and checked for overlap, you can proceed with the matching process. Don't worry – we'll discuss this at length later in this chapter. In short, matching involves pairing individuals from the treatment group with similar individuals from the control group based on their propensity scores.

There are various propensity score matching methodologies, each offering different approaches to balancing treatment and control groups:

- **Stratification matching**: Divides the sample into strata based on propensity scores. Within each stratum, treated and control units are compared. It's implemented by creating quantiles of propensity scores and matching within these quantiles.

- **Optimal matching**: Uses network flow theory to find matches that minimize the overall difference in propensity scores between treated and control units. It's executed using algorithms such as the Hungarian method or network flow optimization.

- **Full matching**: Creates matched sets containing one treated unit and one or more control units (or vice versa). It maximizes the number of matches while minimizing the discrepancy in propensity scores. This is typically implemented using optimization algorithms in statistical software.

- **Kernel and local linear matching**: Uses weighted averages of all controls to construct the counterfactual outcome. Kernel matching applies kernel functions to weight controls, while local linear matching includes a linear term in the weighting. These are implemented using non-parametric regression techniques.

- **Mahalanobis distance matching**: Matches are treated and control units are formed based on the multivariate distance of covariates. It's executed by calculating the Mahalanobis distance between units and matching those with the smallest distances.

- **Coarsened exact matching**: Temporarily coarsens variables into bins, performs exact matching on these bins, and then uses the original (uncoarsened) values for analysis. It's implemented by defining coarsening bins for each variable and then conducting exact matching within these bins.

Each method has its own implementation in statistical software packages, with specific functions or procedures designed for their execution.

These methods offer different tradeoffs: stratification matching is straightforward but can lose data in small strata, while optimal matching minimizes overall score differences but may be computationally intensive. Full matching maximizes data usage but varies in set size and complexity, whereas kernel and local linear matching reduce estimator variance but can be biased in small samples. Finally, Mahalanobis distance matching considers multivariate distances but may not balance covariates effectively, whereas coarsened exact matching improves balance but depends on the choice of coarsening bins and may exclude unmatched data.

After matching, it's important to assess the balance that's been achieved by comparing the distribution of covariates in the matched treatment and control groups. This can be done using standardized mean differences or other balance diagnostics. To assess balance in propensity score matching, various diagnostics are employed: variance ratios gauge variance similarity in covariates, overlap in propensity score distributions indicates group comparability, and covariate balance tables visually compare pre and post-matching distributions. Histograms or density plots, along with QQ plots, offer graphical comparisons of distributions, whereas Love plots display standardized mean differences for all covariates. Statistical tests such as Chi-square, t-tests, and the Kolmogorov-Smirnov test quantitatively compare group distributions, providing a comprehensive balance evaluation.

To estimate causal effects in a balanced matched dataset, several statistical methods are utilized:

- **Difference-in-means test**: This method compares the average outcomes between treatment and control groups to estimate the **average treatment effect** (**ATE**). It's straightforward but assumes perfect balance and doesn't account for residual confounding. For example, if the average outcome in the treatment group is 10 and in the control group is 8, the estimated ATE is 2.

- **Regression analysis (including generalized linear models - GLMs)**: This approach adjusts for any residual imbalances and can incorporate additional covariates or interaction terms. It provides a more nuanced estimate of treatment effects while controlling for other factors. For example, a linear regression model might be $Y = \beta_0 + \beta_1 * Treatment + \beta_2 * Age + \beta_3 * Gender$, where β_1 estimates the treatment effect while controlling for age and gender.

- **Conditional logistic regression**: This method is particularly suited for one-to-one matched pair data with binary outcomes. It accounts for the matched nature of the data by conditioning on the pairs. For example, in a study that involves matching patients based on age and gender, this method would estimate treatment effects on a binary outcome (for example, recovery) within each matched pair.

- **Cox proportional hazards models**: These are used in time-to-event or survival analysis scenarios. They estimate the effect of treatment on the hazard rate of an event occurring over time while accounting for censoring. For example, in a clinical trial, this model might estimate how a new drug affects the time until disease recurrence while accounting for patients who drop out or don't experience recurrence during the study.

- **Mixed effects models**: For data that involves clustering or matched clusters, these models account for random effects associated with these clusters. They're useful when observations within a cluster aren't independent. For example, patients treated in the same hospital may share similar environmental factors or care protocols, which could influence the outcomes. In a multi-center trial, a mixed effects model might estimate treatment effects while accounting for between-center variability.

Each of these methods has its own assumptions and is suitable for different types of data and research questions. The choice depends on the study's design, outcome type, and data structure.

Next, let's learn how to balance variables in this methodology.

Balancing confounding variables

The propensity score's ability to create comparable groups is rooted in probability theory and the conditional independence assumption.

Mathematically, let Y_1 and Y_0 be the potential outcomes, T be the treatment indicator, and X be the vector of observed covariates.

The propensity score, $e(X)$, is defined as follows:

$$e(X) = P(T = 1 \mid X) \qquad (1)$$

The key property of propensity scores is that they satisfy the balancing property: $T \perp X \mid e(X)$. This means that, based on the propensity score, the distribution of observed covariates, X, is the same for treated ($T = 1$) and untreated ($T = 0$) units.

Furthermore, if the strong ignorability assumption holds, $(Y_1, Y_0) \perp T \mid X$, then it also holds conditional on the propensity score: $(Y_1, Y_0) \perp T \mid e(X)$.

By equating the conditional distribution of potential outcomes given the propensity score with that given the covariates, we get the following:

$$P(Y_1, Y_0 \mid T, e(X)) = P(Y_1, Y_0 \mid T, X) \qquad (2)$$

This provides a strong basis for causal inference, allowing for unbiased estimation of average treatment effects by comparing outcomes across treatment groups with similar propensity scores.

Implementing these concepts in R involves using functions from relevant packages (for example, MatchIt, glm, and others) to model the propensity score, check for overlap, and conduct stratification or matching. This programming implementation applies the mathematical concepts in a practical data analysis context.

Oh, a new piece of terminology! **Stratification**; what's that? In propensity score analysis, stratification involves dividing a study sample into subgroups or strata based on propensity scores, ensuring that within each stratum, the distribution of covariates is balanced between treatment and control groups. We will learn more about it in the next section.

Now, let's consider that you're a healthcare company trying to figure out whether a new blood pressure-lowering pill works. You have two groups: patients taking the pill ("super-pill gang") and patients not taking the pill ("no-pill squad"). The challenge is that these groups might differ in age, lifestyle, and health history.

To compare these groups fairly, you use PSM. You assign each patient a score based on their likelihood of taking the pill. Then, you match each super-pill patient with a no-pill patient who has a similar score. This way, you're comparing patients who are alike except for the pill.

If the matched pairs show different blood pressure outcomes, you can attribute it to the pill. However, PSM has limitations: it only matches observed characteristics and may miss hidden factors, potentially leading to biased results. While PSM is a powerful tool, it's not foolproof and needs careful application by a skilled analyst.

Check for confounding using propensity scores

You'll be surprised to learn that you can also detect confounding using propensity scores. This is a two-step process. First, propensity scores are calculated for each individual. The population is then segmented based on these scores (for example, by bucketing based on some thresholds). In the second step, the variation in propensity scores across different population segments is analyzed. If certain segments have higher propensity scores, this suggests that specific covariates are influencing treatment assignment. If these covariates also affect the outcome, this indicates potential confounding. For instance, in the preceding scenario, where we examined the impact of a pro-life policy, if Republicans are more likely to support such policies (as shown by higher propensity scores), and if party affiliation also influences related outcomes (for example, voting behavior), then party affiliation acts as a confounder. Let's check out the code:

```
# Assuming 'prolife_policy' is the dataset, 'treatment' is the
# treatment variable (e.g., support for pro-life policy),
# and 'party_affiliation' along with other covariates are predictors.
# Load necessary library
library(MatchIt)
# Estimate propensity scores
ps_model <- glm(treatment~ party_affiliation + other_covariates,
                data = prolife_policy, family = "binomial")
prolife_policy$propensity_score <- predict(ps_model,
                                           type = "response")


# Analyze the variation in propensity scores
# Here, we can compare the average propensity scores between
```

```
# Republicans and Democrats
mean_score_republicans <- mean(prolife_policy$propensity_score[
  prolife_policy$party_affiliation == "Republican"])
mean_score_democrats <- mean(prolife_policy$propensity_score[
  prolife_policy$party_affiliation == "Democrat"])
```

In this example, `treatment` could be a binary variable indicating support for the pro-life policy, whereas `party_affiliation` is a categorical variable with levels such as `Republican`, `Democrat`, and so on.

The significant difference in mean propensity scores between Republicans (0.5702) and Democrats (0.3547) indicates the presence of confounding since `party_affiliation` appears to influence both treatment assignment and outcomes. These numbers are retrieved when you run the preceding code. This disparity underscores the need for statistical adjustment methods such as matching or weighting to address potential biases in estimating causal effects. The higher propensity scores among Republicans suggest they are more likely to receive the treatment (the binary variable when it's 1) compared to Democrats, given the covariates in the model. Next, let's discuss the challenges in this methodology.

Challenges and caveats

Controlling for propensity scores alone is sufficient to identify the causal effect. This is because propensity scores act as a form of dimensionality reduction, condensing the information contained in multiple confounders into a single score that balances the treated and untreated groups. However, PSM has a few challenges. Let's take a look at a few.

Issue 1 – predictive accuracy versus causal estimation

Suppose that in the preceding voting scenario, you, as a causal researcher, overly focus on refining the model to perfectly predict who supports the policy. You achieve high predictive accuracy, but this doesn't necessarily improve causal estimates. *The highly predictive model might be overfitting the data, capturing noise rather than the underlying causal relationship.* This overfitting can increase the variance in the causal estimates, making them less reliable. A model adept at predicting treatment assignment (policy support) doesn't automatically excel at identifying the causal impact of variables (such as party affiliation) on that support. The model's primary goal should be to balance the groups for a fair comparison, not just to predict who supports the policy.

Issue 2 – inadequate overlap in propensity scores

Now, let's say that in our voting study, you observed that Republicans are more likely to support the policy, whereas Democrats are less likely. A critical issue arises when the propensity score model reveals minimal overlap between the groups, indicating a significant difference in their likelihood of policy support. This lack of overlap leads to an extrapolation problem where causal inferences, such as estimating the impact of party affiliation on policy support, become unreliable due to the fundamental differences between the groups. Reliable causal analysis necessitates sufficient overlap in propensity scores, ensuring the presence of comparable individuals in both groups for an accurate estimation of the causal effect. *Thus, while propensity scores are a powerful tool in observational studies, their effectiveness hinges on ensuring adequate overlap between treatment and control groups.*

Issue 3 – standard error in PSM

A key point in PSM relates to estimating standard errors. When using PSM to align treated and untreated groups in a study, the standard errors that are calculated from these matched samples might not accurately convey the true uncertainty in estimating propensity scores. This limitation is crucial because it affects the reliability of the inferences. If the model for determining propensity scores is complex or the overlap in score distributions between groups is minimal, there's a risk that confidence in the results could be misplaced. The standard errors might either overstate or understate the actual variability in the estimates. So, while PSM is a valuable tool for balancing observed covariates between groups, you must recognize the importance of being cautious with standard error estimation and possibly seek alternative methods to ensure the robustness of any conclusions.

It's essential to balance improving the predictive accuracy of treatment assignment with the goal of causal estimation and to ensure adequate overlap in propensity score distributions to avoid unreliable extrapolation. Next, we'll learn more about stratification.

Stratification and subsampling

Stratification in PSM involves dividing *a study population into strata or subclasses based on their propensity scores.* This involves dividing the study sample into several strata or subclasses based on the distribution of their propensity scores. For example, quintiles are commonly used, dividing the sample into multiple (five) groups with approximately equal numbers of individuals. *The aim is to balance the covariates within each stratum between the treatment and control groups.* By doing so, the method attempts to mimic the conditions of a randomized controlled trial, where treatment assignment would be independent of these covariates.

Theory

After stratifying the data, within each stratum, we can estimate treatment effects independently. Assuming the potential outcome framework, let $Y(1)$ and $Y(0)$ denote the potential outcomes under treatment and control, respectively. The ATE within a stratum can be calculated as follows:

$$ATE_{stratum} = {}^{-}\left(Y(1)_{stratum}\right) - {}^{-}\left(Y(0)_{stratum}\right) \qquad (3)$$

Here, the overall ATE is the weighted average of the ATEs across strata, with weights proportional to the size of each stratum.

Now that we have a better grasp of how PSM works, let's summarize how to compute propensity scores. As aforementioned, we'll use logistic regression to calculate propensity scores. But first, let me provide a brief overview of what logistic regression is. Logistic regression [4] is a statistical method that's used for predicting a binary outcome based on one or more predictor variables. In the context of propensity scores, it's used to estimate the probability of receiving treatment given a set of observed covariates. The logistic regression model is expressed as follows:

$$Logit(e(X)) = ln(e(X) / (1 - e(X))) = \alpha + \beta_1 X_1 + \beta_2 X_2 + ... + \beta_k X_k \quad (4)$$

Here, we have the following:

- $e(X)$ is the propensity score (probability of receiving treatment)
- $X_1, X_2, ..., X_k$ represent the observed covariates
- α is the intercept
- $\beta_1, \beta_2, ..., \beta_k$ are the coefficients for each covariate

The logit function is defined as follows:

$$Logit(p) = ln(p / (1 - p)) \quad (5)$$

Here, p is the probability. Importantly, while the logit can take any real value, the inverse logit function (also known as the sigmoid function) always outputs a value between 0 and 1:

$$e(X) = 1 / (1 + e^\wedge -(\alpha + \beta X)) \quad (6)$$

This property makes logistic regression ideal for estimating propensity scores for several reasons:

- It ensures that the estimated probabilities (propensity scores) are always between 0 and 1.
- It allows for non-linear relationships between the covariates and the probability of treatment.
- It can handle multiple covariates and their interactions.

The output of the logistic regression model gives us the estimated propensity score for each unit, representing the probability of receiving treatment given their observed characteristics. These scores can then be used for matching, stratification, or weighting in subsequent analyses to estimate treatment effects. It's worth noting that while logistic regression is commonly used for propensity score estimation due to its simplicity and interpretability, other methods, such as probit regression, classification trees, or machine learning algorithms, can also be employed, especially when dealing with complex relationships or high-dimensional data.

Now, let's get back to the steps of computing propensity scores:

1. **Estimating propensity scores**: Using logistic regression, estimate the propensity score.
2. **Creating strata**: Divide the dataset into strata based on the estimated propensity scores. This is usually done by dividing the range of propensity scores into equal intervals (for example, quintiles).

3. **Balancing covariates**: Within each stratum, the distribution of covariates should be balanced between the treatment and control groups. This can be checked using standardized mean differences or other balance diagnostics. The term "balanced covariates" refers to achieving a state where the distribution of covariates (the characteristics or variables used in the analysis, such as age, income, education level, and so on) is similar across both the treatment and control groups within each stratum.

4. **Estimating effects**: Estimate the causal effect within each stratum and then aggregate these estimates to obtain an overall effect. This stratified analysis reduces the confounding effect of covariates within each stratum.

Application of propensity scores in R

The key assumption of propensity score stratification is that within each stratum, the treatment assignment is as good as random (conditional independence), and there should be sufficient overlap in the propensity scores across treatment groups. The number of strata can significantly impact the balance that's achieved and the precision of the estimates; *too few strata may fail to adequately control for confounding, while too many can lead to small sample sizes within strata.* When most strata have balanced covariates except for a few, it's essential to address these imbalances to ensure valid causal inferences.

In cases where some strata are imbalanced, you can take several approaches. You may conduct focused analyses on the imbalanced strata to understand the underlying causes and consider additional matching techniques to improve balance. Sensitivity analyses (discussed later in this book) can also be performed to assess the impact of these imbalanced strata on overall results. Alternatively, weighting methods can be applied to reduce the influence of these problematic strata, or covariate adjustment can be used by including the imbalanced covariates as controls in the outcome analysis. It's crucial to remember that propensity score stratification only accounts for observed confounders, leaving unobserved confounding as a potential source of bias in the results.

Let's understand more about stratification by considering an example problem in R. Assume that we aim to understand the effect of education (greater than high school education) on voting behavior (specifically, voting for a certain candidate) while considering potential confounders. The propensity score method will be used to stratify the population into quartiles based on their propensity scores. After, we'll estimate the average potential outcomes within each quartile. Check out the code:

```
# Load necessary libraries
library(MatchIt) # Assuming MatchIt for propensity score estimation
# Synthetic Data Generation
set.seed(123) # For reproducibility
n <- 500 # Number of observations
gssrcc <- data.frame(
  outcome= rbinom(n, 1, 0.5), # Binary treatment indicator
  gthsedu = rnorm(
    n, mean = 12, sd = 2) # A covariate representing education
)
```

The preceding code snippet involves loading the `MatchIt` R library for propensity score estimation and preparing a synthetic dataset. It sets a seed for reproducibility and specifies 500 observations. The dataset features a binary treatment indicator and a continuous covariate representing education level, simulated with specific mean and standard deviation values. This setup is intended for further causal analysis – that is, estimating the treatment effect while controlling for education. Let's run through the code step by step:

1. **Data and propensity score model**: First, the code subsets the data based on the provided IDs. Then, it fits a logistic regression model to estimate the propensity scores (`eb`) for each individual, representing the likelihood of having more than a high school education.

```r
# Propensity Score Estimation Function (Dummy Example)
prop.r <- function(data, ids) {
  ps_model <- glm(outcome ~ gthsedu, data = data[ids, ],
                  family = "binomial")
  e <- predict(ps_model, type = "response")
  return(list(e = e))
}
# Function Definition
equartiles.r <- function(
  data = gssrcc, ids = c(1:nrow(gssrcc))
) {dat <- data[ids, ]

  # Fit the propensity score model using logistic regression
  eb <- prop.r(data, ids)$e

  # Find the quartiles of the propensity score
  quartiles <- quantile(eb, c(0, .25, .5, .75, 1))

  # Assign participants to quartiles
  equartiles <- cut(eb, breaks = quartiles,
                    include.lowest = TRUE)
```

2. **Quartile stratification**: The propensity scores are divided into quartiles. Each individual is assigned to a quartile based on their propensity score. This stratification aims to create comparable groups across different levels of the propensity score:

```r
# Estimate the average potential outcome within each quartile
  out <- glm(outcome ~ gthsedu * equartiles - 1 - gthsedu,
            data = dat)

  # Extract the estimates for E(Y(0)|qk(e)) and E(Y(1)|qk(e))
  EY0 <- out$coef[1:4]
  EY1 <- out$coef[1:4] + out$coef[5:8]
```

3. **Outcome model**: A generalized linear model (`glm`) is fitted to estimate the average potential outcomes for the republican candidates' voting within each quartile. The model includes an interaction term between education (`gthsedu`) and propensity score quartiles (`equartiles`), allowing different effects to occur in each quartile.

4. **Risk difference estimation**: The coefficients from the model (*EY0* and *EY1*) represent the estimated average potential outcomes under control and treatment within each quartile. The risk difference (`RD`) is calculated as the mean difference between these estimated potential outcomes across quartiles.

```
# Estimate the risk difference
  RD <- mean(c(EY1 - EY0))
  return(RD)
}
# Call the function
equartiles.r()
```

This analysis gives a nuanced view of the treatment effect (educational impact on voting behavior) across different propensity score levels, accounting for confounding factors. The risk difference provides an overall measure of the effect size, which in this case is 0.006369. It indicates a modest overall effect of education on voting behavior for Republican candidates when considering different levels of propensity scores. This positive but small risk difference value suggests that higher education is associated with a slightly increased likelihood of voting for a Republican candidate, highlighting a statistically detectable yet not substantial impact. The analysis provides a nuanced understanding by stratifying the treatment effect (educational impact) across propensity score quartiles, effectively accounting for confounding factors. This method offers a comprehensive view of how education influences voting behavior across different segments of the population, as differentiated by their propensity to receive the treatment (education).

Next, we'll discuss the matching technique.

Understanding Propensity Score Matching

Propensity Score Matching (PSM) is a technique that reduces selection bias. How? Simple – by ensuring that treated and untreated groups are comparable in observed characteristics. Consider a study that involves evaluating the effect of a job training program on employment rates. Without PSM, the treatment group might consist of younger, highly motivated individuals with some college education, while the control group includes older, less motivated individuals with only a high school education. This selection bias could lead to an overestimation of the program's effectiveness. However, by using PSM, researchers can calculate propensity scores based on characteristics such as age, motivation, and education level, then match treated individuals with similar untreated counterparts. This approach ensures that both groups are comparable, allowing for a more accurate estimation of the job training program's true impact on employment rates by isolating the effect of the program from other influencing factors.

Selection bias in PSM occurs when the process of selecting subjects for treatment and control groups is influenced by their observed characteristics, leading to a systematic difference in these groups beyond the treatment effect. This bias can distort the estimation of the treatment effect if the characteristics influencing selection are also related to the outcome of interest. PSM aims to mitigate this bias by matching subjects based on their propensity scores, ensuring more comparable groups and thus a more accurate estimation of the causal effect.

Considerations and limitations

Positivity assumption: It's a critical rule that says that in our study, every person, regardless of their characteristics, should have a fighting chance to be in either the treatment or the control group. Imagine that you're throwing a dinner party and want a mix of every kind of guest. The positivity assumption is your guarantee that you don't just end up with only salsa dancers or stamp collectors.

Why is this important? Well, it's all about making fair comparisons. If every type of person is in both the treatment and control groups, you can compare apples to apples. It's like ensuring both vegetarians and meat lovers have something to eat at your party; it keeps everyone happy and the comparisons fair.

Without a good mix, comparing the effects of treatment becomes as reliable as predicting the weather with a horoscope. If, for some combination of traits (such as being a certain age or having a particular hobby), only treated or only untreated people exist, it's impossible to make a solid comparison.

Bias-variance tradeoff: In PSM, there's a balance to be struck between covariate balance and sample representativeness. This balance affects both the bias and variance in the estimation of treatment effects.

Propensity score estimators, in general, tend to have lower variance compared to exact matching algorithms. This is primarily because the size of the matched samples in propensity score matching is usually larger. Larger samples generally lead to lower variance in the estimates. Additionally, within the world of propensity score-based algorithms, methods such as radius and kernel matching make more comprehensive use of the available sample data, which also contributes to lower variance.

Let's expand on what we just said here.

Consider that, in a study that involves evaluating the impact of a health education program on dietary habits, researchers initially employ exact matching to compare participants in the program with a control group. This method matches individuals based on specific confounder values such as age, income, and education level. However, finding perfect matches for each individual proves challenging, particularly when dealing with numeric or continuous confounders. As a result, the exact matching approach yields a small matched sample, which leads to higher variance in the estimated impact of the health program. The limited sample size undermines the stability and reliability of the results, making it difficult to draw definitive conclusions about the program's effectiveness.

Shifting to PSM, you can calculate a propensity score for each individual, reflecting their likelihood of participating in the health program based on their confounders. This approach allows for more flexible pairing than exact matching as it's easier to match individuals with similar propensity scores rather than identical confounder profiles. Consequently, this method results in a larger matched sample, encompassing a broader range of participants. The increased sample size from propensity score matching significantly reduces the variance in the estimate of the program's effect. This reduction in variance leads to more stable and reliable results, enhancing the credibility of the findings regarding the program's influence on dietary habits.

Further enhancing their approach, the researchers explore radius and kernel matching within the propensity score framework. Radius matching matches all treated individuals to control individuals within a certain propensity score radius, not just the closest ones. Kernel matching, on the other hand, weights control individuals based on the closeness of their propensity scores to those of the treated individuals. Both these methods make more comprehensive use of the control group data, effectively increasing the overall sample size used in the analysis. This larger effective sample size from radius and kernel matching further lowers the variance in the estimated impact of the health education program, offering even more precise and reliable insights into its effectiveness.

Thus, the bias-variance tradeoff in PSM involves choosing matching algorithms and parameters that achieve an optimal balance between minimizing bias (through good covariate balance) and reducing variance (by maintaining a sufficiently large and representative sample). This tradeoff is key to ensuring both the accuracy and reliability of causal inferences made using PSM. Next, we'll apply what we've learned to matching in R.

Practical application of PSM in R

The process of matching in R is made simple by the use of R packages such as `MatchIt`, which performs the matching. This package allows various matching methods to be used and provides tools for assessing match quality. After matching, you can use standard statistical methods to estimate the treatment effect.

Let's consider the code:

```
# Loading necessary libraries
library(tidyverse)
library(MatchIt)
library(cobalt)
# Performing nearest neighbor matching
match_data <- matchit(political_campaign ~ age + income +
                      education + party_affiliation,
                  data = data, method = "nearest")

# Viewing match summary
summary(match_data)
```

The `MatchIt` package allows you to create comparable groups, minimizing the differences in covariates between treated and control groups. It's crucial to ensure that the matching process has balanced the covariates across groups:

```
# Checking balance
bal.tab <- bal.tab(match_data, un = TRUE)
print(bal.tab)

# Visualizing balance with Love plot
love.plot(bal.tab, stars = "raw")
```

Here, `bal.tab` and `love.plot` provide a quantitative and visual assessment of balance, respectively, ensuring that the matching process was effective.

In PSM, assessing the balance between treatment and control groups is a critical step to ensure the validity of causal inferences. *The `cobalt` package in R is particularly useful for this purpose due to its comprehensive and user-friendly tools for balance assessment.* However, other methods and tools are available for this purpose as well.

Balancing methods

There are a few alternative methods you can use to seek balance:

- **Standardized mean differences (SMD)**: This measure indicates the difference in means of covariates between treatment and control groups, standardized by the standard deviation. It's a straightforward, non-parametric measure to assess balance, that doesn't rely on the units of the covariates.

- **Visualization**:

  - **Love plots**: These are commonly used for visualizing balance and are available in both `cobalt` and other packages such as `MatchIt`. They provide a clear visual representation of the standardized mean differences before and after matching.

  - **Histograms and box plots**: These can be used to compare the distribution of covariates in the treatment and control groups.

- **Variance ratios**: Another approach to assess balance is to compare the ratios of variances of covariates between groups. Ideally, these ratios should be close to 1.

- **Overlap in propensity score distributions**: Assessing the overlap in the distributions of propensity scores between groups can be done using density plots. Adequate overlap is crucial for valid comparisons.

- **Kolmogorov-Smirnov test**: This non-parametric test can be used to compare the distribution of covariates between groups. This is outside the scope of this book, but feel free to read more; please refer to [2] in the *References* section at the end of this chapter.

Now, let's compare the voting behavior between the matched groups to estimate the effect of the campaign:

```
# Extracting matched data
matched_data <- match.data(match_data)
# Comparing outcomes between groups
effect <- with(matched_data, t.test(
  vote[political_campaign == 1],
  vote[political_campaign == 0]))
```

This step analyzes whether the political campaign had a significant impact on voting behavior.

Sensitivity analysis

We perform sensitivity analysis to check the robustness of our findings against unmeasured confounding:

```
library(sensitivitymv)
sen <- senmv(data)
```

Sensitivity analysis evaluates how the results might change under different scenarios or assumptions. This is particularly important in observational studies where randomization isn't present, and some level of bias is always a possibility due to unobserved variables.

Let's delve into what we can learn from the output of a typical sensitivity analysis in PSM.

The primary output of sensitivity analysis in PSM is an indication of how much an unmeasured confounder could change your estimated treatment effect. Essentially, it tells you how sensitive your results are to potential variables that weren't included in your model. Sensitivity analysis often provides a threshold level of an unmeasured confounder's effect that would be required to invalidate your findings. This could be in terms of the strength of the confounder's relationship with both the treatment and the outcome.

If the sensitivity analysis shows that only a very strong unmeasured confounder could overturn your findings, you gain confidence in the robustness of your results. Conversely, if the analysis indicates that a relatively weak unmeasured confounder could change the conclusions, it suggests that your results should be interpreted with caution.

In our code, the output from the senmv function indicates a highly significant result in the context of your causal analysis. The p-value of 0 suggests strong statistical evidence against the null hypothesis, implying that the observed effect is unlikely to be due to chance. The deviate value of 26.49545 and the statistic value of 254.9124, both significantly higher than the near-zero expectation, reinforce this finding, suggesting a substantial deviation from what would be expected under the null hypothesis. The variance of 92.56342 indicates the variability of the test statistic, which should be considered when assessing the reliability and stability of these results. Overall, this output points to a significant effect or association, but the specific interpretation should be contextualized within the details of your study and the assumptions of the analysis. Next, let's learn how we can visualize our results.

Visualizing the results

Finally, we'll visualize the average voting behavior in the matched groups:

```
# Analyzing and plotting the data
matched_data %>%
  # Grouping data by 'political_campaign'
  group_by(political_campaign) %>%
  # Calculating the average vote for each campaign
  summarise(Average_Vote = mean(vote, na.rm = TRUE)) %>%
  # Creating a bar plot
  ggplot(aes(
    x = factor(political_campaign, labels = c("No", "Yes")),
    y = Average_Vote, fill = factor(
      political_campaign, labels = c("No", "Yes")))) +
  # Using identity stat for pre-summarized data:
  geom_bar(stat = "identity") +
  # Adding labels and titles
  labs(x = "Political Campaign",
       y = "Average Vote", fill = "Campaign",
       title = "Average Vote by Political Campaign") +
  # Applying a minimalistic theme with increased text size
  theme_minimal() +
  theme(
    text = element_text(size = 14),  # Increase base text size
    plot.title = element_text(size = 14, face = "bold"),  # Larger,
bold title
    axis.title = element_text(size = 16),  # Larger axis titles
    axis.text = element_text(size = 12),  # Larger axis text
    legend.title = element_text(size = 14),  # Larger legend title
    legend.text = element_text(size = 12)  # Larger legend text
  )
```

The preceding code generates a single bar chart where each bar corresponds to a different political campaign, with the bar's height reflecting the average vote for that campaign. This visual representation facilitates an easy comparison of average votes among various campaigns (*refer to Figure 6.2*).

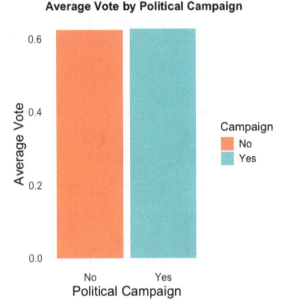

Figure 6.2 – The average vote statistics over political campaigns

With this, we have come to the final topic that will be discussed in this chapter: weighting.

Weighting in PSM using R

Inverse probability weighting (**IPW**) is adopted to adjust for confounding in causality assessment in observational studies. It is grounded in the framework of potential outcomes and relies on the use of – you guessed it – propensity scores.

Once the propensity scores are estimated, each individual is assigned a weight. *This is achieved by reweighting each data point using propensity scores, $e(H)$, which are the probabilities of receiving the treatment given the confounders. Treated individuals are reweighted by $1/(e(H)$ and untreated ones by $1/(1 - e(H))$. The ATE is then estimated by calculating the difference in expected outcomes between treated and untreated groups in this adjusted pseudo-population, essentially balancing out the influence of confounding variables to reveal the treatment's true impact.*

These weights are designed to create a synthetic sample in which the distribution of covariates is independent of treatment assignment, mimicking a randomized experiment. After assigning weights, the causal effect is estimated by applying these weights to the outcome analysis. In its simplest form, this involves calculating the weighted average of the outcomes for the treated and untreated groups and then computing the difference between these averages for the ATE. Alternatively, weighted regression models can be used, where the weights are applied in the regression analysis.

Why do we need this?

Specifically, IPW allows causal effects to be estimated in the presence of confounding by balancing the distribution of covariates between treatment groups. It's particularly useful when we're dealing with high-dimensional covariate spaces. This method aims to create "pseudo-populations," where the association between treatment and outcome mimics causation by adjusting for confounding. The goal is to reweight the data to make the treatment assignment mechanism, $P(T \mid W)$, independent of the confounders, W.

As useful as it is, IPW can be sensitive to the specification of the propensity score model. *Mis-specification can lead to biased estimates.* When propensity scores are close to 0 or 1, the weights can become extremely large, leading to high variance in the estimates. This is known as the problem of extreme weights.

Several variants of IPW exist to address its limitations, such as truncating the weights or using stabilized weights. These approaches aim to reduce the variance of the estimates while preserving the unbiasedness property. Feel free to read more about them by referring to the *References* section at the end of this chapter [3]. For now, let's check the implementation in R:

```r
# Load necessary libraries
library(MASS)
library(ipw)
library(dplyr)


# Synthetic Data Generation
set.seed(123) # For reproducibility
n <- 500 # Number of observations
df <- data.frame(
  treatment = rbinom(n, 1, 0.5), # Binary treatment indicator
  outcome = rnorm(n), # Continuous outcome variable
  confounders = matrix(rnorm(n * 3),
                       ncol = 3) # Matrix of confounders
)


# Estimate propensity score
ps_model <- glm(
  treatment ~ confounders.1+confounders.2+confounders.3,
  family = "binomial", data = df)
df$propensity_score <- predict(ps_model, type = "response")


# Calculate weights
df$weights <- ifelse(
  df$treatment == 1, 1 / df$propensity_score,
  1 / (1 - df$propensity_score))


# Estimate ATE
```

```
ate <- with(
  df, sum(weights * treatment * outcome) /
    sum(weights * treatment) - sum(
      weights * (1 - treatment) * outcome) /
    sum(weights * (1 - treatment)))
print(ate)
```

The estimated ATE of 0.1132704, which is derived using propensity score weighting, indicates that, on average, the treatment increases the outcome variable by approximately 0.113 units compared to the absence of treatment. This positive value suggests that the treatment has a beneficial effect on the outcome. Here, the assumption is that all relevant confounders have been adequately accounted for in PSM. Furthermore, there are no unmeasured confounders affecting the analysis.

Additionally, be aware that the **conditional average treatment effect** (CATE) is an advancement of ATE that focuses on estimating treatment effects within specific subpopulations defined by certain covariates. Unlike IPW, which estimates an overall effect, CATE provides a more detailed analysis by evaluating how the treatment effect varies across different groups based on these covariates. However, this detailed approach often leads to higher variance in the estimates due to the reduced sample sizes in these targeted subpopulations.

There are a few other pointers to learn about IPW. One is that extreme weights (when propensity scores are near 0 or 1) can cause instability in estimates. In such cases, the common practice is to trim or cap the weights to reduce variance, though this can introduce bias. Next, the effectiveness of IPW hinges on accurately specifying the propensity score model, ensuring that it accounts for the variables influencing treatment assignment correctly. Additionally, IPW operates under the assumption that there are no unmeasured confounders while acknowledging that extreme propensity scores can result in estimates with high variance.

Summary

In this chapter, we provided an in-depth exploration of propensity score techniques in causal inference, highlighting its crucial role in adjusting for confounding variables in observational studies. We began with an introduction to the concept of propensity scores, emphasizing their importance in transforming observational studies so that they resemble randomized trials. Then, we shifted to practical applications using R, where we presented various methods such as matching, stratification, and weighting, each with its unique features and implementation strategies. After, we introduced the theoretical foundations of propensity scores, such as balancing confounding variables and understanding their underlying assumptions and limitations, and also guided you through the practical aspects of estimating and applying these scores using R programming.

Furthermore, we covered different methods of propensity score matching, including nearest-neighbor, caliper, and full matching, and discussed weighting techniques such as IPW and stabilized weights, both of which were complemented by relevant R code examples. We addressed the challenges of implementing these methods, such as the bias-variance tradeoff and issues related to the predictive accuracy of treatment assignment models versus their ability to estimate causal effects. Advanced topics such as sensitivity analysis and handling complex data structures were also explored to enhance your understanding of the nuanced application of propensity score methods in real research scenarios.

In the next chapter, we'll look at a few regression approaches for causal inference.

References

1. Rosenbaum, P.R., & Rubin, D.B. (1983). *The Central Role of the Propensity Score in Observational Studies for Causal Effects.* Biometrika, 70(1), 41-55.

2. Frank J. Massey Jr. (1951) *The Kolmogorov-Smirnov Test for Goodness of Fit*, Journal of the American Statistical Association, 46:253, 68-78, DOI: 10.1080/01621459.1951.10500769

3. Chesnaye NC, Stel VS, Tripepi G, Dekker FW, Fu EL, Zoccali C, Jager KJ. *An introduction to inverse probability of treatment weighting in observational research.* Clin Kidney J. 2021 Aug 26;15(1):14-20. doi: 10.1093/ckj/sfab158. PMID: 35035932; PMCID: PMC8757413.

4. Bewick V, Cheek L, Ball J. Statistics review 14: *Logistic regression.* Crit Care. 2005 Feb;9(1):112-8. doi: 10.1186/cc3045. Epub 2005 Jan 13. PMID: 15693993; PMCID: PMC1065119.

7
Employing Regression Approaches for Causal Inference

We have learned a lot till now. This chapter will further lead you into the deep roots of regression-based methods to discern causality. We will keep our pace pretty much the same as other chapters, which means we will first start with theory and then get to practice the learned theoretical models in real use cases in R, using provided coding scripts and synthetic datasets.

Choosing the right model is an art as much as it is a science, influenced by the nature of the data at hand and the specific causal relationships under investigation. In this chapter, we go deep into model selection, providing you with the insights needed to make informed decisions in choosing the best model for the job. We'll tackle model diagnostics to assess and address the assumptions that underpin the models you deploy. In this, we will discover a wide range of regression models, spanning linear and non-linear versions. We will learn about the assumptions and constraints these models must adhere to provide a systematic outlook on causality. OK, let's get started!

Here are the topics that we will cover:

- Role of regression in causality
- Choosing the appropriate regression model
- Linear regression for causal inference
- Model diagnostics and assumptions
- Non-linear regression for causal inference
- Important considerations in regression modeling

Technical requirements

You can find the code examples for this chapter in this book's GitHub repository: `https://github.com/PacktPublishing/Causal-Inference-in-R/tree/main/chap_07`.

Role of regression in causality

Let's start with a bit of background on regression. A historical understanding of regression may enrich your appreciation for the methodological nuances and widespread application in contemporary causal research. The foundation of regression analysis can be traced back to Sir Francis Galton, who, in the late 19th century, introduced the concept of regression . Galton's work was initially concerned with understanding hereditary characteristics, but his developed principles laid the groundwork for statistical regression [1]. Karl Pearson and Yule further developed regression analysis in the early 20th century, focusing on correlation and linear regression models. Their work emphasized the mathematical relationships between variables, setting the stage for the later use of regression in causal analysis [2,3].

A significant milestone in using regression methods for causal inference was the development of the **Rubin Causal Model** (**RCM**) [4] and the potential outcomes framework in the 1970s. Donald Rubin, along with Paul Holland, introduced the concept of potential outcomes to define causal effects formally [4,5]. This framework shifted the focus from association to causation, emphasizing the importance of counterfactual reasoning in causal inference.

Simultaneously, econometricians such as Trygve Haavelmo and James Heckman's work on the identification of causal models laid the theoretical foundations for using statistical models to infer causal relationships [6]. Heckman further developed these ideas, particularly focusing on selection bias and using instrumental variables to estimate causal effects [7]. Recent developments have further expanded the toolbox for causal inference, including regression discontinuity designs, difference-in-differences methods, and the use of **machine learning** (**ML**) techniques to improve causal estimation [8, 9]. Feel free to look up these works, but we give you a picture early on, to establish that what you are about to learn next is a deep and profound toolset to establish or discern causation.

Choosing the appropriate regression model

As you will see in this chapter, there are a plethora of regression models (many available in R). Selecting the appropriate regression model for causal inference is a critical decision that significantly influences the accuracy of causal effect estimates, the clarity of results interpretation, and the robustness of analytical conclusions. This process demands a thoughtful assessment of various factors, including the type of outcome variable, the relationship dynamics between covariates, adherence to model assumptions, and the need to account for confounding and interaction effects (we will look at this in a bit). Through a systematic review of these elements, we will choose a regression model that not only aligns with your data and business objectives but also strengthens the validity and reliability of the causal findings. In this section, let's familiarize ourselves with what it entails to deploy a regression model.

Understanding the nature of the outcome variable

Let's begin by saying that the type of the outcome variable plays a crucial role in model selection. Linear regression models excel in capturing linear relationships with continuous outcomes, while logistic regression is optimal for binary outcomes by modeling their occurrence probabilities. For count data, Poisson or negative binomial regression models are utilized due to their discrete distribution handling, whereas Cox proportional hazards models are applied for time-to-event data, addressing censoring and time-varying covariates. Each of these models will be explored in detail within the R programming environment next.

Continuous outcomes using linear regression

Let's try to predict house prices based on their size:

```
library(ggplot2)
# Simulated data
set.seed(123)
house_data <- data.frame(size = runif(100, 1000, 5000),
                         price = runif(100, 200000, 500000))
# Linear regression model
model <- lm(price ~ size, data = house_data)
summary(model)
ggplot(house_data, aes(x=size, y=price)) +
  geom_point() +
  geom_smooth(method="lm", col="blue")
```

The output of `lm` provides coefficients indicating the average effect of a one-unit increase in house size on the house price. If the coefficient of `size` is positive and statistically significant (p-value less than a threshold, usually 0.05), it suggests that larger houses tend to be more expensive, with the coefficient value quantifying the average price increase per square unit of size.

Now, in R, let's estimate the probability of passing an exam based on study hours. We choose a logistic regression model as the outcome is a binary variable – 1 for pass and 0 for fail:

```
set.seed(123)
exam_data <- data.frame(study_hours = runif(100, 0, 20),
                        passed = rbinom(100, 1, prob=0.5))

# Logistic regression model
model <- glm(passed ~ study_hours, data=exam_data,
             family="binomial")

# Summary of the model
```

```
summary(model)

# Predicting probabilities
exam_data$predicted_probability <- predict(
  model, type = "response")
```

A logistic regression model tool uses something called log odds, which tells us how likely we are to pass or fail based on how much we study. It turns out that the more hours we spend studying, the better our chances of passing. This is shown by the positive number we get when we study more, meaning more study time leads to higher odds of success. We store the predictions from this model in the `predicted_probability` column, which predicts our chances of passing the exam based on how much we study.

Count outcomes using a Poisson regression model

Now, we will try modeling the number of daily website sign-ups based on advertising spend:

```
library(MASS)
set.seed(123)
ad_data <- data.frame(ad_spend = runif(100, 100, 1000),
                      sign_ups = rpois(100, lambda=20))
# Poisson regression model
model <- glm(sign_ups ~ ad_spend, data=ad_data, family="poisson")
summary(model)
```

In Poisson regression, the coefficient for `ad_spend` tells us how the log of the expected count of sign-ups changes with each additional unit of advertising spend. A positive coefficient suggests that higher advertising spending is associated with an increase in the number of sign-ups.

Time-to-event outcomes using a Cox proportional hazards model

Now, how about we analyze the time-to-equipment failure based on maintenance frequency?

```
library(survival)
set.seed(123)
maintenance_data <- data.frame(
  maintenance_freq = runif(100, 1, 12),
  failure_time = runif(100, 0, 24),
  event = rbinom(100, 1, 0.5))
# Cox proportional hazards model
surv_obj <- Surv(time = maintenance_data$failure_time,
                 event = maintenance_data$event)
model <- coxph(surv_obj ~ maintenance_freq,
               data = maintenance_data)
summary(model)
```

The Cox model provides the hazard ratio associated with maintenance frequency, where a hazard ratio greater than 1 indicates that higher maintenance frequency is associated with a higher risk of equipment failure. The significance of the `maintenance_freq` coefficient helps us understand whether the maintenance frequency significantly affects the time-to-equipment failure, controlling for other factors in the model. Here, we provide a conceptual overview of what is available in the regression space. Feel free to explore these models and test them in your use cases.

Now that we know about a diverse set of regression models, let's dig deeper into what goes around the variables that we choose to use in these models.

Consideration of confounding and interaction effects

You might know already by now that models should be chosen and specified in a way that allows for the control of confounding variables, which could otherwise bias causal effect estimates. If interaction effects between variables are expected, the chosen model must be capable of accommodating these interactions to accurately capture their impact on the outcome. Now, what are interaction effects?

Interaction effects occur in statistical models when the effect of one independent variable on the dependent variable depends on the level of another independent variable. In other words, the presence of an interaction effect indicates that the relationship between two variables changes across the levels of another variable. This concept is crucial in regression analysis and also in other statistical modeling techniques, as it can reveal complex dynamics between variables that would not be apparent when considering each variable in isolation.

Let's consider a hypothetical example to illustrate interaction effects. Think of a scenario where we're studying the effect of a new teaching method (treatment: traditional versus innovative) on student performance (outcome), and we're also interested in how the effect of the teaching method varies by the students' initial interest level in the subject (interest level: high versus low).

Without considering interaction, we might find that the innovative teaching method improves student performance compared to the traditional method, on average. However, this analysis might miss how the effect varies among students with different levels of initial interest.

With interaction, an interaction term between the teaching method and interest level in the model allows us to see that perhaps the innovative method significantly improves performance for students with high initial interest but has little to no effect on students with low initial interest. This interaction tells us that the effectiveness of the teaching method depends on the student's initial interest level. Later in this chapter, we will see it in action in R. Next, let's learn other considerations to think of while choosing the right model for the task.

Model complexity, parsimony, and assumptions

The **principle of parsimony** (or **Occam's razor**) suggests choosing the simplest model that adequately describes the relationships between variables. Overly complex models may overfit the data, performing well on the sample data but poorly on new data. However, complex models may be necessary to capture the true nature of the data but require larger sample sizes and more robust validation to ensure generalizability.

Overfitting occurs when a model learns the training data too well. That includes any noise or random fluctuations, leading to poor generalization on new, unseen data. On the other hand, underfitting happens when a model is naive or too simple to capture the true pattern in the data. To identify overfitting, you should look for a model that performs significantly better on training data than on validation or test data. An underfit model performs poorly on training and validation data. You can use several techniques to detect these issues, including cross-validation, learning curves (plotting model performance against training set size), and regularization methods. In practice, it's a trade-off. Balancing model complexity with performance on both training and validation data is crucial to avoiding both overfitting and underfitting.

Another aspect that needs attention is that *each regression model comes with its own set of assumptions(explained later in the chapter). Violating these assumptions can lead to biased or incorrect estimates.* Before selecting a model, it is essential to consider whether the data meets these assumptions or if adjustments are necessary.

Of course, the chosen model should have good predictive performance, indicated by chosen metrics such as R-squared, **root mean squared error** (**RMSE**), or **area under the curve** (**AUC**) for classification problems. That being said, the other considerations mentioned are very important too. Next, let's dive deeper into the theory of linear regression.

Linear regression for causal inference

Linear regression models are employed to estimate the causal effect of one or more independent variables (your treatments or interventions) on a dependent variable (the outcome). This section delves into the application of linear regression for causal inference, highlighting its assumptions, methodologies, and practical considerations.

The theory

Linear regression models predict the value of a dependent variable based on the linear combination of one or more independent variables. Here is the equation for this model:

$$Y = \beta_0 + \beta_1 X_1 + \beta_2 X_2 + \dots\dots + \beta_n X_n + \epsilon \qquad (1)$$

In this context, Y is the dependent variable we want to predict. The independent variables are $X_1, X_2, \dots\dots X_n$. The intercept β_0 represents the value of Y when all X values are zero.

The coefficients $\beta_1, \beta_2, \dots\dots \beta_n$ show how changes in each X affect Y, keeping other variables constant. Each coefficient indicates the impact of a unit change in its corresponding X on Y.

These coefficients are crucial as they quantify the relationship between variables. For example, a positive β means an increase in X leads to an increase in Y, while a negative β means an increase in X leads to a decrease in Y. The β values provide insight into causality, assuming we have controlled for confounding variables. They measure how changes in independent variables affect the dependent variable, guiding decisions and strategies.

Application of regression modeling in R

For a comprehensive example of linear regression, let's explore the use case of studying the impact of screen exposure (mobile or TV) on adolescents, focusing on an outcome variable such as academic performance. This example will involve generating synthetic data, considering potential confounders, and examining interaction effects. The analysis aims to justify the application of a linear model through detailed code and visualizations in R.

To ensure the synthetic data exhibits a linear relationship between screen time (mobile and TV) and academic performance and includes interaction effects and confounders with a clear linear relationship, let's adjust our data generation approach. We will create a dataset where the impact of screen time on **grade point average (GPA)** is linear, including the effects of confounders such as physical activity, sleep quality, and **socioeconomic status (SES)**, ensuring these relationships align more closely with linear assumptions:

```
set.seed(123) # Ensure reproducibility
n <- 500
# Generating synthetic data with a clearer linear relationship
data <- data.frame(
  mobile_hours = runif(n, 1, 5), # Mobile screen time in hours
  tv_hours = runif(n, 1, 5), # TV screen time in hours
  # Physical activity in hours per week:
  physical_activity = runif(n, 0, 10),
  # Sleep quality on a scale of 1 to 10:
  sleep_quality = sample(1:10, n, replace = TRUE),
  SES = sample(c("low", "medium", "high"),
               n, replace = TRUE), # Socioeconomic status
  age = sample(12:18, n, replace = TRUE) # Age in years
)
# Convert SES to a numeric scale for simplicity
data$SES_numeric <- as.numeric(factor(
  data$SES, levels = c("low", "medium", "high")))
# Adjusting academic performance to ensure a linear relationship
data$GPA <- with(
  data, 3 +
    -0.2 * mobile_hours + #Negative impact from mobile screen time
    -0.15 * tv_hours + # Negative impact from TV screen time
    0.05 * physical_activity + # Positive impact from physical
activity
    0.1 * sleep_quality / 10 + # Positive impact from sleep quality
```

```
    0.1 * SES_numeric + # Positive impact from higher SES
    -0.02 * age + # Slight negative impact from older age
    rnorm(n, mean = 0, sd = 0.25) # Random noise
)
```

To confirm the linear relationships between screen time, confounders, and GPA, we can visualize these relationships using scatter plots with linear regression lines (refer to *Figures 7.1* and *7.2*):

```
library(ggplot2)
# Visualizing GPA against mobile hours with SES as color
ggplot(data, aes(x = mobile_hours, y = GPA, color = SES)) +
  geom_point(alpha = 0.5) +
  geom_smooth(method = "lm", se = FALSE) +
  theme_minimal() +
  labs(title = "GPA vs. Mobile Screen Time by SES",
       x = "Mobile Screen Time (hours)",
       y = "GPA") +
  theme(
    text = element_text(size = 14),  # Increase base text size
    plot.title = element_text(size = 18, face = "bold"),  # Larger,
bold title
    axis.title = element_text(size = 16),  # Larger axis titles
    axis.text = element_text(size = 12),  # Larger axis text
    legend.title = element_text(size = 14),  # Larger legend title
    legend.text = element_text(size = 12)  # Larger legend text
  )
```

We get the following visualization:

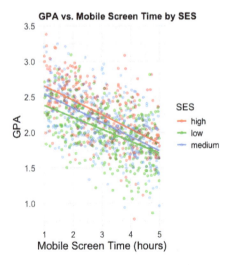

Figure 7.1 – Linear relationship between GPA and mobile screen time in hours

Let's now visualize GPA against TV hours:

```
ggplot(data, aes(x = tv_hours, y = GPA, color = SES)) +
  geom_point(alpha = 0.5) +
  geom_smooth(method = "lm", se = FALSE) +
  theme_minimal() +
  labs(title = "GPA vs. TV Screen Time by SES",
       x = "TV Screen Time (hours)",
       y = "GPA") +
  theme(
    text = element_text(size = 14),  # Increase base text size
    plot.title = element_text(size = 14, face = "bold"),  # Larger,
bold title
    axis.title = element_text(size = 16),  # Larger axis titles
    axis.text = element_text(size = 12),  # Larger axis text
    legend.title = element_text(size = 14),  # Larger legend title
    legend.text = element_text(size = 12)  # Larger legend text
  )
```

We get the result as follows:

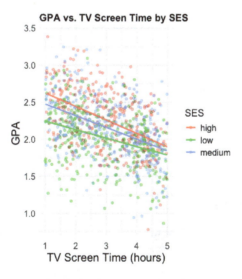

Figure 7.2 – Linear relationship between GPA and TV screen time in hours

These visualizations demonstrate the linear relationships we're assuming between screen time and GPA, differentiated by SES levels. Through this approach, we ensure our synthetic dataset is well suited for illustrating the linear regression analysis, including examining the impact of screen time on academic performance while considering the effects of various confounders.

Before fitting the model, let's visualize the relationships between screen time, confounders, and academic performance.

Start off by loading the `ggplot2` library, which is a versatile package in R for creating data visualizations. It's based on the grammar of graphics, and it allows you to create complex plots from data in a data frame in a declarative manner.

First, we are interested in exploring the relationship between the number of hours students spend on their mobile devices and their GPA, considering their SES as well. To do this, we use the `ggplot()` function, specifying the data frame as `data` and mapping the aesthetics (`aes`) such that `mobile_hours` is on the *x* axis and GPA is on the *y* axis:

```
library(ggplot2)

# Plotting GPA against mobile and TV hours
ggplot(data, aes(x = mobile_hours, y = GPA)) +
  geom_point(aes(color = SES), alpha = 0.5) +
  geom_smooth(method = "lm", se = FALSE, color = "blue") +
  facet_wrap(~SES) +
  theme_minimal() +
  labs(
    title = "Impact of Mobile Hours on GPA by Socioeconomic Status",
    x = "Mobile Screen Time (hours)",
    y = "GPA")
```

To visualize the data points, we will be adding `geom_point()` to the plot, coloring the points based on SES to see if there's a pattern across different SESs. We set the opacity of these points to 0.5 with `alpha = 0.5` for better visualization, especially where points overlap:

```
ggplot(data, aes(x = tv_hours, y = GPA)) +
  geom_point(aes(color = SES), alpha = 0.5) +
  geom_smooth(method = "lm", se = FALSE, color = "red") +
  facet_wrap(~SES) +
  theme_minimal() +
  labs(title = "Impact of TV Hours on GPA by Socioeconomic Status",
       x = "TV Screen Time (hours)",
       y = "GPA")
```

Now, let's fit a linear regression model to assess the impact of screen time on academic performance, including interaction effects and controlling for confounders:

```
# Fitting the linear model
model <- lm(GPA ~ mobile_hours + tv_hours + physical_activity +
                  sleep_quality + SES + age + mobile_hours:tv_hours,
              data = data)
summary(model)
```

The model includes an interaction term (`mobile_hours:tv_hours`) to explore how combined screen time impacts GPA. It also controls for confounders such as physical activity, sleep quality, SES, and age.

The justification for using a linear model lies in the initial exploratory analysis, showing a roughly linear relationship between screen time and GPA, and the assumption that the combined effect of mobile and TV screen time on academic performance can be captured through additive and interaction terms:

```
ggplot(data, aes(x = mobile_hours + tv_hours, y = GPA)) +
  geom_point(aes(color = SES), alpha = 0.5) +
  geom_smooth(method = "lm", formula = y ~ x + I(x^2),
              se = FALSE, color = "green") + theme_minimal() +
  labs(title = "Model Fit: Impact of Total Screen Time on GPA",
       x = "Total Screen Time (hours)",
       y = "GPA")
```

This visualization, alongside the regression summary, shows insights into the strength and direction of the relationships, including how confounders modify the impact of screen time on academic performance. The choice of a linear model is supported by the assumption of linearity in these relationships, confirmed through **exploratory data analysis** (**EDA**) and the statistical significance of model coefficients, which indicate the relative influence of each predictor, including interaction effects, on GPA. Now, let's focus on the distinction between kinds of regression.

Single versus multivariate regression

Let's discuss single-variable linear regression first. It expresses the impact of one variable upon another. In this, we use the following formula:

$$Y = \alpha + \beta X + \varepsilon \qquad (2)$$

Here, the following applies:

- Y is the outcome we want to predict
- X is the predictor variable
- α is the intercept (Y-value when $X = 0$)
- β is the coefficient (change in Y for one unit change in X)
- ε is the error term

β represents the relationship between X and Y, calculated using their covariance and X's variance.

In multivariate linear regression, we extend this to multiple predictors:

$$Y = \alpha + \beta_1 X_1 + \beta_2 X_2 + ... + \beta_k X_k + \varepsilon \qquad (3)$$

Each β_i shows the effect of X_i on Y, controlling for other variables.

The **Frisch-Waugh-Lovell (FWL)** theorem [12] helps isolate each variable's effect (explained later in the chapter), similar to conditions in a randomized trial. This approach is crucial in causal inference, allowing us to estimate the impact of a treatment while accounting for other factors.

Let's keep learning more in this direction.

Treatment orthogonalization

Let's get right to the gist of it. The technique here basically transforms the treatment variable such that it becomes orthogonal (that is, uncorrelated) to the confounding variables. Why do we do it? The purpose is to isolate the variation in the treatment that is independent of these confounders, which can then be used to estimate the causal effect of the treatment on the outcome more accurately.

Let's define it mathematically. Let T be the treatment variable, X be the matrix of confounders, and Y be the outcome. The treatment orthogonalization process can be represented as follows:

- Regress T on X: $T = X\beta + \varepsilon$
- Calculate residuals: $T\perp = T - X\beta$

Here, $T\perp$ represents the orthogonalized treatment variable.

The methodological underpinnings of treatment orthogonalization are further illuminated by the FWL theorem, which delineates a framework for decomposing multivariate regression into distinct steps, one of which crucially encompasses a debiasing phase. In this phase, the treatment variable is regressed against the confounders, yielding residuals that represent the purified treatment effect, stripped of the influence of these confounding factors. These residuals are then utilized in subsequent analyses, ensuring that the estimation of the treatment effect is not compromised by the confounders.

The FWL theorem states that in the model, the following applies:

$$Y = T\alpha + X\beta + \varepsilon \qquad (4)$$

The **ordinary least squares (OLS)** estimate of α is equivalent to the estimate obtained from the regression:

$$Y\perp = T \perp \alpha + u \qquad (5)$$

Here, we have the following:

$$Y\perp = Y - X(X'X)^{-1}X'Y \text{ (residuals from regressing } Y \text{ on } X)$$

$$T\perp = T - X(X'X)^{-1}X'T \text{ (residuals from regressing } T \text{ on } X)$$

This theorem demonstrates that by orthogonalizing both the treatment and outcome variables with respect to the confounders, we can obtain an unbiased estimate of the treatment effect α, effectively controlling for the confounding variables. Treatment orthogonalization often involves two steps:

1. **Modeling the treatment assignment**: First, we fit a model to predict the treatment assignment based on the observed confounders. This model can take various forms, such as a linear regression model or a logistic regression model for binary treatments. The goal here is to capture how the confounding variables influence the likelihood of receiving the treatment.

2. **Residualizing the treatment**: The fitted model from the first step is then used to calculate the residuals for each observation, which represent the part of the treatment variable that cannot be explained by the confounders. These residuals are essentially the treatment variable that has been orthogonalized with respect to the confounders. The orthogonalized treatment variable is then used in the outcome model to estimate the causal effect.

This approach is particularly useful in the context of linear models and has been extended to more complex models and settings. Treatment orthogonalization helps in reducing bias in causal effect estimation, especially in observational studies where randomized control trials are not feasible and confounding is a significant concern.

One notable implementation of treatment orthogonalization in the context of causal inference with R programming is through the use of instrumental variables(explained in *Chapter 10*) and **two-stage least squares** (**2SLS**) regression, where the first stage can be seen as an effort to orthogonalize the treatment with respect to the confounders, and the second stage estimates the effect of the orthogonalized treatment on the outcome.

Example of the FWL theorem

Let us illustrate the FWL theorem using the treatment (T), confounders (X), and outcome (Y) framework. Consider a scenario where we want to estimate the causal effect of a treatment T on an outcome Y while controlling for confounders X. The FWL theorem provides a step-by-step approach to isolate the treatment effect:

1. Orthogonalize the treatment with respect to confounders:

   - Fit a model: $T = \alpha X + \varepsilon$

   - Calculate residuals: $T\perp = T - \hat{\alpha}X$. Here, $T\perp$ represents the part of T that cannot be explained by X.

2. Orthogonalize the outcome with respect to confounders:

    - Fit a model: $Y = \beta X + \eta$

    - Calculate residuals: $Y\perp = Y - \beta X$. Here, $Y\perp$ represents the part of Y that cannot be explained by X.

3. Estimate the treatment effect:

    - Fit the final model: $Y\perp = \gamma T\perp + \xi$

The coefficient γ in this final model represents the causal effect of T on Y, controlling for X. This process effectively does the following:

- Removes the influence of confounders from both the treatment and outcome.

- Estimates the relationship between the parts of T and Y that are not explained by the confounders.

By following these steps, we isolate the true effect of the treatment on the outcome, free from confounding influences. This method is particularly useful when we have a binary treatment and want to control for multiple confounders, as it provides a clear path to estimating the treatment effect while accounting for these other variables.

By partitioning the data and analyzing the residuals, we can understand the true nature of influence and causality. This method allows us to see the true structure of relationships, guiding us through the complexities of linear regression.

Next, we will learn about model diagnostics.

Model diagnostics and assumptions

Linear regression requires certain assumptions to be met to ensure accurate results:

- **Linearity**: The first assumption is linearity, which means there should be a straight-line relationship between the independent and dependent variables. For example, if you plot study hours against exam scores and find a straight line that fits well, then the linearity assumption is satisfied. However, if the data forms a curve, it indicates that the linearity assumption might be violated, which could affect the accuracy of the regression results. In other words, the relationship between independent and dependent variables should be linear.

- **Assumption of independence**: In this assumption, each data point should be independent of others. For example, in a clinical trial, each participant's outcome should not be influenced by another's. Violating this assumption can lead to incorrect results.

- **Homoscedasticity**: It means that the variance of residuals (also known as errors) should be consistent across all levels of the independent variables. Residuals are the differences between observed and predicted values in a regression model. If this condition is not met, it indicates heteroscedasticity (opposite of homoscedasticity), which suggests that the model might be missing important information.

- **The normality of error terms**: This assumption is crucial for reliable statistical inferences. The errors, or the differences between predicted and actual values, should ideally follow a normal distribution. If they don't, the reliability of T-tests and F-tests is compromised.

- **Multicollinearity**: Lastly, no perfect multicollinearity means that independent variables should not be highly correlated with each other. If they are, it becomes difficult to determine the individual effect of each variable on the dependent variable. To assess multicollinearity, one can calculate the Pearson correlation between independent variables. If high correlations are present, various methods can be employed to address the issue, such as removing one of the correlated variables, combining variables, or using regularization techniques.

As you can see, regression analysis relies on several key assumptions, and careful consideration is crucial to ensure accurate results. Here's a straightforward guideline to follow when you apply regression models in your causal problems:

- **A large sample size** is essential for reliable results. A small sample size may lead to imprecise estimates and statistical insignificance. Ensure your sample is sufficiently large to capture the true relationships in your data. Here, sample size is subjective and can vary depending on the domain. Different fields may have different standards for what constitutes a "sufficiently large" sample size due to variations in data characteristics and research goals.

- **Omitted variable bias**: Missing important variables can lead to incorrect conclusions.

- **Irrelevant variables**: Including unnecessary variables can introduce noise.

- **Multicollinearity**: Highly correlated independent variables can make it difficult to isolate individual effects. Use the **Variance Inflation Factor** (**VIF**) to detect multicollinearity.

- **Outliers**: Identify and assess the impact of outliers. To identify and assess outliers in regression analysis, use scatter plots, residual plots, and standardized residuals to flag potential outliers. Investigate whether outliers are due to errors or represent valid observations, and decide whether to keep, remove, or apply robust regression techniques. Finally, rerun the analysis to compare results with and without outliers for a comprehensive understanding.

- **Handling missing values**: Real data is often incomplete and has missing values. Addressing them in your dataset is essential before conducting regression analysis. Including missing values significantly impacts the accuracy and reliability of your results. There are several methods to handle missing data. One approach is **complete case analysis** (**CCA**), which involves removing any observations with missing values. While this method is straightforward, it can lead to a loss of valuable information. Instead, you can use various imputation techniques to estimate and replace missing values. For instance, methods such as the use of mean, median, or mode of the variable can be used to fill in missing values. In addition, you can do regression imputation based on other variables and multiple imputation, which creates several plausible datasets with different estimates for the missing values.

Now, which methodology should you adopt? Good question. To answer this question, you should investigate the extent of missingness, the mechanism behind the missing data (**Missing Completely At Random – MCAR, Missing At Random – MAR**, or **Missing Not At Random – MNAR**), and the specific requirements of your analysis. Tools such as R's `mice` package for multiple imputation or the `missForest` package for random forest imputation can assist in effectively managing missing values. Remember that ensuring that your dataset is as complete and accurate as possible will enhance the validity of your regression analysis. Perform diagnostic checks to ensure your model meets the assumptions. To understand, let's run through a linear regression application example in R.

Consider a scenario where we evaluate the impact of a job training program on wages, with prior work experience as a confounding variable.

Let's first create a dataset in which we include variables such as training program participation, prior work experience, and wages. We'll then build a linear regression model including the confounding variable (prior work experience) to control for its effect:

```
n <- 1000
experience <- rnorm(n, mean=5, sd=2)
# Probability of training increases with experience
training_prob <- plogis(0.5 * experience)
training <- rbinom(n, 1, training_prob)
# Simulating wages, influenced by both training and experience
wages <- 30 + 5 * training + 2 * experience + rnorm(n)
data <- data.frame(wages, training, experience)
```

Now, we'll fit a linear regression model to estimate the effect of the job training program on wages, controlling for prior work experience:

```
model <- lm(wages ~ training + experience, data=data)
summary(model)
```

From the `summary` output, we'll focus on the coefficient for `training`. This coefficient estimates the average effect of participating in the job training program on wages, controlling for the effect of prior work experience.

Our simulation used a sample size of 1,000, which is generally adequate for linear regression analysis. However, the required sample size can vary depending on the effect size, the number of predictors, and the desired power of the study. We included `experience` as a control variable because of its potential to confound the relationship between `training` and `wages`. Careful consideration of which variables to include is crucial to avoid omitted variable bias and reduce noise. If we had additional variables, we would need to check for multicollinearity, possibly using VIF analysis.

VIF measures the extent of multicollinearity among independent variables in a regression model, indicating how much the variance of a coefficient is inflated due to linear dependence on other predictors. A VIF value of 1 suggests no correlation, values between 1 and 5 indicate moderate correlation considered acceptable, while a value of 5 or above signals problematic multicollinearity that might require remedial measures such as removing or combining correlated variables. High VIF values compromise the reliability of regression estimates, making it essential to check VIF in model diagnostic procedures. The following code shows how you can do that in R using the `car` package:

```
# Package required for VIF analysis
if (!require(car)) {
  install.packages("car")
  library(car)
}
# Perform VIF analysis
vif_results <- vif(model)
print(vif_results)

# Interpret VIF results
# VIF > 5 or 10 is often considered problematic

# If multicollinearity is detected, you can mitigate it by:
# 1. Removing one of the highly correlated variables
model_reduced <- lm(wages ~ training, data=data)
summary(model_reduced)
```

However, in our simple model, this concern is minimal. After fitting the model, it's essential to check for assumption violations:

```
# Checking for non-linearity and heteroscedasticity
plot(model, which=1:6)
# Conducting the Breusch-Pagan test
library(lmtest)
bptest(model)
```

This diagnostic process helps ensure the reliability of our causal inference. *By examining the residuals, we can detect potential problems such as non-linearity, heteroscedasticity, or outliers that might invalidate our model assumptions. If such issues are present, further steps might include transforming variables, adding interaction terms, or using robust regression techniques.*

The first part, `plot(model, which=1:6)`, generates diagnostic plots for checking non-linearity and heteroscedasticity in the residuals, which are essential for assessing the linearity of the relationship between predictors and the outcome, as well as the consistency of variance across predictions. *Figure 7.3* shows a plot of residuals randomly scattered around zero, indicating that the linearity assumption is reasonably met. The second part, using `bptest(model)` from the `lmtest` package, conducts the Breusch-Pagan test to formally test for heteroscedasticity, or unequal variance in the error terms, across the levels of the independent variables.

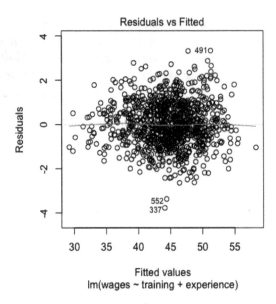

Figure 7.3 – Residual versus fitted plots to verify linearity

This example demonstrates how linear regression, supplemented with careful consideration of practical issues and diagnostic checks, can be a powerful tool for causal inference in R. OK, let's now get our hands dirty with non-linear regression.

Non-linear regression for causal inference

Regression's capability extends beyond linear associations by leveraging parameter linearity rather than necessitating linearity in the data itself. Parameter linearity in regression means that the model is linear in its coefficients (parameters), regardless of whether the relationship between variables is linear or non-linear.

This flexibility allows the modeling of non-linear relationships through data transformations and the inclusion of polynomial or interaction terms. Specifically, interaction terms, which model the multiplicative effects between variables, are incorporated by adding a product term to the regression equation. For example, to model the interaction between variables X_1 and X_2, we introduce the term $X_1 \times X_2$ into the model. Additionally, non-linear dependencies can be quantitatively assessed using entropy-based metrics such as mutual information, providing a robust framework for understanding complex variable relationships beyond the linear model's scope.

On the other hand, when relationships are non-linear, polynomial regression, spline models, or **generalized additive models (GAMs)** [10] can capture more complex dynamics.

In causal science, non-linear regression models stand out for their ability to elucidate complex relationships between variables that linear models might oversimplify or misinterpret. These models expand the analytical possibilities by accommodating various types of outcome variables and the intricate dynamics between predictors and outcomes. Logistic regression is particularly pivotal when dealing with binary outcomes, such as success versus failure or yes versus no scenarios. It excels by modeling the probability of an outcome, offering interpretable results in the form of odds ratios. This feature makes logistic regression invaluable in fields such as medicine and social science, where binary outcomes are common. However, its reliance on the linearity assumption in the log odds scale and potential biases from model misspecification or omitted variables are notable limitations.

Other types of non-linear models

Poisson and negative binomial regressions are specialized models for analyzing count data – data where the outcome is a non-negative integer (such as the number of accidents or disease cases). Poisson regression is simpler and assumes the mean equals the variance of the counts. It's useful when this assumption holds true in your data. Negative binomial regression is an extension of Poisson that allows for greater variability in the data. It's helpful when counts are more spread out than Poisson would predict. Note that the Poisson model assumes equidispersion (mean equals variance). If your count data is overdispersed (variance > mean), the Poisson model may not be the appropriate choice, and a negative binomial regression could be a better option. Both models are commonly used in fields such as epidemiology and insurance, where counting events is crucial.

You can do it in R, as seen here:

```
# Simulated count data for Poisson and Negative Binomial regression
(accidents)
set.seed(123)
data <- data.frame(
  training = sample(0:1, 100, replace = TRUE),
  experience = rnorm(100, 5, 2),
  accidents = rpois(100, lambda = exp(
    1 + 0.5 * sample(0:1, 100, replace = TRUE) +
      0.3 * rnorm(100, 5, 2))),
  wages = rnorm(100, 50000, 10000)
)


# Poisson Regression with count data
model_poisson <- glm(accidents ~ training + experience,
                     family = "poisson", data = data)
summary(model_poisson)

# Check for overdispersion
dispersion <- sum(residuals(
  model_poisson,
```

```
      type = "pearson")^2) / df.residual(model_poisson)
dispersion  # If > 1, it suggests overdispersion

# If overdispersion is detected, fit Negative Binomial regression
if (dispersion > 1) {
  model_negbin <- glm.nb(accidents ~ training + experience,
                         data = data)
  summary(model_negbin)
}
```

For time-to-event data, proportional hazards models (or Cox regression) offer a robust framework. These models are instrumental in survival analysis, enabling researchers to understand how covariates influence the hazard rate without specifying the baseline hazard function. When dealing with time-to-event data, it is often important to simulate **censoring**, where some observations don't experience the event by the end of the study period. Despite their flexibility and ability to include time-varying covariates, they assume proportional hazards, which may not always hold, and interpreting hazard ratios can be less straightforward than direct effect sizes:

```
# Simulated time-to-event data for Cox regression
# (failure_time and event)
set.seed(123)
maintenance_data <- data.frame(
  maintenance_freq = runif(100, 1, 12),
  failure_time = rexp(100, rate = 0.1), # Time-to-event data
  event = rbinom(100, 1, 0.7)           # Event occurrence
)

# Create censoring (if failure_time > 20, it's censored)
maintenance_data$event <- ifelse(
  maintenance_data$failure_time > 20, 0, 1)

# Cox Proportional Hazards Model
surv_obj <- Surv(time = maintenance_data$failure_time,
                 event = maintenance_data$event)
model_cox <- coxph(surv_obj ~ maintenance_freq,
                   data = maintenance_data)
summary(model_cox)

# Check proportional hazards assumption
cox.zph(model_cox)  # Proportional hazards diagnostic
```

GAMs represent a leap in modeling flexibility, adeptly handling non-linear relationships through smooth functions. This adaptability makes GAMs suitable for a wide range of applications, from ecological to financial data, where non-linearity is a given. The trade-off, however, lies in the complexity of model interpretation and the critical need for careful selection of smoothing parameters to prevent overfitting or underfitting:

```
# Generalized Additive Model (GAM) to handle non-linear relationships
# Treat `training` as a factor (binary) and apply smoothing to
# `experience` (continuous)
gam_model <- gam(wages ~ training + s(experience), data = data)
summary(gam_model)

# Visualization of the GAM model with smoothing on `experience`
ggplot(data, aes(
  x = experience, y = wages, color = as.factor(training))) +
  geom_point() +
  geom_smooth(method = "gam", formula = y ~ s(x), col = "green")
```

Lastly, **multilevel models** (or **hierarchical models**) address the challenges of grouped or clustered data, such as students nested within schools. By accounting for the data's hierarchical structure, these models offer nuanced insights into both the fixed effects of predictors and the random effects of groups. Although powerful, their application is marred by complex estimation and interpretation processes, especially with extensive levels or groups, and demands sufficient within-group data to accurately estimate group-level effects:

```
# Simulated data for multilevel model (grouped wages)
set.seed(123)
multilevel_data <- data.frame(
  group = factor(rep(1:10, each = 10)),  # 10 groups
  training = sample(0:1, 100, replace = TRUE),
  experience = rnorm(100, 5, 2),
  wages = rnorm(100, 50000, 10000)
)

# Multilevel Model with random intercepts for groups
multilevel_model <- lmer(wages ~ training + experience + (
  1 | group), data = multilevel_data)
summary(multilevel_model)

# Updating dataset: adding effect of training and experience on wages
multilevel_data$wages <- multilevel_data$wages + 5 *
  multilevel_data$training + 2 * multilevel_data$experience +
  rnorm(100, 0, 5000)
summary(multilevel_model)
```

Updating the dataset in the previous example was meant to illustrate how the effects of training and experience on wages could be simulated, ensuring that the model captures relationships accurately. However, if you already have a dataset and don't need to simulate the effects, you can skip this step.

In conclusion, non-linear regression models are indispensable in causal inference. They help tackle a broad spectrum of research questions across various disciplines. The choice among these models hinges on the specific characteristics of the data at hand, the nature of the outcome variable, and the intricacies of the underlying relationships. By carefully weighing each model's strengths and limitations, researchers can select the most fitting approach, paving the way for deeper insights and more robust causal inferences.

Application of a non-linear regression problem in R

Let's consider how we can model a non-linear regression problem in R. Let's think that we wish to explore the impact of traffic patterns on people's work-life balance and stress. In this, we will use a synthetic dataset that simulates this scenario. We will create a non-linear regression model to capture the complex relationships and potential interaction effects between traffic conditions, work-life balance, and stress levels. We'll also consider confounders such as working hours, mode of commute, and personal factors such as flexibility in work schedule.

First, we'll generate synthetic data that reflects non-linear relationships and includes potential confounders:

```r
n <- 1000
data <- data.frame(
  working_hours = runif(n, 6, 12), # Total working hours per day
  commute_mode = factor(sample(c(
    "Car", "Public Transport", "Bike", "Walk"),
    size = n, replace = TRUE)), # Mode of commute
  # Traffic level on a scale from 0 (no traffic) to 10 (heavy traffic)
  traffic_level = runif(n, 0, 10),
  # Flexibility in work schedule (
  #0 = no flexibility, 1 = high flexibility)
  work_flexibility = runif(n, 0, 1),
  stress_level = rep(0, n) # Placeholder for stress level
)
# Non-linear relationship + interaction + confounding
data$stress_level <- with(
  data, 2 + (0.5 * working_hours) + (traffic_level^2)/10 -
    (work_flexibility * 2) + ifelse(
      commute_mode == "Car", traffic_level * 0.4,
      ifelse(commute_mode == "Public Transport",
             traffic_level * 0.2, 0)) +
    rnorm(n, mean = 0, sd = 2))
# Visualize the non-linear relationship between traffic level and
```

```
stress level
library(ggplot2)
ggplot(data, aes(x = traffic_level, y = stress_level,
                 color = commute_mode)) +
  geom_point(alpha = 0.5) +
  geom_smooth(method = "loess", formula = y ~ x, linewidth = 1) +
  labs(
    title = "Impact of Traffic Level on Stress Level by Commute Mode",
    x = "Traffic Level",
    y = "Stress Level") +
  theme_minimal()
```

The synthetic data includes a non-linear term for traffic level (`traffic_level^2`) to simulate the increasing stress levels at a non-linear rate as traffic worsens. The interaction effect between `traffic_level` and `commute_mode` represents how different modes of commute modify the impact of traffic on stress. The `work_flexibility` variable acts as a confounder, influencing both the stress level and, potentially, the choice of commute mode.

The visualization will show the non-linear relationship between traffic levels and stress (refer to *Figure 7.4*), differentiated by commute modes, providing an intuitive justification for employing a non-linear model:

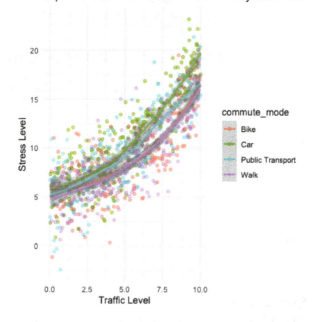

Figure 7.4 – Non-linear relationship between stress and traffic

For the non-linear regression, we will use the `nls` function in R, which is designed for **non-linear least squares** (**NLS**) estimation. We'll model the stress level as a function of traffic level, commute mode, working hours, and work flexibility, including non-linear terms and interactions:

```
# Initial guess for parameters
start_vals <- list(a = 2, b = 0.5, c = 0.1, d = -2)
# Fitting the non-linear model
non_linear_model <- nls(
  stress_level ~ a + b * working_hours + c *
    I(traffic_level^2) + d * work_flexibility,
  data = data,
  start = start_vals)
summary(non_linear_model)
```

The `nls` model summary provides estimates for the parameters governing the relationship between stress level and its predictors. A significant `c` parameter for the `traffic_level^2` term confirms the non-linear impact of traffic on stress. The model also estimates the effects of working hours and work flexibility, accounting for their influence on stress levels.

This example underscores the importance of considering non-linear relationships and interaction effects in modeling complex real-world phenomena. The synthetic data and subsequent analysis illustrate how non-linear regression can capture these dynamics, offering deeper insights than linear models. Through careful consideration of confounders and the use of visualizations, we can better understand the suitability of non-linear models for specific research questions.

Important considerations in regression modeling

This section outlines a set of discussion points that we should bring up. You will face them as you model regression equations to support finding evidence of causality. Let's start!

Which covariates to consider in the model?

The question of whether to control for all available covariates in causal analysis is nuanced, reflecting deep considerations around the potential for confounding, the relevance of variables, and the structural relationships among them. The primary concern is that while controlling for variables (covariates) aims to mitigate confounding, this strategy does not uniformly lead to more accurate causal conclusions. The rationale often cited for extensive control is to yield more conservative hypothesis tests. However, this can backfire, as controlling for the wrong variables may introduce spurious effects or even reverse the signs of effects, leading to erroneous conclusions about causal relationships.

What are these effects? Spurious effects refer to observed associations between two variables that are not causally related but are due to a third variable or chance. Reversal of effect signs occurs when the inclusion or exclusion of relevant variables from a model changes the direction of an estimated effect from positive to negative or vice versa. This is often related to omitted variable bias.

Let's consider a hypothetical example in the context of economic research, where the goal is to analyze the impact of education level X_1 on income Y, but other variables are included that may not be relevant or could introduce spurious relationships and potentially reverse the sign of the effect. Suppose we initially hypothesize that higher education levels should positively correlate with higher income, based on the theory that better education provides better employment opportunities.

Initially, we fit a simple linear regression model:

$$Y = \beta_0 + \beta_1 X_1 + \epsilon \qquad (6)$$

Here, the following applies:

- Y = Income
- X_1 = Years of education
- β_0 = Intercept
- β_1 = Effect of education on income
- ϵ = Error term

In this simple model, suppose we find $\beta_1 > 0$, confirming our hypothesis that higher education levels are associated with higher income.

Next, we decide to add more variables to the model, including the number of hours spent watching TV per week X_2 and the number of hours spent on physical activities per week X_3, leading to the following:

$$Y = \alpha_0 + \alpha_1 X_1 + \alpha_2 X_2 + \alpha_3 X_3 + \epsilon \qquad (7)$$

Unexpectedly, in this over-specified model, we find that $\alpha_1 < 0$, suggesting that higher education levels are associated with lower income, which contradicts our initial findings and common sense.

This may result in the following:

- **Spurious effect**: The inclusion of X_2 (TV watching) and X_3 (physical activities) might introduce spurious relationships. For example, if both X_2 and X_3 are negatively correlated with Y due to reasons unrelated to education (for example, people with lower income might spend more time watching TV because it's a cheaper form of entertainment and might work in more physically demanding jobs), their inclusion without proper theoretical justification can create a misleading model.

  Including X_2 (TV watching) and X_3 (physical activities) can cause spurious effects if they correlate with the outcome Y for unrelated reasons, such as lower-income individuals watching more TV due to its affordability and having more physically demanding jobs. These correlations, unrelated to education, can mislead the model by attributing changes in Y to X_2 and X_3 instead of the true causes. Proper theoretical justification is crucial to avoid such misleading relationships.

- **Reversed sign effect:** The sign reversal for α_1 from positive to negative could result from omitted variable bias or the accidental control of mediators or colliders in the pathway from education to income. For instance, suppose there's a variable Z (such as job market conditions) that we failed to include in our model. Z could have a significant impact on the relationship between education and income. By including X_2 and X_3, we might inadvertently control for factors that are affected by Z, distorting the original relationship observed between X_1 and Y.

A nuanced approach recommended by Becker et al. [11] suggests exercising discretion in the selection of control variables, advising against the use of uncertain variables as controls and advocating for conceptually meaningful choices. This approach is underlined by the recommendation to conduct comparative tests and review results with and without the inclusion of control variables, thereby gauging their impact on the analysis.

Knowledge of the causal graph significantly aids in deciding which variables to control for, simplifying the task when the causal structure is known. Yet, without such knowledge, the decision becomes inherently complex. I would argue while there's no universal solution to the dilemma of whether to control for all covariates, a deep understanding of causality can vastly improve decision-making in this context.

The decision to control for all available covariates is not straightforward and requires careful consideration of the causal structure, relevance, and potential impact of each variable on the analysis. The guidance provided leans toward a strategic approach, prioritizing theoretical knowledge and empirical evidence over broad heuristics, thereby fostering more accurate and reliable causal inference. In the world of variables, let's learn about a new type next.

Dummy variables? What are they?

Creating dummy variables allows for the inclusion of categorical data into regression models, enabling the analysis of how different groups affect the outcome variable. This approach extends the flexibility and applicability of regression analysis, making it a powerful tool for exploring and interpreting the effects of categorical predictors on a continuous outcome.

To be more specific, a dummy variable, also known as an indicator variable, is a numerical variable used in regression analysis to represent subgroups of the sample in your analysis. In essence, it's a way to include categorical data in a regression model, which requires numerical input. Each dummy variable can only take the value of 0 or 1, indicating the absence or presence of some categorical effect that may be expected to shift the outcome variable.

Dummy variables are crucial for controlling for categorical predictors in regression models, allowing for the estimation of separate constant terms (intercepts) for different groups. For example, if you have a categorical variable such as gender (male/female), you can create a dummy variable that takes the value of 1 if the individual is female and 0 if male (or vice versa). When dealing with a large number of dummy variables resulting in sparse data, techniques such as regularization (for example, Lasso or Ridge regression), dimensionality reduction methods (for example, **Principal Component**

Analysis or **PCA**), or feature selection algorithms can help manage computational complexity and potential overfitting issues. These methods can effectively handle high-dimensional sparse data while maintaining model interpretability and performance [13].

Imagine we want to model the salary of individuals based on their gender and years of experience. The dataset contains the following columns: `Salary` (continuous), `Gender` (categorical with `Male` and `Female`), and `YearsExperience` (continuous).

First, we need to create a dummy variable for the `Gender` categorical variable:

```
data <- data.frame(
  Salary = c(50000, 55000, 57000, 61000, 67000, 58000),
  Gender = c("Male", "Female", "Male",
             "Female", "Male", "Female"),
  YearsExperience = c(1, 2, 3, 4, 5, 6)
)
# Creating dummy variable for Gender
data$GenderFemale <- as.numeric(data$Gender == "Female")
```

With the dummy variable in place, we can now include it in our regression model to see how gender, along with years of experience, affects salary:

```
# Linear regression model with dummy variable
model <- lm(Salary ~ YearsExperience + GenderFemale, data=data)
# Summary of the model to see coefficients
summary(model)
```

The `GenderFemale` coefficient will tell us the difference in salary between females and males, holding years of experience constant. A positive coefficient for `GenderFemale` indicates that females, on average, have a higher salary than males when years of experience are accounted for, while a negative coefficient would suggest the opposite.

Next, let's understand the true effect of using a regression approach in causality.

Orthogonalization effect in R

As we discussed earlier, orthogonalization in regression isolates the treatment's effect from covariates, making it independent of their influence. This ensures that the treatment's unique contribution to the outcome is accurately captured. In this discussion, let's visualize this concept to understand it.

This orthogonalization happens through the projection of the treatment variable onto the space spanned by the other covariates and then considering the component of the treatment variable that is perpendicular (orthogonal) to this space. The regression coefficient of the treatment variable then represents the effect of a one-unit change in the treatment on the outcome, holding all other covariates constant:

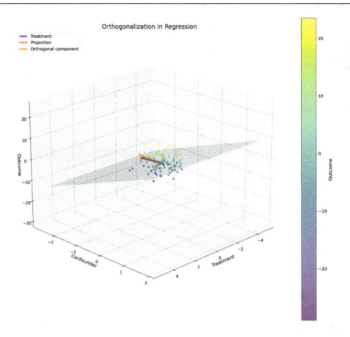

Figure 7.5 – Orthogonalization effect in regression modeling

The concept is grounded in the geometry of least squares regression. In a multiple regression context, the estimation of the coefficient for the treatment variable can be thought of as first projecting the treatment variable onto the subspace defined by the other covariates (thus accounting for their effects) and then assessing the residual variation in the treatment variable that is orthogonal to this subspace (see *Figure 7.5*). The regression coefficient for the treatment effect is estimated based on this orthogonal component, ensuring that it captures the effect of the treatment that is not explained by the other covariates. *Figure 7.5* illustrates orthogonalization in regression using a 3D plot. The purple arrow represents the total treatment effect, the red arrow shows its projection onto the confounder plane (gray), and the orange arrow displays the orthogonal component. This orthogonal component represents the isolated effect of the treatment after accounting for the confounder, demonstrating how regression separates the unique contribution of variables.

This property of regression is particularly important in causal inference because it implies that if the model is correctly specified and includes all relevant confounders, the coefficient for the treatment variable can be interpreted as an estimate of the causal effect of the treatment on the outcome. By orthogonalizing the treatment with respect to the confounders, regression analysis provides a way to estimate causal effects even in observational data, assuming that there are no unmeasured confounders.

To illustrate the concept of orthogonalization in regression through R code, let's consider a simple example. We'll simulate data where we have a treatment variable, a confounder, and an outcome variable. The goal is to demonstrate how regression can isolate the effect of the treatment on the outcome, making the treatment effect orthogonal to the confounder.

First, let's simulate some data. We'll create a scenario where both the treatment and the outcome are influenced by a confounder:

```
set.seed(123)
n <- 100 # Number of observations

# Simulate a confounder
confounder <- rnorm(n, mean = 50, sd = 10)

# Simulate a treatment variable that is also influenced by the
confounder
treatment <- 0.5 * confounder + rnorm(n)

# Simulate an outcome variable influenced by both the treatment and
the confounder
outcome <- 2 * treatment + 0.3 * confounder + rnorm(n)
# Combine into a data frame
data <- data.frame(outcome, treatment, confounder)
```

Next, let's fit a model without controlling for the confounder, to see the biased estimation of the treatment effect:

```
model_without_confounder <- lm(outcome ~ treatment, data = data)
summary(model_without_confounder)
```

Now, let's fit a regression model that includes the confounder, to demonstrate orthogonalization:

```
model_with_confounder <- lm(outcome ~ treatment + confounder, data =
data)
summary(model_with_confounder)
```

The coefficient of `treatment` in `model_with_confounder` will show the effect of the treatment on the outcome, adjusted for the confounder. This adjustment makes the treatment effect orthogonal to the confounder within the model, meaning it isolates the effect of the treatment from the influence of the confounder.

To explicitly show the orthogonalization process, we can decompose the treatment variable into components that are orthogonal and not orthogonal to the confounder:

```
treatment_on_confounder <- lm(treatment ~ confounder, data = data)

# Extract the residuals, which represent the part of
# treatment orthogonal to the confounder
data$orthogonal_treatment <- residuals(treatment_on_confounder)
# Fit a model using the orthogonal component of the treatment
model_orthogonal_treatment <- lm(
```

```
    outcome ~ orthogonal_treatment + confounder, data = data)
summary(model_orthogonal_treatment)
```

In `model_orthogonal_treatment`, the coefficient for `orthogonal_treatment` represents the effect of the treatment on the outcome, independent of the confounder. This process demonstrates how regression orthogonalizes the treatment with respect to the confounder, allowing us to estimate the unique contribution of the treatment to the outcome. This principle is foundational in causal inference, where the goal is to discern the causal impact of treatments or interventions while accounting for potential confounders.

Summary

In this chapter, we discussed regression methods to unearth causality from intricate datasets, beginning with the basics of regression's role in causality and advancing toward choosing the right models, including both linear and non-linear approaches. We emphasized hands-on application, leveraging R scripts and synthetic datasets to marry theory with practical execution. This journey through various regression techniques such as linear, logistic, and poisson regression equipped you with the necessary skills for model diagnostics and covariate selection, with a special focus on the creation and use of dummy variables to highlight the effects of categorical predictors.

Furthermore, we highlighted the crucial process of orthogonalization in regression analysis, which isolates the unique contributions of treatment variables, independent of confounders. Through practical examples and R code demonstrations, we illustrated the precision of regression models in estimating causal effects. The chapter also delved into critical considerations for causal inference, such as strategic covariate selection and the interpretation of model coefficients, culminating in a rich understanding of regression approaches in causal analysis. This synthesis of practical insights and foundational theory effectively bridges the gap between statistical concepts and their real-world applications, equipping you with a robust framework for causal inference. Now, we are perfectly placed to learn real-world experimental settings, specifically A/B testing and controlled experiments, in the next chapter.

References

1. Stigler, S. M. (1986). *The history of statistics: The measurement of uncertainty before 1900*. Harvard University Press.

2. Pearson, K. (1896). *Mathematical contributions to the theory of evolution. III. Regression, heredity, and panmixia*. Philosophical Transactions of the Royal Society of London. Series A, Containing Papers of a Mathematical or Physical Character, 187, 253-318.

3. Yule, G. U. (1909). *The theory of correlation*. Journal of the Royal Statistical Society, 72(4), 721-730.

4. Rubin, D. B. (1974). *Estimating causal effects of treatments in randomized and nonrandomized studies*. Journal of Educational Psychology, 66(5), 688-701.

5. Holland, P. W. (1986). *Statistics and causal inference.* Journal of the American Statistical Association, 81(396), 945-960.

6. Haavelmo, T. (1943). *The statistical implications of a system of simultaneous equations.* Econometrica, 11(1), 1-12.

7. Heckman, J. J. (1979). *Sample selection bias as a specification error.* Econometrica, 47(1), 153-161.

8. Imbens, G. W. & Lemieux, T. (2008). *Regression discontinuity designs: A guide to practice.* Journal of Econometrics, 142(2), 615-635.

9. Angrist, J. D. & Pischke, J.-S. (2009). *Mostly harmless econometrics: An empiricist's companion.* Princeton University Press.

10. Wood, S.N. (2017). *Generalized Additive Models: An Introduction with R, Second Edition (2nd ed.).* Chapman and Hall/CRC. https://doi.org/10.1201/9781315370279

11. Becker, Thomas E., Guclu Atinc, James Breaugh, Kevin D. Carlson, Jeffrey R. Edwards, and Paul E. Spector. *Statistical control in correlational studies: 10 essential recommendations for organizational researchers".* Journal of Organizational Behavior 37 (2016): 157-167.

12. Frisch, Ragnar; Waugh, Frederick V. (1933). *Partial Time Regressions as Compared with Individual Trends.* Econometrica. 1 (4): 387–401. doi:10.2307/1907330. JSTOR 1907330.

13. Eric Chu, Jennifer Beckmann, and Jeffrey Naughton. 2007. *The case for a wide-table approach to manage sparse relational data sets.* In Proceedings of the 2007 ACM SIGMOD international conference on Management of data (SIGMOD '07). Association for Computing Machinery, New York, NY, USA, 821–832. https://doi.org/10.1145/1247480.1247571

8

Executing A/B Testing and Controlled Experiments

In this chapter, we are excited to discuss A/B testing and experimental designs to uncover causal relationships. Here, the main goal is to equip you with a comprehensive understanding of how to effectively plan, execute, and analyze controlled experiments.

These experimental techniques are extremely vital because they provide a robust framework for isolating the impact of specific factors and identifying causal effects. The topics covered in this chapter will help you know the key stages of the experimental process, from hypothesis formulation and sample size planning to randomization, group assignment, and advanced statistical analysis.

By the end of the chapter, you'll have a solid grasp of the tools and techniques needed to design and interpret high-quality A/B tests and controlled experiments. This will empower you to make more informed and impactful business decisions.

The following topics are covered:

- Designing and conducting A/B tests
- Controlled experiments and causal inference
- Common pitfalls and challenges
- Implementing A/B test analysis in R

Technical requirements

You can find the code examples for this chapter in this book's GitHub repository: `https://github.com/PacktPublishing/Causal-Inference-in-R/tree/main/chap_08`.

Designing and conducting A/B tests

A deep dive into A/B testing and experimental design helps us establish a credible link between cause and effect. As you know by now, at the core of causal inference is the challenge of distinguishing between simple association and actual causation, which requires both statistical skills and an acute, insightful mind. This section will guide you through this process step by step.

Concepts

First, let's understand a few key concepts:

- **Potential outcomes framework**: This framework posits that for any individual and any intervention, potential outcomes correspond to each possible action. The causal effect is the difference between these outcomes, typically unobservable for the same unit under both states.

- **Randomization**: In this setting, randomization is an important concept. By randomly assigning subjects to treatment or control groups, randomization ensures that confounding variables are evenly distributed across groups, isolating the effect of the treatment.

Now, let's focus on defining A/B testing.

A/B testing is a method where two or more different versions of an artifact such as a web page or app are created—let's say Version A (the current version) and Version B (a modified version). These versions are randomly shown to different users, and their performance is measured to see which one achieves better results, such as higher click rates or conversions to purchase of membership, and so on. This data-driven approach helps businesses make informed decisions about what works best for their audience. Now, to ensure the reliability and validity of an A/B test, several key principles of experimental design must be adhered to.

At the heart of our experiment, *we start with randomization, a clever trick to ensure our results are due to our intervention, not hidden factors.* We keep a control group that doesn't get the treatment, to compare results and see what difference our intervention makes. It's crucial that our experiment can be repeated with the same results, proving its reliability. We also try to blind participants and researchers to who gets what treatment to avoid bias. Ensuring both our treated and control groups are similar at the start makes our comparison fair. While randomization usually does the job, we still check and adjust to make sure everything's balanced, much like ensuring both sides of a scale are equally weighted before measuring.

These guardrails for A/B testing help ensure that the conclusions drawn from the data are as accurate and unbiased as possible, providing a solid foundation for making informed decisions based on the results. Next, let's go into the details of planning your A/B test.

Planning your A/B test

Here, we'll specifically focus on formulating hypotheses, understanding test statistics, determining sample size, and discussing various randomization techniques.

Formulating hypotheses

Formulating a hypothesis in A/B testing starts with crafting a precise question. This first step sets the stage for comparing two hypotheses: the control group (group A) and the treatment group (group B). Each hypothesis takes an opposing view on the performance of these groups.

The **null hypothesis** (H_0) states that there is no real difference in performance between the control and treatment groups. In other words, any differences we see are just due to random chance. We use statistical significance to test this—if our p-value is less than 0.05, we reject H_0 and conclude that the difference is likely real and not just a fluke. The p-value is the probability of obtaining results as extreme as observed, assuming the null hypothesis is true. A p-value below 0.05 (5% chance) is often used to indicate statistical significance, suggesting the observed difference is unlikely due to chance alone. This threshold balances false positives and negatives. When $p < 0.05$, we typically reject the null hypothesis, inferring a real effect or relationship between variables. The p-value helps gauge evidence strength against the null hypothesis.

The **alternative hypothesis** (H_1) argues that the treatment does cause a significant change in the performance metric. We use hypothesis testing to figure out if the differences we observe are real or just due to random chance. When we combine two independent normal distributions, their sum or difference also forms a normal distribution. In this scenario, the mean of the combined distribution is the sum (or difference) of the individual means, and the total variance is the sum of the individual variances. This helps us understand and predict the behavior of combined data from different groups.

The following plot (*Figure 8.1*) explains this concept well. It shows the normal distributions of two independent groups (blue and green) and their combined distribution (red), reflecting the difference in means and the sum of variances. It helps illustrate how hypothesis testing predicts the behavior of combined data, supporting the alternative hypothesis that a treatment or factor significantly affects the performance metric:

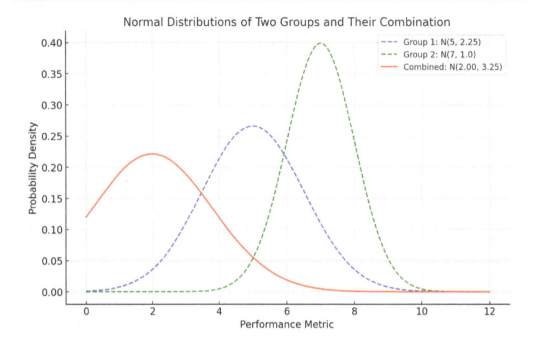

Figure 8.1 – Hypothesis testing predicts the behavior of combined
data, supporting the alternative hypothesis

This principle extends beyond being merely intriguing; it serves as a vital tool. By leveraging computer simulations and coding, we can determine whether observed differences are statistically significant or simply due to random variation. This process ensures we focus on uncovering true effects rather than noise.

Next, let's learn about test statistics.

Understanding test statistics

In hypothesis testing, understanding normal distributions is fundamental. Normal distributions are bell-shaped probability distributions that are symmetric around their mean, with the mean representing the average of the dataset and variance measuring the spread of data points around that mean. When normally distributed variables are added or subtracted, the result remains normally distributed; the means add or subtract directly, while the variances combine. To facilitate comparison, we often standardize our data by subtracting the mean and dividing by the standard deviation, transforming any normal distribution into a standard normal distribution with a mean of 0 and a variance of 1. The **test statistic** serves as a standardized measure of your sample data, allowing you to assess how far your observed results deviate from what would be expected under the null hypothesis. This structured approach clarifies the role of the test statistic in quantifying support for or against the null hypothesis in hypothesis testing.

To be precise, a test statistic is a numerical value calculated from sample data that plays a crucial role in hypothesis testing by helping us make decisions about a hypothesis. It acts as a measuring tool that indicates how different our observed data is from what we would expect if the null hypothesis were true. By converting raw data into a standardized form, the test statistic allows for easy comparison to known probability distributions. It quantifies the evidence against the null hypothesis, with larger values generally indicating stronger evidence. We can then compare the calculated test statistic to critical values from statistical tables; if the test statistic is more extreme than the critical value, we typically reject the null hypothesis. This process suggests that our data is significantly different from what we would expect under the null hypothesis. In essence, the test statistic serves as a bridge between our actual data and the theoretical world of probability distributions.

We often utilize a test statistic to examine the question: Is there sufficient data to dispute the null hypothesis H_o? The structure of a test statistic is such that its increasing values indicate a growing disagreement with H_o, helping us decide whether to reject or accept H_o. One commonly used test statistic for this purpose is the t-statistic, calculated as follows:

$$\frac{(\bar{X}_D - \mu_{o)}}{SE(\bar{X}_D)} \qquad (1)$$

Here, $\bar{X}_D$ specifies the mean of the participants, while μ_o is the expected mean under the null hypothesis H_o.

If H_o holds true, we would find μ_o unchanged, making the numerator's value meaningless.

The denominator of the t-statistic, $SE(\bar{X}_D)$, measures the variability in the data.

The mathematical formula for the **standard error** (**SE**) of the difference between two means $\left(\bar{X}_D \right)$ is the following:

$$SE\left(X_D\right) = \sqrt{\left[\left(\tfrac{s1^2}{n1}\right) + \left(\tfrac{s2^2}{n2}\right)\right]} \qquad (2)$$

Here, the following applies:

- $\left(\tfrac{s1^2}{n1}\right) + \left(\tfrac{s2^2}{n2}\right)$ are the sample variances of the two groups
- $n1$ and $n2$ are the sample sizes of the two groups

This formula is used when comparing two independent samples. It measures the variability or precision of the difference between two sample means. The formula combines the variances of both samples, accounting for their respective sample sizes. Taking the square root of the sum gives us the SE of the difference between the means. This SE is crucial in hypothesis testing, particularly in t-tests, as it forms the denominator of the t-statistic. It quantifies the uncertainty in the estimate of the difference between two population means based on sample data.

Simply put, $\bar{X}_D$ represents the actual difference between our study groups, while mu_o represents the expected difference if there were none, according to H_o. The t-statistic helps us determine if this difference is significant or just due to random chance by accounting for data variability.

If the calculated t-statistic surpasses critical values, it suggests evidence against the null hypothesis H_o. In simpler terms, if our t-statistic exceeds the commonly accepted thresholds (such as ±1.96 for a 95% confidence level), it indicates a significant result, hinting that we might reject H_o in favor of the alternative hypothesis:

$$t_{stat} = \frac{(diff_\mu - 0)}{diff_{se}} \tag{3}$$

Here, we have the following:

- $diff_\mu$: The mean of the differences
- $diff_{se}$: The SE of the differences

To repeat, the t-statistic helps us examine the validity of the null hypothesis H_o, using p-values to gauge the significance of our results. Next, let's talk about sample sizes.

Determining sample size

Selecting the appropriate sample size is crucial for ensuring the reliability of A/B testing outcomes. The primary objective is to optimize the statistical **power** of the test, which refers to the test's capacity to accurately detect a genuine significant effect resulting from the intervention. A power level of 0.8, or 80%, is typically targeted, indicating a strong likelihood (80% chance) of identifying true differences if they exist. This benchmark ensures a high probability of detecting actual effects, making it essential for researchers to carefully calculate sample size to achieve this power level. Thus, the process involves a meticulous blend of statistical reasoning and strategic planning to ensure that experiments are both powerful and efficient, aiming for a robust ability to discern true effects within the data.

Now, think of a car dealership considering a move to stock a new car model. They decide to run an A/B test to see if this new addition could indeed rev up their sales. The might of this A/B test, or its "power," lies in its talent for recognizing if the new model is a true tarmac-blazer, sales-wise.

To pull this off, the dealership needs to factor in a few things to ensure they're betting on the right car: the oomph they believe this new model will add to their sales, the line they don't want to cross with a bad guess (also known as the level of **significance**), and the tendencies of sales variability. The name of the race is all about the number of models they need to sell—bigger sales nets provide clearer distinctions, but that chews a lot into resources and time.

Conducting an A/B test requires careful balancing to accurately assess the impact of a new feature or strategy, ensuring significant insights are derived without overextending resources. The goal is to optimize the test to capture valuable data effectively, avoiding unnecessary complexity or resource drain. In essence, it's about strategically focusing efforts to identify actionable insights and make informed decisions based on solid evidence. This process involves a meticulous design to isolate and measure the true effect of changes introduced, distinguishing genuine improvements from statistical noise. Ultimately, the aim is to discern whether the new strategy significantly enhances outcomes, guiding future directions with confidence based on reliable data.

Let's list down the factors that influence the determination of sample size:

- **Effect size:** The smallest difference in outcomes between groups that is of practical significance to the researcher. Smaller effect sizes require larger sample sizes to detect.

- **Significance level (α):** The probability of rejecting the null hypothesis when it is true (Type I error). This is commonly set at 0.05.

- **Power** ($1 - \beta$): The probability of correctly rejecting the null hypothesis when it is false. A higher power requires a larger sample size. This is commonly set at 0.8 or 0.9. Here, β is a Type II error.

- **Variability in the data:** Greater variability requires a larger sample size to detect a given effect size.

Calculating the sample size typically involves statistical formulas or software (such as R) that considers these factors, ensuring the test is adequately powered to detect meaningful differences.

Randomization in depth

In the setting of randomized control trials, the framework whereby an association is inferred to reflect causation is rigorously justified. This is accomplished by exploiting the core advantage of randomization, which inherently counteracts confounding elements, thereby providing a mechanism to quantifiably attribute the divergence in the outcome, captured by the mathematical equation $E[Y(1)] - E[Y(0)]$, to the attested effect of the intervening treatment through the measurement of the observable judicatures $E[Y \mid T = 1] - E[Y \mid T = 0]$. Let's dive deeper next.

Math in randomization

In an ideal experiment, the only difference between the intervention and control groups is the intervention itself, ensuring that any outcome differences stem from the intervention. This concept, known as **covariate balance**, means both groups are similar across all potential influencing factors, both seen and unseen, typically achieved through random assignment. Mathematically, this is expressed as $P(X \mid T = 1) = P(X \mid T = 0)$, assuming random allocation to treatments T ensures an even distribution of pre-treatment characteristics X. While randomization can show covariate balance, establishing a causal link requires showing that the outcome probability under intervention, $P(y \mid do(t))$, equals the outcome probability given the treatment, $P(y \mid t)$, implying treatment T and factors X are independent.

Mathematically, covariate balance allows us to attribute differences in outcomes Y directly to the treatment, supporting the principle that in well-controlled experiments, observed associations signify causation.

Exchangeability

Here, we will discuss the topic of exchangeability for a bit. Consider choosing who gets to buy a car not by chance but by drawing names from a hat. If you are in treatment $T = 1$, you're in line for a car; if not $T = 0$, you're not. *The idea of exchangeability suggests that if we could swap people between buying or not buying without messing up the overall results, then the way we picked who gets to buy a car (randomly, like pulling names from a hat) really does make the two groups comparable except for who ends up with a new set of wheels.* Putting it simply, exchangeability means the following:

- Whether everyone gets a car or no one does, we expect the outcomes to be the same across both groups

- If nobody buys a car, those outcomes should still match up between groups

This concept shows us that by using random selection, we can make sure any connections we see between getting a car and whatever outcomes we're measuring are due to the car itself, not just by chance. It's a neat trick that helps researchers turn observations into solid evidence of cause and effect, all thanks to the power of randomization in setting up fair comparisons.

Backdoor paths

In our exploration of causality within randomized experiments, we have discussed graphical causal models, highlighting how randomization clears the path between treatment T and outcome Y, free from confounders. Randomization, sometimes comparable to a lottery, ensures T is not influenced by past causes, thus maintaining a direct causal link to Y, a process perfectly captured by the backdoor criterion:

$$P(Y \mid do(T = t)) = P(Y \mid T = t) \tag{4}$$

This principle underscores that observed relationships in randomized experiments genuinely reflect causality. Key to this are concepts such as covariate balance, ensuring groups are comparable; exchangeability, making groups interchangeable without affecting outcomes; and the prevention of backdoor paths, safeguarding against hidden biases. Together, these ensure the observed causal effects in randomized experiments are both genuine and reliable, providing a clear understanding of the causal mechanisms at play.

Randomization techniques

As you may know by now, randomization plays a crucial role in the integrity of A/B testing by ensuring treatment and control groups are equivalent, controlling for both observable and unobservable confounders.

We discuss several randomization methods next:

- **Simple randomization** offers each participant an equal chance of assignment to either group, though it may result in group size imbalances in smaller samples

- **Block randomization** groups participants into blocks to guarantee equal numbers in each treatment within blocks, requiring predetermined sample sizes

- **Stratified randomization** categorizes participants by key characteristics such as age or gender before random assignment, promoting balance across significant variables

- **Cluster randomization** is applied when participants naturally form clusters (such as groups in schools), assigning these clusters to treatments, which helps account for similarities within each cluster but necessitates larger samples

These strategies are designed to reduce bias in assigning participants to treatment conditions, thereby strengthening the causal interpretations of A/B testing outcomes. Implementing an A/B test with meticulous planning, from hypothesis creation and sample size calculation to choosing the right randomization technique, bolsters the study's reliability and validity. This careful approach supports making well-informed decisions based on the test results. But you would not know it unless we implement an A/B test next.

Implementation details

The detailed technical aspects of implementing an A/B test involve extensive planning and execution across several critical areas: setting up control and treatment groups, ensuring test validity through blinding and the use of placebo controls, and adopting appropriate data collection methods. Let's look into each step in this section.

Setting up control and treatment groups

At the heart of A/B testing is a simple yet powerful idea: take two groups, introduce a new element to one (the treatment group), and leave the other as is (the control group). Think of the control group as your baseline or the "before" picture, and the treatment group as the "after" picture where you've made a change to see what happens.

Here are some points to consider:

- **Starting on the same page**: It's vital that both groups begin the experiment under similar conditions. This level playing field allows us to confidently say that any differences we see later on are because of the change we introduced, not because the groups were different to begin with.

- **Mixing it up**: How we decide who goes into which group is where the magic of randomization comes in. By randomly assigning people, we make sure that both the stuff we know about and the things we don't are evenly spread out across both groups. This way, we can trust that any outcomes we observe are fair and square the result of our intervention.

Ensuring test validity – blinding and placebo controls

Blinding is a method used to minimize bias in experiments by keeping participants and, sometimes, researchers in the dark about who's in the control or treatment group. It can take one of the following forms:

- **Single-blind**: Participants don't know their group, helping to avoid bias in how they report their experiences or act.

- **Double-blind**: Neither the participants nor the researchers know who's in which group, which helps prevent bias in how the experiment is conducted and its results are interpreted.

For instance, if a car dealership tests a new model with some salespeople (treatment) and not others (control), knowing their group might change how salespeople act, affecting sales data. To avoid this, the dealership could hide which cars are part of the test when showing them to customers, ensuring sales approaches and customer experiences are consistent, leading to fairer test results.

Placebo controls are used when treatments have subjective effects (such as pain relief). They look like the real treatment but have no active element, helping to measure the treatment's true effect beyond just the belief in being treated.

In the next section, let's proceed to learn about data collection methodologies.

Adopting appropriate data collection methods

In preparing for an experiment, selecting the appropriate data collection methodology is crucial, aligning with both the nature of the data sought and the experiment's specific requirements:

- **Quantitative data collection**: When the experiment demands objective metrics, such as **click-through rates (CTRs)** or sales volumes, leveraging automated digital tools and software facilitates efficient and accurate data capture.

- **Qualitative data collection**: For insights into subjective experiences, such as user satisfaction or product usability, employing surveys, interviews, or observational studies provides depth and nuance to the data collected.

- **Ensuring data integrity**: Maintaining uniformity in data collection across both experimental and control groups is essential for the validity of comparative analyses. The selection of reliable and validated instruments for data collection underpins the accuracy of the measurements.

The temporal aspect of data collection—its timing and duration—is pivotal. Synchronizing data collection across groups and defining an appropriate timeframe are vital to capture the intervention's impact accurately while avoiding the influence of extraneous variables.

Consider, for instance, a car dealership conducting an A/B test on a new car model's introduction. By integrating quantitative tracking of sales data with qualitative feedback from customers, the dealership can form a comprehensive understanding of the new model's market performance. This dual-faceted approach ensures a holistic assessment, combining objective sales metrics with subjective customer experiences.

Adherence to consistent data collection practices and the utilization of validated methodologies fortify the experiment's integrity, allowing for a robust analysis of the new model's success. Moreover, ethical considerations, particularly regarding participant consent and data privacy, are paramount, underscoring the ethical conduct required in experimental research. Such meticulous planning and execution of A/B testing enhance the reliability of the results, facilitating informed decision-making based on a confluence of empirical evidence and consumer sentiment.

In the next section, we delve into the crucial role that controlled experiments play in the field of causal inference.

Controlled experiments and causal inference

Here, we will be outlining how controlled experiments can enhance our understanding and determination of causal relationships. Moreover, we discuss advanced experimental designs beyond traditional A/B testing, such as multi-armed bandit tests and factorial designs, which offer sophisticated ways to explore and identify causality within complex systems. Understanding these concepts and methodologies is crucial for implementing causal inference in statistical programming environments such as R. Various R packages provide powerful tools for conducting causal inference analysis based on experimental data, allowing researchers and analysts to estimate causal effects, perform sensitivity analyses, and extend causal inference to observational studies.

Enhancing causal inference

Randomized controlled trials (RCTs) are the gold standard for establishing causal relationships by carefully manipulating one variable and measuring the impact on another variable. RCTs balance both known and unknown factors between treatment and control groups, allowing researchers to isolate the effect of the intervention. However, RCTs can be impractical, unethical, or unrealistic in certain real-world scenarios. Despite their strengths, RCTs may still struggle with selection bias, highlighting the limitations of these methods. Controlled experiments aim to observe causal effects by standardizing all factors except for the experimental change, ensuring that any observed differences can be attributed to the manipulation.

Beyond A/B testing – multi-armed bandit tests and factorial designs

A/B testing is your go-to for straightforward causal inference, but the experimental design world is vast, filled with options for more nuanced inquiries. When the question at hand involves multiple treatments or the interplay between variables, it's time to look toward more complex experimental designs.

Multi-armed bandit tests

Multi-armed bandit tests extend A/B testing by evaluating several options at once and adjusting resource allocation based on real-time performance. This approach is particularly effective in scenarios requiring rapid adaptation, such as enhancing online user engagement, by immediately applying the best-performing strategies without waiting for the experiment's end.

These tests are dynamic optimization algorithms that balance exploration and exploitation in real time. Unlike traditional A/B tests, which maintain fixed allocation ratios throughout the experiment, these tests continuously adjust the traffic allocation based on performance. They start by distributing traffic equally among all variants, then gradually shift more traffic to better-performing options. This approach allows for faster learning and immediate application of insights, making it particularly useful in fast-paced environments such as e-commerce or digital advertising. Multi-armed bandit tests can significantly reduce opportunity costs associated with showing underperforming variants to users.

However, they may be less suitable for experiments requiring a fixed sample size or when the goal is to understand the precise effect size of each variant.

Let's consider a real-world example of using multi-armed bandit tests for an app that sells specially designed mattresses. The company wants to optimize its marketing messages to increase user engagement and sales during a promotional campaign.

The mattress app has three different promotional messages (A, B, and C) that they want to test simultaneously. Each message emphasizes different features of their mattresses:

- Message A highlights comfort
- Message B emphasizes durability
- Message C focuses on eco-friendliness

Let's discuss the observations:

1. **Initial distribution**: At the start of the campaign, the app shows each message to an equal number of users (33% each). This allows the company to gather initial data on how each message performs.

2. **Real-time performance monitoring**: As users interact with the app, the algorithm continuously monitors the CTRs and conversion rates for each promotional message.

3. **Dynamic traffic allocation**: After a few hours, Message B starts to show a higher CTR (5%) compared to Message A (3%) and Message C (4%). The algorithm adjusts the traffic allocation, increasing the percentage of users who see Message B to 50% while reducing Messages A and C to 25% each.

4. **Ongoing adjustments**: Throughout the campaign, the algorithm continues to analyze user interactions. If Message C begins to perform better after a few days, traffic allocation will shift again to favor Message C.

5. **Final outcome**: By the end of the promotional period, the app has dynamically allocated more traffic to the best-performing message(s), maximizing user engagement without waiting for a fixed testing period to conclude.

Using this technique, the app can quickly respond to user preferences in real time, ensuring that more users see the most effective promotional messages. By minimizing exposure to underperforming messages, the company maximizes potential sales during the campaign. This approach allows for ongoing adjustments, meaning that even as user preferences change over time, the marketing strategy can adapt accordingly. Next, let's study one more advanced method for experimentation.

Factorial designs

Factorial designs allow for the analysis of multiple variables and their interactions simultaneously, providing a deeper understanding of how different elements influence outcomes together. This method can reveal complex relationships and combined effects of variables, such as the synergistic impact of various marketing strategies, which single-variable experiments might not capture.

Factorial designs combine all levels of each factor, creating comprehensive experimental conditions. For instance, a 2x2 design with two factors at two levels each yields four conditions. This method is useful when factors may interact. It's more efficient than separate experiments, requiring fewer participants for equal statistical power. Factorial designs reveal complex interactions between variables, offering deeper insights. However, too many factors or levels can complicate interpretation and increase false-positive risks.

Let's consider the mattress app experiment using three factors, each with two levels:

- Message type (A: comfort versus B: durability)

- Offer type (X: discount versus Y: free trial)

- **Call-to-action (CTA)** placement (P: top of screen versus Q: bottom of screen)

A full factorial design would test all possible combinations:

- **A-X-P**: Comfort message, discount offer, CTA at top

- **A-X-Q**: Comfort message, discount offer, CTA at bottom

- **A-Y-P**: Comfort message, free trial offer, CTA at top

- **A-Y-Q**: Comfort message, free trial offer, CTA at bottom

- **B-X-P**: Durability message, discount offer, CTA at top

- **B-X-Q**: Durability message, discount offer, CTA at bottom

- **B-Y-P**: Durability message, free trial offer, CTA at top

- **B-Y-Q**: Durability message, free trial offer, CTA at bottom

Initially, the app would distribute traffic equally among all eight combinations. The app would then collect data on user engagement (e.g., CTRs, time spent on the page) and conversion rates for each combination. After a set period or number of interactions, statistical analysis would be performed to determine the following:

- **Main effects**: How each factor independently affects the outcome

- **Interaction effects**: How factors influence each other

The analysis might reveal the following, for example:

- The durability message (B) performs better overall than the comfort message (A)

- There's a strong interaction between message type and offer type (for example, the durability message works best with a free trial offer)

- CTA placement has little effect on its own but interacts with the message type

Based on these insights, the app can implement the best-performing combination and potentially explore further refinements. Some advantages we can see in this approach are the following:

- **Efficiency**: Tests multiple factors simultaneously, saving time and resources
- **Interaction insights**: Reveals how different elements work together, which single-factor tests miss
- **Comprehensive understanding**: Provides a fuller picture of what drives user engagement and conversions
- **Reduced sample size**: Requires fewer total users than testing each factor individually

Factorial design in this mattress app scenario would provide deeper insights into how different marketing elements interact, allowing for more nuanced and effective optimization of the app's user experience and marketing strategy.

These sophisticated experimental designs peel back the layers of complex causal relationships, allowing you to explore the rich tapestry of interactions that influence outcomes. Whether you're optimizing a digital platform with the fluid agility of multi-armed bandit tests or dissecting the interdependencies between strategies with factorial designs, these methods expand the horizon of causal inference. They not only provide clarity on what influences outcomes but also how different elements come together to shape those outcomes, offering a comprehensive understanding that paves the way for informed decision-making and innovative solutions.

Ethical considerations

Ethical considerations in experimental design are essential for upholding research integrity and protecting participants' rights. *They emphasize the importance of informed consent, privacy, and considering the study's moral impact. These principles play a pivotal role in ensuring participant welfare and promoting ethical research practices, marrying scientific rigor with respect for human dignity, and ensuring research processes honor the individuals involved.*

Respect for persons turns the spotlight on informed consent and autonomy, making sure everyone involved understands the stakes and retains their right to bow out gracefully.

Beneficence acts as our moral compass, guiding us to weigh the scales of risk and reward thoughtfully, ensuring our pursuit of answers doesn't inadvertently harm those who help us seek them.

Justice demands that no group unfairly bears the burdens or is denied the benefits of the research.

Ethical practices in research not only protect participants but also enhance the research process, turning data collection into a joint venture based on trust and respect. By integrating these ethical principles, researchers ensure their work is both scientifically robust and deeply respectful of the personal stories that fuel our scientific inquiries.

Informed consent

Informed consent in experimental research is the ethical linchpin that ensures participants enter the study with eyes wide open, fully briefed on the study's aims, methods, potential risks, and benefits, along with their unequivocal rights. It's a threefold pact of disclosure: making all aspects of the study transparent; comprehension, ensuring information is accessible and clear, taking into account diverse linguistic, literacy, and cultural backgrounds; and voluntariness, guaranteeing participants' freedom to join or exit the study without pressure or penalty. This process elevates informed consent from mere procedure to a profound commitment to respect and partnership, ensuring participants are not just subjects but active, valued collaborators in the research narrative, thereby enriching the study with integrity, trust, and human dignity.

Privacy concerns

In experimental research, privacy protection is a cornerstone ethical principle and a methodological necessity. Stringent data safeguards—encryption, anonymization, and restricted access—form the foundation of this commitment, ensuring participant information is rigorously shielded. The principle of data utilization, bound by informed consent, mandates using participant data solely for agreed purposes, maintaining transparency about its use, sharing, and future implications.

Navigating diverse regulatory landscapes, from the **Health Insurance Portability and Accountability Act (HIPAA)** in the US to the **General Data Protection Regulation (GDPR)** in the EU, underscores a profound commitment to participant privacy and rights. This approach surpasses mere legal compliance, elevating the research process's ethical integrity. By integrating these principles into experimental design and execution, researchers not only meet ethical and legal standards but also cultivate an environment where participant welfare and data protection are most important. This practice embodies the essence of responsible, respectful scientific inquiry, reinforcing trust between researchers and participants while ethically advancing knowledge.

Common pitfalls and challenges

In conducting A/B testing and controlled experiments, you may have to be careful about biases and external factors that can compromise the findings. Challenges such as selection bias, attrition, nonresponse, and external variables pose risks to data integrity and accuracy. Mastery in experimental design involves identifying and mitigating these risks to preserve data purity. By doing so, you must create robust experiments capable of yielding precise and meaningful insights, thereby informing decisions with reliable evidence.

In the scientific process, researchers face critical challenges that can compromise findings and ethical integrity. Selection bias, where participant groups don't accurately represent the broader population, can skew results and misrepresent intervention effects. Strategies such as randomization, stratification, and matching help ensure a representative sample, maintaining result integrity.

Attrition and nonresponse, where participants drop out or fail to provide data, threaten research completeness. Methods such as follow-up efforts, **intention-to-treat** analysis, and sensitivity analysis mitigate these effects, ensuring all participant data contributes to the final analysis. External factors can falsely influence outcomes, but careful experiment design, statistical corrections, and thoughtful participant grouping protect studies from such influences.

In A/B testing and controlled experiments, researchers must navigate methodological challenges alongside ethical responsibilities. This process demands precise experimental design, adherence to ethical standards, and advanced statistical analysis to extract valid insights. Addressing incomplete data, mitigating spill-over effects, and implementing adaptive experimentation are crucial for ensuring reliability and validity.

Strategies for dealing with incomplete data

When data goes missing due to nonresponse or participants dropping out, it risks skewing the results and diminishing the study's effectiveness. Tackling this issue head-on involves several strategic approaches:

- **Multiple imputation**: This method crafts multiple versions of the dataset, filling in the blanks based on the patterns found in the existing data. By analyzing these completed datasets individually and then pooling the findings, researchers can account for uncertainty introduced by gaps in their data.

- **Maximum likelihood estimation (MLE)**: MLE offers a way to deduce the most probable parameters for your model, even with pieces missing, under the assumption that the data's absence is purely random. It leverages all the information at hand, sidestepping the need to guess missing values.

- **Sensitivity analysis**: This technique tests the resilience of your conclusions by exploring how they might shift in different scenarios regarding the missing data. It's a way of asking, "What if the data went missing for reasons beyond randomness?" and ensures that the study's findings are not just accurate under ideal conditions but robust across various plausible situations.

Through these methods, you can mitigate the impact of incomplete data, ensuring findings remain both valid and reliable, a crucial step in painting a true picture of the phenomena under investigation.

Mitigating spill-over effects

Spill-over effects, where an intervention affects not just the target group but also others, can muddy the results of a study. In other words, spill-over effects occur when an intervention or treatment intended for one group unintentionally influences another group, typically the control group, in an experiment or study. This unintended influence can compromise the validity of the results by blurring the distinction between treatment and control groups. For example, in a study testing a new educational app in a school, students in the control group might learn about the app from their friends in the treatment group, potentially altering their behavior and skewing the results. Understanding and addressing spill-over effects is crucial for accurately assessing the impact of an intervention.

To mitigate this, researchers can do the following:

- Separate treatment and control groups either by space or time to prevent unintended influence.

- Use cluster randomization to confine the intervention's effects within predefined groups.

- Apply statistical adjustments during analysis to correct for any spill-over, ensuring the study accurately captures the intervention's true impact. These measures help clarify the intervention's actual effects, bolstering the study's validity.

Adaptive experimentation – when and how to adjust your experiment

Adaptive experimentation is a flexible approach that lets researchers adjust their experiments in response to early findings, enhancing efficiency and relevance, especially in fast-changing or complex scenarios. Let's look at some forms this can take:

- **Pre-planned adaptations**: Setting out adaptation criteria beforehand allows for strategic adjustments, such as shifting resources based on performance or ending the experiment early for clear outcomes.

- **Sequential analysis**: This method involves checking the data at intervals, requiring careful statistical corrections to avoid false-positive results from multiple data examinations.

- **Bandit algorithms**: For experiments testing multiple options, these algorithms smartly divert resources to the most promising options, balancing the need to explore different treatments against concentrating on the most effective ones.

Adopting these approaches necessitates a keen eye on ethical considerations, especially when tweaking the experiment as it unfolds. The primary goal is to maintain the study's integrity and participant safety. Through judicious use of adaptive strategies, researchers can tackle the intricacies of experimental design, ensuring their results are both robust and relevant. Now, let's get our hands on R programming in this context.

Implementing A/B test analysis in R

We're going to take a closer look at how A/B testing plays out in a car dealership, examining the effects of introducing a new sales strategy on monthly sales. Our case study will craft synthetic data to mirror the intricate dynamics of actual sales settings, incorporating elements such as the type of sales strategy employed (A versus B), the experience level of the sales team, various customer demographics, and the financing options on offer. The aim here is straightforward: we want to figure out if the novel sales strategy (B) can truly surpass the performance of the conventional method (A) in boosting sales figures.

Step 1 – Generating synthetic data

We'll begin by creating a synthetic dataset representing two groups: one that employs the traditional sales strategy (A) and another that uses the new sales strategy (B). Other variables will include salesperson experience (years), customer age, customer income, whether financing was used (yes/no), and the number of cars sold:

```
set.seed(123) # Ensure reproducibility
n <- 200 # Number of observations per group
# Generating data for Group A
sales_A <- data.frame(
  strategy = rep('A', n),
  # Avg 5 years experience, SD 2:
  salesperson_experience = rnorm(n, 5, 2),
  customer_age = sample(25:65, n, replace = TRUE),
  # Avg income 60k, SD 10k:
  customer_income = rnorm(n, 60000, 10000),
  financing = sample(c('yes', 'no'), n, replace = TRUE,
                     prob = c(0.7, 0.3)),
  # Avg 3 cars sold, poisson distribution:
  cars_sold = rpois(n, 3)
)
# Generating data for Group B
sales_B <- data.frame(
  strategy = rep('B', n),
  # Same distribution for fairness
  salesperson_experience = rnorm(n, 5, 2),
  customer_age = sample(25:65, n, replace = TRUE),
  customer_income = rnorm(n, 60000, 10000),
  financing = sample(c('yes', 'no'), n, replace = TRUE,
                     prob = c(0.7, 0.3)),
  # Hypothesizing an improvement with strategy B:
  cars_sold = rpois(n, lambda = 4)
)
# Combining both datasets
sales_data <- rbind(sales_A, sales_B)
```

Step 2 – Exploratory data analysis (EDA)

Before conducting A/B testing, it's crucial to explore the data for insights and ensure there are no anomalies:

```
library(ggplot2)
# Plotting distribution of cars sold by strategy
```

```
ggplot(sales_data, aes(x = strategy, y = cars_sold)) +
  geom_boxplot() +
  labs(title = "Distribution of Cars Sold by Strategy",
       y = "Cars Sold", x = "Sales Strategy")
```

This plot (*Figure 8.2*) will help visualize the central tendency and dispersion of cars sold under each strategy, hinting at the effectiveness of strategy B over A. Feel free to perform more exploration of the dataset in R:

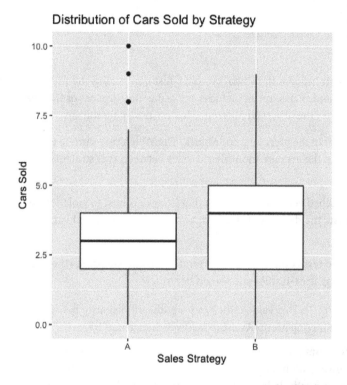

Figure 8.2 – Distribution of cars sold by strategy (A and B)

Step 3 – Statistical testing

Now, we'll conduct a t-test to evaluate if the difference in the number of cars sold between the two strategies is statistically significant:

```
t_test_result <- t.test(cars_sold ~ strategy, data = sales_data)
print(t_test_result)
```

The result will look like this :

```
Welch Two Sample t-test
data:  cars_sold by strategy
t = -4.9877, df = 395.94, p-value = 9.16e-07
alternative hypothesis: true difference in means between group A and
group B is not equal to 0
95 percent confidence interval:
 -1.3174825 -0.5725175
sample estimates:
mean in group A mean in group B
         3.025           3.970
```

This code performs an independent two-sample t-test comparing the mean number of cars sold between strategies A and B. The result includes the t-statistic, degrees of freedom, and the p-value, which indicates whether the observed difference is statistically significant.

Let's interpret the result of the preceding code block. The **Welch two-sample t-test** reveals a significant statistical difference in the average monthly car sales between two strategies, identified as group A and group B.

Welch's t-test is a modified version of the t-test that is more robust to violations of the equal variance assumption. It adjusts the degrees of freedom to account for unequal variances, providing more accurate p-values.

The breakdown of the test's output gives us a nuanced understanding (the numbers seen next are retrieved upon running the t-test R code shown before):

- **t-statistic** (-4.9877): This value indicates a significant disparity between the groups, with the negative sign implying group A's mean is lower than group B's.

- **Degrees of freedom** (df = 395.94): Adjusted for unequal variances (Welch correction), this figure shapes the t-distribution for p-value calculation, reflecting the samples' size and variance. This value indicates a relatively large sample size and unequal variances between groups. Neither a high nor low value is inherently better; it depends on the context and assumptions of the t-test.

- **p-value** (9.16e-07): Its minuscule size, well below the 0.05 threshold, robustly challenges the null hypothesis (asserting no difference), suggesting the observed mean difference is not by chance.

- **95% confidence interval** (-1.32 to -0.57): Excluding 0, it confidently estimates the true mean difference (favoring group B) lies within this range, indicating a substantial and consistent outperformance by group B.

- **Sample estimates** (mean in group A = 3.025, mean in group B = 3.970): Directly observed means underscore group B's superior sales performance.

This detailed analysis underscores strategy B's statistically validated superiority over strategy A in enhancing car sales, providing a solid basis for potentially shifting the dealership's sales approach to capitalize on strategy B's efficacy.

Step 4 – Multivariate analysis

Considering multiple variables might affect the outcome, we perform a regression analysis to control for these factors.

The code converts the `financing` variable from text to numeric format for use in the regression analysis. It then performs a multiple linear regression, with the number of cars sold as the dependent variable and sales strategy, salesperson experience, customer age, customer income, and financing as the independent variables. This allows the analysis to control for the effects of these multiple factors and isolate the impact of the sales strategy. The summary of the regression model (*Figure 8.3*) provides important statistics to assess the overall fit and significance of the model:

```
# Converting 'financing' to numeric for regression analysis
sales_data$financing_numeric <- ifelse(
  sales_data$financing == 'yes', 1, 0)
lm_result <- lm(
  cars_sold ~ strategy + salesperson_experience +
    customer_age + customer_income + financing_numeric,
  data = sales_data)
summary(lm_result)
```

Here is the result that we see:

```
Residuals:
    Min      1Q  Median      3Q     Max
-4.2997 -1.3747 -0.0528  1.1535  6.7354

Coefficients:
                        Estimate Std. Error t value Pr(>|t|)
(Intercept)            2.989e+00  7.146e-01   4.183 3.55e-05 ***
strategyB              9.334e-01  1.888e-01   4.944 1.13e-06 ***
salesperson_experience 5.233e-02  4.859e-02   1.077  0.28224
customer_age          -2.190e-02  8.154e-03  -2.686  0.00753 **
customer_income        1.138e-05  9.679e-06   1.175  0.24060
financing_numeric      1.595e-01  2.020e-01   0.790  0.43014
---
Signif. codes:  0 '***' 0.001 '**' 0.01 '*' 0.05 '.' 0.1 ' ' 1

Residual standard error: 1.88 on 394 degrees of freedom
Multiple R-squared:  0.08249,   Adjusted R-squared:  0.07084
F-statistic: 7.084 on 5 and 394 DF,  p-value: 2.33e-06
```

Figure 8.3 – Summary output of our regression model

Incorporating a comprehensive suite of variables, the regression model evaluates the influence of the sales strategy on car sales volumes, while factoring in the experience level of salespersons, customer demographic profiles, and available financing options.

The multivariate linear regression analysis dissects the relative contributions of multiple predictors—namely, the sales strategy, salesperson experience, customer age, customer income, and the chosen financing option—toward the total car sales. This approach elucidates the interplay among these variables, offering a detailed perspective on their individual and collective impact on sales outcomes.

Residuals

The residuals give an idea of the distribution of errors between the observed and predicted values: the range of residuals from -4.2997 to 6.7354 suggests variability in the model's error, indicating that while the model fits well for many observations, there are outliers or instances where the prediction deviates significantly from the actual value.

Coefficients

The regression model articulates the relationship between sales strategy and car sales volume, adjusting for variables such as salesperson experience, customer demographics, and financing options, quantified as follows:

- `Intercept` (2.989): The model posits an initial expectation of approximately 2.989 car sales in scenarios where contributory variables are absent, serving primarily as a mathematical anchor rather than a real-world scenario.

- `strategyB` (0.9334): This led to an increase of approximately 0.933 cars sold compared to strategy A, with a highly significant p-value of 1.13e-06. This strong evidence supports the superiority of strategy B over strategy A.

- `salesperson_experience` (0.05233): An additional year under a salesperson's belt translates to a marginal sales increase of 0.052 cars, although its statistical insignificance (p-value = 0.28224) hints at a tenuous link between experience and sales performance in this context.

- `customer_age` (-0.0219): An increase in customer age correlates with a decrement of about 0.022 cars sold, with statistical significance (p-value = 0.00753) suggesting a preference trend or varying purchase propensities among different age cohorts.

- `customer_income` (1.138e-05): A unit rise in customer income marginally elevates car sales, yet the lack of statistical significance (p-value = 0.24060) indicates a weak or non-existent relationship between income levels and car purchasing behavior in this dataset.

- `financing_numeric` (0.1595): The availability of financing options is linked to a nominal sales increase of 0.159 cars, but the statistical non-significance (p-value = 0.43014) implies that financing availability may not be a pivotal factor in driving car sales according to this model's analysis.

This detailed breakdown provides a mathematical and statistical lens through which the dynamics of car sales can be understood, highlighting strategy B's effectiveness while also shedding light on the nuanced contributions of salesperson experience, customer demographics, and financing options to the sales equation.

Model fit

In this section, we will deal with a metric called **R-squared (R^2)**. Let's discuss it first.

R^2 represents the proportion of variance in the dependent variable that is predictable from the independent variable(s). It ranges from 0 to 1, where, $R^2 = 1$ indicates that the regression model explains all the variability in the dependent variable; $R^2 = 0$ indicates that the model explains none of the variability. Mathematically, R^2 is calculated as follows:

$$R^2 = 1 - (SSR / SST) \qquad (5)$$

Here, we have the following:

- SSR = sum of squared residuals (unexplained variance)
- SST = total sum of squares (total variance)

Adjusted R^2 is a modified version of R^2 that adjusts for the number of predictors in the model. It increases only if the new term improves the model more than would be expected by chance. It's always lower than R^2. The formula for adjusted R^2 is the following:

$$Adjusted\ R^2 = 1 - [(1 - R^2) * (n - 1) / (n - k - 1)] \qquad (6)$$

Here, we have the following:

- n = number of observations
- k = number of predictors

A high R^2 (close to 1) suggests that the model explains a large portion of the variance in the dependent variable. A low R^2 (close to 0) indicates that the model doesn't explain much of the variance. Adjusted R^2 helps prevent overfitting by penalizing the addition of variables that don't significantly improve the model. When interpreting these values, consider the following:

- The nature of your data and field of study (in some fields, lower R^2 values are acceptable)
- The difference between R^2 and adjusted R^2 (a large difference may indicate overfitting)
- Other model diagnostics such as residual plots and p-values for a comprehensive evaluation

The model's **residual SE (RSE)** of 1.88 suggests predictions are, on average, within 1.88 cars of actual sales, indicating the model's accuracy level. The multiple R^2 value at 0.08249, with an adjusted R^2 value of 0.07084, shows the model explains only about 8.25% of the variance in car sales, pointing to the potential influence of unaccounted factors. The significant F-statistic (7.084) with a p-value of 2.33e-06 confirms the model's overall validity, highlighting that certain predictors meaningfully affect car sales, yet suggesting room for model refinement by including more variables to improve explanatory power.

Step 5 – Interpreting results

The model confirms that the sales strategy significantly impacts car sales, with strategy B outperforming strategy A. Customer age also plays a role, with younger customers associated with higher sales. However, variables such as salesperson experience, customer income, and financing options do not show a significant impact on sales within this dataset. The relatively low R^2 values suggest the need for exploring additional variables or interactions that might better explain the variability in car sales.

The analysis includes both a t-test and a regression analysis. The low p-value from the t-test (less than 0.05) suggests the new sales strategy had a significant impact on sales volumes. The regression analysis provides more detailed insights into how different factors contribute to the success of the sales strategy.

This case study showcases a holistic method for conducting A/B testing using R, spanning from synthesizing data to executing statistical tests and interpreting their results. Such a thorough examination furnishes critical insights, aiding strategic decision-making within the automotive sales domain.

To elevate the analysis of the A/B test's efficacy in R, it's possible to venture into more advanced statistical methodologies and analytical techniques. Next are further steps and snippets of R code designed to enrich the interpretation of the experimental findings.

Step 6 – Checking assumptions of the t-test

Before fully trusting the t-test results, it's crucial to verify its assumptions, such as the normality of the distribution of the differences and equality of variances. Here, we see code for the **Shapiro-Wilk normality test** that provides an F test to compare two variances.

It helps ensure the data meets the necessary assumptions for the t-test, as violations of these assumptions can compromise the validity and reliability of statistical conclusions drawn from the analysis:

```
# Checking for normality
shapiro.test(sales_data$cars_sold[sales_data$strategy == 'A'])
shapiro.test(sales_data$cars_sold[sales_data$strategy == 'B'])
# Checking for equality of variances
var.test(cars_sold ~ strategy, data = sales_data)
```

The F test's outcome, when assessing the variance in car sales between two strategies, offers detailed insights:

- **F = 0.86558**: This F statistic quantifies the variance ratio between the groups. An F value of 1 signals equal variances; here, a value below 1 indicates the first group's variance is smaller than the second's.

- **Degrees of freedom**: With num df = 199 and denom df = 199, reflecting sample sizes, these values shape the distribution critical for evaluating the F statistic.

- **p-value = 0.3094**: This suggests the observed variance ratio (or more extreme) under the null hypothesis (no variance difference) is not rare, with a p-value over 0.05 indicating no significant variance difference between strategies.

- **95% confidence interval (0.6550559 to 1.1437511)**: Encompassing 1, this interval supports no significant variance difference, consistent with the p-value's implication.

- **Variance ratio = 0.8655755**: Suggests one strategy's sales variance is about 86.56% of the others, but this is not statistically significant per the p-value and confidence interval.

This analysis concludes no significant variance difference in car sales between strategies, important for equal variance assumptions in further tests such as the t-test, and indicates mean differences aren't due to variance disparities.

Step 7 – Effect-size calculation

Understanding the magnitude of the difference between groups, in addition to its statistical significance, can provide insights into the practical significance of the experiment's outcome. We will use **Cohen's d statistics**. Cohen's d is a standardized measure of the effect size, which quantifies the difference between two groups in terms of standard deviation units. It is calculated as the difference between the means of the two groups divided by the pooled standard deviation:

```
library(effsize)
effect_size <- cohen.d(
  sales_data$cars_sold ~ sales_data$strategy, pooled = TRUE)
print(effect_size)
```

The Cohen's d statistic of -0.4987737 indicates a small to medium effect size, aligning with established benchmarks (0.2 for small, 0.5 for medium, 0.8 for large), and signifies that the average performance of strategy A is lower than that of strategy B by roughly half a standard deviation in terms of car sales. This difference, underscored by the negative value, suggests a directional effect where strategy A's sales lag behind strategy B's.

The confidence interval for Cohen's d, stretching from -0.6984013 to -0.2991462 and not encompassing zero, solidifies the effect's statistical significance and directional consistency—strategy A's sales figures are reliably lower. *This outcome, both statistically significant and of practical relevance, underscores a modest yet meaningful advantage of strategy B's strategy over strategy A's, marking the impact as significant enough to warrant consideration in strategic decision-making without categorizing it as a transformative change.*

Step 8 – Power analysis

Conducting a power analysis can help determine if the sample size was sufficient to detect a significant effect, which is crucial for interpreting null results or planning future experiments. Power analysis is a statistical technique used to determine the minimum sample size required to detect an effect of a given size with a desired level of statistical power. Statistical power refers to the probability of correctly rejecting the null hypothesis when it is false; that is, the ability to detect an effect if it truly exists. Basically, power analysis can guide researchers in designing more informative studies with an appropriate sample size to have sufficient statistical power.

Take a look at the following code:

```
library(pwr)
pwr_result <- pwr.t.test(d = effect_size$estimate,
                         sig.level = 0.05,
                         power = 0.8,
                         alternative = 'two.sided')
print(pwr_result)
```

The power calculation for a two-sample t-test outlines the sample size needed to identify a specific effect size with set statistical power and significance levels:

- **Sample size** (n = 64.07467): To detect the target effect size with the desired power, each group must consist of approximately 65 participants (rounding up from 64.07467 for practicality).

- **Effect size** (d = 0.4987737): The Cohen's d value of about 0.5 signifies a small to medium difference between groups, measured in standard deviations, aiming to capture a moderate disparity in the study.

- **Significance level** (sig.level = 0.05): The chosen alpha level of 0.05 allows for a 5% chance of a Type I error, a standard threshold indicating a 5% risk of incorrectly finding a difference.

- **Power** (0.8): Setting the test's power to 80% ensures an 80% probability of accurately rejecting the null hypothesis if a true effect of size d exists, striking a balance between minimizing Type II errors and practical sample size considerations.

- **Test type** (alternative = two.sided): A two-sided test aims to detect variances in either direction, assessing if group A significantly differs from group B without presupposing which group might exhibit higher values.

This calculation framework supports precise planning for experimental designs, ensuring that studies are adequately equipped to discern anticipated effects within established error margins.

This power analysis informs the planning stage of an experiment or study, indicating that to reliably detect a moderate effect size with 80% power and at a 5% significance level, at least 65 individuals in each group are necessary. This ensures the study is adequately powered to discern meaningful differences between the two strategies or treatments being compared, reducing the risk of both Type I and Type II errors.

Step 9 – Post-hoc analyses

If multiple comparisons or subgroup analyses are needed (for example, analyzing by salesperson experience level or customer income brackets), adjusting for multiple testing and conducting stratified analyses can reveal more nuanced insights:

```
# Stratified analysis example by financing option
lm_finance_yes <- lm(cars_sold ~ strategy, data = sales_data,
                     subset = (financing == 'yes'))
lm_finance_no <- lm(cars_sold ~ strategy, data = sales_data,
                    subset = (financing == 'no'))
summary(lm_finance_yes)
summary(lm_finance_no)
# Adjusting for multiple testing if necessary
p.adjust(c(lm_finance_yes$coefficients[2],
           lm_finance_no$coefficients[2]),
        method = 'bonferroni')
```

When we break down how sales strategies influence car sales, taking into account whether customers used financing, we find some intriguing insights:

- **With financing** (`lm_finance_yes`): Starting with strategy A, dealers sell an average of about three cars. Switching to strategy B, sales jump by almost one car (0.8995, to be exact), a significant boost (with almost zero chance of it being random, thanks to a p-value of less than 0.0001). However, the model isn't perfect—it misses the actual sales mark by about 1.866 cars and only explains a tiny slice (5.522%) of why sales vary, though it's clear the strategy plays a significant role (again, $p < 0.0001$).

- **Without financing** (`lm_finance_no`): Here, starting sales are a tad lower at roughly 2.9 cars for strategy A. But strategy B shines even brighter, increasing sales by over one car (1.0609, to be precise), with a low chance (p-value = 0.00269) of this finding being due to luck. As with its counterpart, this model also isn't a bullseye, deviating from real sales by about 1.963 cars and explaining just under 7% (6.871%) of sales variance, yet it significantly confirms strategy B's effectiveness (p = 0.002693).

In layman's terms, whether customers finance their purchase or not, switching to strategy B tends to sell more cars. The effect is slightly stronger when no financing is involved, even though both scenarios show that a lot about what influences car sales remains a mystery beyond strategy alone.

Both models affirm that switching to strategy B significantly increases the number of cars sold, regardless of financing. However, the effect is slightly more pronounced when no financing is involved. Despite the statistical significance of the strategy's impact, the low R^2 values in both cases indicate that the sales strategy, while important, is just one of many factors that could affect car sales. This analysis underscores the necessity of considering financing options as a potential moderator in the effectiveness of sales strategies and suggests further investigation into other variables that might influence sales outcomes.

Let's take the following output:

```
strategyB strategyB
        1         1
```

The code applies a Bonferroni correction to the p-values associated with the `strategyB` coefficient in two separate regression models, one for the `finance_yes` group and one for the `finance_no` group. The resulting adjusted p-values are both 1, indicating that the effect of the `strategyB` strategy is not statistically significant in either group after accounting for multiple comparisons, suggesting the need to re-evaluate the effectiveness of the two strategies or consider alternative analytical approaches.

If the original p-values were indeed small (as indicated by the significance stars in the model summaries), applying the Bonferroni correction (especially with just two tests) should not inflate the p-values to 1 unless there was an error in the calculation or input values. A p-value of 1 would suggest no evidence against the null hypothesis, even without adjustment.

For a more accurate interpretation, ensure the input to the `p.adjust` function correctly references the p-values from the regression model coefficients. If you're adjusting for multiple comparisons, the correct approach would be something like this, assuming `p1` and `p2` are the p-values from the respective models:

```
adjusted_p_values <- p.adjust(c(p1, p2), method = 'bonferroni')
```

The expected output should provide adjusted p-values reflecting the corrected significance levels of the coefficients, typically increasing the p-values but not to the extent of equating them to 1 unless the original values were very high or an error occurred in the process.

Step 10 – Visualizing interaction effects

Visualizing interaction effects between the sales strategy and other variables (for example, customer income or salesperson experience) can provide deeper insights into how the strategy performs under different conditions. Have a look at the following code:

```
# Interaction effect of strategy and salesperson experience
interaction_model <- lm(
  cars_sold ~ strategy * salesperson_experience,
  data = sales_data)
summary(interaction_model)
```

```
library(ggplot2)
ggplot(sales_data, aes(x = salesperson_experience,
                       y = cars_sold, color = strategy)) +
  geom_point(size = 3) +
  geom_smooth(method = 'lm', size = 1.5) +
  labs(title = "Strategy/Experience on Car Sales",
       x = "Salesperson Experience (years)",
       y = "Cars Sold") +
  theme(text = element_text(size = 14),
        axis.text = element_text(size = 12),
        legend.text = element_text(size = 12),
        plot.title = element_text(size = 16))
```

The following plot (*Figure 8.4*) shows an interaction effect between sales strategy (A or B) and salesperson experience, where the effectiveness of the strategies depends on the experience level of the salespeople. For less experienced salespeople (up to around 6 years), strategy B appears to generate higher car sales, but for more experienced salespeople (6 years and above), strategy A becomes more effective. This suggests the optimal sales strategy may need to be tailored based on the experience profile of the sales team in order to maximize performance:

Figure 8.4 – Interaction effect of strategy and salesperson experience in years

These additional analyses provide a more rounded and comprehensive understanding of the A/B testing experiment's success or failure, offering actionable insights and guiding future experimental designs.

Summary

In this chapter, we walked through the process of A/B testing and controlled experiments, essential for identifying causal relationships. Starting with hypothesis formation and sample size determination, we aimed to lay the groundwork for effective experimentation. Emphasizing randomization helped ensure a solid foundation for participant allocation, crucial for mitigating confounders. The execution phase, including setting up control and treatment groups and implementing blinding and placebo controls, was detailed to maintain experimental validity. Utilizing R for statistical analysis, we explored various methods to distill causal insights from data while also addressing the ethical responsibilities toward participants. Concluding with challenges and adaptability in experimental design, I hope to have provided a comprehensive guide for conducting robust and ethical A/B testing, readying readers to unveil the causal mechanisms influencing our observations. In the next chapter, we'll cover implementing doubly robust estimation in R.

9

Implementing Doubly Robust Estimation

This chapter focuses on **doubly robust** (**DR**) estimation. This technique, when used in causal analysis, uses two models: one for treatment and one for the outcome. The strength of DR estimation lies in its ability to provide reliable results, even if one of the models isn't perfect. We'll break down the theory behind this method and show you how to apply it using R. Starting with the basics, we'll move on to more advanced techniques and compare DR estimation with other methods, delineating why it's such a dependable approach. By the end of this chapter, you'll know how to use DR estimation to get accurate and robust results in your data analysis.

We'll cover the following topics:

- What is doubly robust estimation?
- Doubly robust estimation in R
- Discussing doubly robust methods

Technical requirements

You can find the code examples for this chapter, in this book's GitHub repository: `https://github.com/PacktPublishing/Causal-Inference-in-R/tree/main/chap_09`.

What is doubly robust estimation?

Like many statistical methods, DR estimation addresses the challenge of obtaining unbiased estimates when faced with potential model misspecification, offering protection against errors in either the outcome model or the propensity score model. In causal inference, DR estimation uses two models: the exposure/treatment model, $e(H)$, and the outcome model, $E(Y \mid H, T)$, to ensure accurate results. DR provides reliability even if one model is incorrect.

Let's dive deeper and learn more about them.

The DR estimator leverages two strengths: accurately modeling the outcome based on covariates (outcome model) and predicting treatment distribution (exposure/treatment model). Remarkably, only one needs to be correct for a reliable causal estimate (we'll explain why later). This makes the DR estimator a failsafe tool in causal analysis. This dual-model approach is particularly beneficial in complex data analysis, providing a safety net that enhances the reliability of causal inferences. So, the question is, how do we apply DR in practice? What's the theory behind it? Let's take a look.

An overview of DR estimation

In DR estimation, the art of propensity score weighting and the science of regression analysis are combined so that biases can be removed from the modeling side.

Let's examine the core mathematical formula of DR estimation, which provides insights into counterfactual outcomes:

$$\widehat{\mu}_t^{DR} = \frac{1}{N} \sum_{i=1}^{N} \left(\widehat{\pi} X_i + \frac{T_i}{\widehat{e} X_i} \left[Y_i - \widehat{m} X_i \right] \right) \tag{1}$$

Let's break down each component and explain how they interact:

- $\widehat{\mu}_t^{DR}$: This is the DR estimate of the average treatment effect
- N: The total number of individuals in the study
- $\widehat{\pi} X_i$: This represents the estimated propensity score for individual i, which is the probability of receiving treatment given their covariates, X_i
- T_i: The treatment indicator for individual i. It's typically 1 if the individual received treatment and 0 otherwise
- $\widehat{e} X_i$: This is the inverse of the propensity score, $\widehat{1/\pi} X_i$
- Y_i: The observed outcome for individual i
- $\widehat{m} X_i$: The predicted outcome for individual i based on their covariates, X_i, regardless of treatment status

The DR estimator combines two key elements:

- **Propensity score weighting**: The term $T_i/e^{\widehat{}}(X_i)$ weights each observation by the inverse of its probability of receiving the treatment it received. This helps balance the treated and untreated groups.
- **Outcome regression**: The term $[Y_i - m^{\widehat{}}(X_i)]$ represents the difference between the observed outcome and the predicted outcome based on covariates alone.

The estimator works as follows:

1. For each individual, it calculates $\hat{\pi} X_i$, which is the estimated probability of receiving treatment given their characteristics.

2. Then, it adds the inverse propensity score $(e^{\hat{}}(X_i))$ and the difference between the observed outcome (Y_i) and the predicted outcome $(m^{\hat{}}(X_i))$ to the product of the treatment indicator (T_i).

3. This calculation is performed for every individual in the study.

4. Finally, it takes the average of these values across all individuals.

The "double robustness" property comes from the combination of propensity score weighting and outcome regression. If either the propensity score model $(\pi^{\hat{}}(X_i))$ or the outcome model $(m^{\hat{}}(X_i))$ is specified correctly, the estimator will provide consistent estimates of the average treatment effect. This approach effectively combines the strengths of propensity score methods and regression adjustment, providing a safeguard against misspecification in either model. It allows you to leverage the benefits of both approaches while mitigating the risks associated with relying solely on one method.

In our analysis, $\hat{m}X$ predicts outcomes from the covariates, X, with great precision. Simultaneously, $\hat{e}X$ guides us through treatment probabilities accurately. As mentioned previously, the strength of the DR estimator lies in providing trustworthy results, so long as either the outcome predictions or the treatment probability estimations are correct. You can say it's very similar to having a backup system that ensures our conclusions about causal relationships are solid, even if one part of our analysis fails.

For example, if our propensity score model, $\hat{e}X$, is incorrect, $\hat{m}X$ will still ensure our analysis is accurate. Conversely, if $\hat{m}X$ is wrong but $\hat{e}X$ is correct, our DR estimator still works due to the accurate treatment probability estimation. The term "doubly robust" means the estimator can handle uncertainties in model specification and still provide accurate results, even when there are potential pitfalls.

At the heart of this approach is calculating the **average treatment effect** (ATE), where we compare the estimated counterfactual outcomes for both treated and untreated groups:

$$ATE = \hat{\mu}_1^{DR} - \hat{\mu}_0^{DR} \qquad (2)$$

This represents the ATE, where $\hat{\mu}_1^{DR}$ and $\hat{\mu}_0^{DR}$ are the estimated average outcomes for the treated and control groups, respectively. The DR superscript indicates these estimates are derived using DR methods, protecting model misspecification.

While we use various models, linear and logistic regression are often preferred for their ease of interpretation. The process begins in R with the creation of a covariate matrix, X, possibly using the `dmatrix` function in R. Logistic regression then estimates the propensity score, $\hat{e}X$, followed by linear regression for each treatment to determine $\hat{m}X$.

The approach tailors models to specific data segments by dividing the control group into meaningful subgroups, fitting separate regression models for each segment, and then aggregating these models to create a comprehensive outcome model estimate ($\hat{\mu}_0$). This method allows for a more precise estimation of outcomes within subgroups, captures nuanced relationships between covariates and outcomes, and accommodates heterogeneity in the control group, ultimately leading to a more accurate overall estimate for causal inference studies.

From this, we can see that the DR estimator combines propensity score modeling and outcome regression, providing a safeguard against misspecification in either model. This approach enhances the precision and reliability of causal inference in practical applications.

Technique behind DR

At this point, you might be wondering, how does DR work? The core idea behind this technique is that it uses complementary strengths of different approaches. Propensity score methods excel at balancing treatment groups, while outcome modeling captures the relationship between covariates and outcomes. By combining these approaches, DR estimation leverages the strengths of both. The propensity score component helps reduce bias from confounding variables, while the outcome model adjusts for remaining imbalances in covariates. When both models are combined, they work together to provide a more accurate estimate of the treatment effect, even if one model is misspecified.

Propensity scores are often used for matching. This process involves the following aspects:

- Calculating propensity scores for all subjects
- Matching treated subjects with control subjects who have similar propensity scores
- Comparing outcomes between these matched groups

This matching helps to create comparable groups, reducing bias in the treatment effect estimate.

The DR estimator remains consistent if either the propensity score model or the outcome model is specified correctly. This is a very interesting aspect and it happens because, if the propensity score model is correct, it weights the observations properly, effectively balancing the treatment groups. If the outcome model is correct, it accurately predicts counterfactual outcomes for each subject. Even if one model is misspecified, the other model compensates, providing a "double" chance of getting the correct estimate. That's why the DR estimator uses both models in its formula.

The propensity scores are used to weight observations, giving more weight to underrepresented treatment assignments. The outcome model predictions are used to estimate counterfactual outcomes. The difference between observed and predicted outcomes is then weighted by the inverse propensity scores. This combination allows the estimator to benefit from both the balancing properties of propensity scores and the predictive power of outcome models, resulting in a more robust estimate of the treatment effect.

Now that we know what DR estimation is, let's check out some other estimation methods.

Comparison with other estimation methods

The DR estimation technique notably accommodates advanced machine learning models, offering a significant edge in scenarios with high-dimensional data where conventional models falter. Specifically, DR estimation shines when the precise modeling of outcomes or treatment assignments is uncertain.

Yet, applying the DR estimation process is not without its challenges. The complexity of accurately deploying this method demands meticulous attention to the selected models and their underlying assumptions. Moreover, the presence of misspecification in both models could still skew the DR estimator, introducing bias into the analysis.

Due to this, it's nice to be aware of other approaches in this space. Here's a brief overview of some alternatives. Please feel free to learn more about these.

- **Conditional outcome modeling (COM)**: COM involves fitting a single model to the conditional expectation of the outcome given covariates and treatment. It's a straightforward approach that combines all data into a single model, including treatment as a predictor alongside other covariates. To use COM, fit a single model while predicting the outcome using all available data. This method is particularly useful when treatment effects are expected to be similar across subgroups as it simplifies the modeling process and can capture treatment-covariate interactions efficiently. COM's strength lies in its simplicity and ability to provide a unified framework for estimating treatment effects.

- **Grouped conditional outcome modeling (GCOM)**: GCOM involves training separate models for each treatment group but may not utilize all available data, potentially leading to inefficiencies. To implement GCOM, split your data into treated and control subsets, then fit distinct models for each group. These models are then used to predict counterfactual outcomes. GCOM is particularly valuable when treatment effects are heterogeneous or when the treatment substantially alters the relationship between covariates and outcomes. By allowing for different covariate effects in each group, GCOM can capture nuanced treatment impacts that might be missed by simpler methods.

- **Treatment-Agnostic Representation Network (TARNet)**: An extension of GCOM, TARNet uses a shared representation for both treatment groups but still treats the groups separately for the final estimation. It builds upon GCOM by leveraging neural networks to learn a shared representation of covariates, followed by separate output layers for treated and control groups. To use TARNet, implement a neural network architecture with shared lower layers and group-specific output layers. This method is particularly useful for high-dimensional data where traditional methods might struggle as it can capture complex, non-linear relationships while still allowing for group-specific estimation. TARNet combines the benefits of shared learning and group-specific estimation, making it a powerful tool for modern causal inference challenges.

- **X-learner**: X-learner builds upon GCOM estimators by using all of the data for both models in the estimator. This method follows a three-step process that aims to utilize data more efficiently and is particularly suited for estimating **individualized treatment effects (ITEs)**. X-learner is an advanced method that builds upon GCOM estimators by utilizing all available data more efficiently. To implement X-learner, first fit separate models for treated and control groups, then use these models to estimate individual treatment effects, and finally fit a model on these estimated effects. This three-step process is particularly effective in settings with unbalanced treatment groups and can provide more accurate estimates of individual treatment effects. X-learner's strength lies in its ability to leverage information from both treatment groups to improve overall estimation accuracy.

- **Inverse probability weighting (IPW)**: Along similar lines, IPW is a different strategy that creates pseudo-populations by reweighting the observed data to adjust for confounding. This method relies on the propensity score to balance the distribution of covariates across treatment groups, thus creating a dataset where association more closely resembles causation. IPW is a method that creates pseudo-populations by reweighting the observed data to adjust for confounding. To use IPW, first estimate propensity scores for all units, then weight observations by the inverse of their propensity scores, and finally estimate treatment effects on this weighted sample. IPW is particularly useful when the propensity score model is well-specified but the outcome model might be challenging. It creates a pseudo-population where treatment assignment is independent of covariates, allowing for more accurate causal inference. IPW can also be combined with outcome regression for DR estimation, proving to be a versatile tool in your causal inference toolkit.

In comparison, DR estimation offers a more robust alternative. These methods rely solely on either the treatment model or the outcome model being correct, while DR estimation provides a safety net by combining both. This makes DR estimation particularly valuable in empirical studies where the true model isn't known and where it's essential to account for potential misspecifications in the modeling process.

In summary, DR estimation is a powerful tool in your causal inference toolkit that allows you to achieve consistent estimates even in the presence of model misspecification. Its capacity to integrate machine learning models makes it adaptable to a wide range of applications, although it requires careful implementation and an understanding of the underlying assumptions. Next, we'll learn how to apply DR estimations in R.

Implementing doubly robust estimation in R

Let's consider a scenario consisting of a floral mega-corporation evaluating the influence of packaging on flower sales during the festive season. In this domain, details matter, and numerous variables interplay to drive sales – factors such as the weather, festival proximity, store capacity, and competitors' pricing strategies.

In this example, we'll consider two models: one predicting sales based on packaging and other factors, and another estimating the likelihood of a store choosing better packaging. If the sales prediction model is accurate, but the packaging choice model isn't, DR estimation still provides reliable results. This is because the accurate sales prediction compensates for the inaccurate packaging choice estimate. Conversely, if the packaging choice model is accurate but the sales prediction model isn't, DR estimation remains reliable because the accurate estimate of packaging choice compensates for the sales prediction.

This dual reliability is why DR estimation is superior to other methods, which may require both models to be accurate simultaneously. DR estimation ensures that our conclusions are robust, even if one model is wrong, making it a highly dependable method for causal inference.

Additionally, DR is highly adaptable and flexible, which means we can include machine learning techniques, something that's particularly handy for dealing with high-dimensional data. This flexibility is crucial because it allows us to capture complex, non-linear relationships that might otherwise be missed.

In essence, employing DR estimation improves robustness and ensures that our causal conclusions are as reliable and informative as possible.

Going back to our florist example, how do we model DR in R? Essentially, it involves creating a covariate matrix so that we can include all relevant predictors. It uses logistic regression to estimate propensity scores, representing the probability of receiving the treatment. Then, linear regression predicts outcomes based on these covariates and treatment interaction. By combining the propensity score model and outcome model, DR estimation ensures reliable treatment effect estimates, even if one model is misspecified or under-specified. In the next section, you'll learn how R's statistical tools and diagnostic functions facilitate thorough robustness checks for building a DR model.

Preparing data for DR analysis

To implement DR estimation in R for our florist scenario, we must prepare the data appropriately. The preparation phase involves creating a dataset with relevant variables and then defining a treatment indicator and outcome variable, which in our case would be whether the flower packaging was improved (treatment) and the resultant sales (outcome). In your specific use cases, you can utilize your own data. For demonstration purposes, we'll create synthetic data to model a real-world example.

In this case study, we aim to evaluate the influence of packaging on flower sales during the festive season. Our data will include the following aspects:

- `packaging`: A binary treatment variable (0 = standard packaging, 1 = enhanced packaging)
- `sales`: The outcome variable representing flower sales
- Covariates that might affect sales, such as `weather`, `festival_proximity`, `store_capacity`, and `competitors_pricing`

Let's walk through the R code for generating synthetic data and preparing it for DR analysis.

We start by setting a seed for reproducibility and defining the number of observations as 500. Variables are then generated: packaging for treatment assignment, weather conditions, festival_ proximity indicating whether the period of sales is close to a festival or not, store_capacity, and competitors_pricing:

```
# GENERATE DATA
set.seed(123) # For reproducibility
n <- 500 # Number of observations
packaging <- rbinom(n, 1, 0.5) # Treatment assignment
weather <- runif(n, 5, 30) # Weather conditions
festival_proximity <- rbinom(
  n, 1, 0.5) # Whether close to a festival or not
store_capacity <- rnorm(n, 100, 20) # Store capacity
competitors_pricing <- runif(
  n, 50, 150) # Competitors' pricing strategies
```

The outcome variable, sales, is generated using a linear model with added noise, incorporating the effect of these variables. All these variables are combined into a DataFrame called data:

```
# Outcome model (True effect + noise)
sales <- 200 + 50 * packaging + 10 * weather - 15 *
  festival_proximity + 0.5 * store_capacity - 0.25 *
  competitors_pricing + rnorm(n, 0, 50)

data <- data.frame(packaging, weather,
                   festival_proximity, store_capacity,
                   competitors_pricing, sales)
```

Implementing basic DR estimators

The DR estimator capitalizes on two models: a **propensity score model** and an **outcome model**. As explained previously, if either model is specified correctly, the estimator is consistent, which means it will converge to the true average treatment effect as the sample size grows. This is particularly useful in practical scenarios like ours, where the outcome model might be difficult to specify due to the multitude of influencing factors.

Now, let's check out the R implementation.

Here, the code estimates propensity scores using logistic regression. The propensity scores, representing the probability of receiving the treatment given the covariates, are predicted and added to the DataFrame:

```
# ESTIMATE PROPENSITY SCORES
# Logistic regression for propensity scores
propensity_model <- glm(
  packaging ~ weather + festival_proximity +
```

```
      store_capacity + competitors_pricing,
   family = binomial(), data = data)
data$propensity_score <- predict(propensity_model,
                                  type = "response")
```

An outcome regression model is then fitted using linear regression to predict sales based on the covariates and treatment:

```
# OUTCOME REGRESSION MODEL
outcome_model <- lm(
  sales ~ packaging + weather + festival_proximity +
    store_capacity + competitors_pricing,
  data = data)
```

Calculating weight

For DR estimation, the **inverse probability of treatment weighting** (**IPTW**) method is used to calculate weights based on the propensity scores.

The following line creates weights for each observation based on whether they received the treatment (`packaging == 1`) or not. It uses the `ifelse` function to assign weights: if treated, the weight is 1 / `data$propensity_score`; if not, the weight is 1 / (1 - `data$propensity_score`). These weights balance the treated and control groups to account for the different probabilities of receiving the treatment and making them more comparable:

```
# DOUBLY ROBUST ESTIMATION
# Inverse Probability of Treatment Weighting (IPTW)
weights <- ifelse(
  data$packaging == 1, 1 / data$propensity_score,
  1 / (1 - data$propensity_score))
```

This line predicts the sales for each observation as if everyone received the treatment (`packaging = 1`). It uses the `predict()` function with an existing outcome model and temporarily sets the `packaging` variable to 1 for all observations to simulate a treated scenario. The predicted values are stored in `predicted_sales_treated`, which represents the expected sales under treatment. This approach helps us estimate the impact of the treatment by providing a basis for comparison with the untreated scenario.

Here, we predict the sales while assuming everyone received the treatment to estimate the potential impact of the treatment on the outcome. By simulating a scenario where all observations are treated, we can directly compare these predicted outcomes with those from a scenario where no one is treated. This comparison allows us to estimate the treatment's effect, isolating the influence of the treatment from other factors.

This approach is crucial in causal inference as it provides insights into how the treatment changes the outcome, which is essential for making informed decisions about implementing the treatment in real-world settings:

```
# Predicted sales under treatment and control, using the outcome model
predicted_sales_treated <- predict(
  outcome_model, newdata = transform(data, packaging = 1))
predicted_sales_control <- predict(
  outcome_model, newdata = transform(data, packaging = 0))
```

Crafting the DR estimator

We built two models for DR estimation:

- **Estimate propensity scores**: A logistic regression model that estimates the probability of each store opting for enhanced packaging based on covariates.

- **Outcome regression model**: This linear model predicts sales based on the treatment and covariates. It's used to estimate the potential outcome under both treatment and control conditions.

Now, let's explore the steps involved in crafting the DR estimator, along with the code and a technical interpretation of the results. For DR estimation, we combine the propensity score weights and the outcome regression model to estimate the ATE:

```
# DR Estimation
dr_estimate <- mean(weights * (
  data$packaging * data$sales + (1 - data$packaging) *
    predicted_sales_treated - data$packaging *
    predicted_sales_control - (1 - data$packaging) * data$sales))
```

The DR estimation is then computed using these weights and the predicted sales values.

In this code, we're calculating the DR estimate to determine the average treatment effect. The formula combines both observed and predicted outcomes, weighted by the inverse of the propensity scores. Specifically, it multiplies the observed outcomes (`sales`) and the predicted outcomes (`predicted_sales_treated` and `predicted_sales_control`) by the weights and adjusts based on whether the treatment was received (`packaging`). The result, which is stored in `dr_estimate`, represents a more accurate estimate of the treatment effect by leveraging both the outcome model and propensity score model. This dual approach ensures robustness, providing reliable results even if one of the models is misspecified.

Next, let's apply the `AER` and `sandwich` packages for variance estimation in the context of DR estimation in R. This is crucial for assessing the uncertainty of our treatment effect estimate. While DR estimation gives us a point estimate, we also need to gauge its reliability. Using these packages allows us to calculate robust standard errors while accounting for potential heteroskedasticity or model misspecification. This provides a more accurate measure of variance, leading to more reliable confidence intervals and significance tests, thus enhancing the robustness of our causal inference.

Discussing doubly robust methods

In this section, we'll look into advanced concepts regarding DR estimation so that we can leverage its full potential for more accurate and reliable causal inference results.

Estimating variance

The **AER** package, which stands for **Applied Econometrics with R**, is commonly used for instrumental variable estimations and other econometric models. The `sandwich` package, on the other hand, provides robust covariance matrix estimators, which are essential for reliable standard error estimation in the presence of heteroscedasticity or other model misspecifications.

But why do we need variance estimation? In the DR estimation method within R, variance estimation is crucial for a few key reasons.

First, accurate variance estimation allows for reliable standard error calculation, which is fundamental for hypothesis testing and constructing confidence intervals around causal effect estimates. Without proper variance estimation, we might either overstate or understate the precision of our estimates, leading to incorrect inferences about the presence and size of causal effects. Thus, incorporating `AER` and `sandwich` into the DR estimation process in R enhances the robustness and credibility of our causal inference analysis.

When applying the `AER` and `sandwich` packages for this purpose, we aim to achieve a more nuanced and accurate understanding of our estimations' reliability and precision.

It's worth noting that while `AER` isn't directly used for DR estimation, it's still relevant because it helps us analyze the variance and distribution of estimators, which is crucial for assessing the reliability of treatment effect estimates in econometric analysis.

The `sandwich` package, on the other hand, is directly relevant to DR estimation for providing robust covariance matrix estimators. This is particularly important because DR methods combine models (for example, propensity score models and outcome models) that may have different variance properties. Robust covariance matrix estimators help in accurately estimating standard errors, even in the presence of heteroscedasticity (that is, when the variance of the error term isn't constant across observations) or other model misspecifications.

Let's adjust the R code so that it incorporates these packages for variance estimation. First, ensure that the `AER` and `sandwich` packages are installed and load them:

```
# Install packages if not already installed
if (!require("AER")) install.packages("AER")
if (!require("sandwich")) install.packages("sandwich")
```

```
# Load packages
library(AER)
library(sandwich)
```

Here, as we did previously, we use the lm function for our outcome model but adjust how we calculate the standard errors using the sandwich package. The vcovHC function from the sandwich package allows us to compute robust standard errors. This step is crucial for ensuring our model's variance estimation is resilient to heteroscedasticity:

```
robust_standard_errors <- sqrt(diag(vcovHC(outcome_model, type =
"HC1")))
summary(outcome_model, robust = robust_standard_errors)
```

The final step remains similar to the initial DR estimation, but we incorporate robust variance estimation for more accurate standard errors. This adjusted approach emphasizes the importance of robust variance estimation in causal inference, ensuring our conclusions about the impact of packaging on sales during the festive season are as reliable as possible.

The final step in the DR estimation process usually involves summarizing these estimates and providing a clear interpretation of their implications. Based on this context, let's demonstrate how to report and interpret the results from the outcome model using robust standard errors:

```
# Calculate robust standard errors using vcovHC from the sandwich
package
robust_se <- sqrt(diag(vcovHC(outcome_model, type = "HC1")))
# Update the summary of the outcome model with robust standard errors
coefs <- summary(outcome_model)$coefficients
robust_summary <- cbind(
  coefs[,1:2], Robust.Std.Error = robust_se,
  t = coefs[,1] / robust_se,
  `Pr(>|t|)` = 2 * (1 - pt(abs(coefs[,1] / robust_se),
                         df = outcome_model$df.residual)))
# Print the updated summary table with robust standard errors
print(robust_summary)
```

The preceding code snippet calculates robust standard errors for the linear model coefficients using the vcovHC function from the sandwich package. Then, it constructs a new summary table that includes these robust standard errors alongside the original model coefficients, their standard errors, and their t-values and p-values:

- **Robust standard errors**: The sandwich package's vcovHC function provides heteroscedasticity-consistent standard errors, which are crucial for reliable inference in the presence of heteroscedastic errors – a common issue in observational data.

- **Outcome model interpretation**: By incorporating robust standard errors, we can more confidently interpret the statistical significance and reliability of the estimated treatment effects (in this case, the impact of packaging on sales).

In conclusion, while the direct DR estimation code remains unchanged, the integration of AER and sandwich for variance estimation significantly enhances our confidence in the robustness of the causal effect estimates derived from the outcome model. This step is crucial for ensuring the reliability of causal inferences in complex real-world scenarios, such as the one described for the floral mega-corporation. Next, let's learn about some more advanced concepts in DR estimation.

Advanced DR techniques (using the tmle and SuperLearner packages)

Let's try to do a deeper analysis of the DR space. To do so, we'll be using the tmle and SuperLearner packages in R. We can harness these tools to enhance our estimation's precision and adaptability. The **tmle** package allows for targeted maximum likelihood estimation, a method that refines estimates by adjusting them iteratively to minimize bias. Meanwhile, **SuperLearner** is an ensemble method that combines predictions from multiple models to improve prediction accuracy. Together, these packages offer a robust framework for estimating causal effects with high-dimensional data, leveraging the strengths of machine learning models while maintaining desirable statistical properties.

Here's how you might proceed with its implementation in R. First, select a suite of algorithms for SuperLearner to consider. This could include logistic regression, random forests, and more, depending on the problem at hand and the available predictors:

```
if (!require("tmle")) install.packages("tmle")
if (!require("SuperLearner")) install.packages("SuperLearner")
library(tmle)
library(SuperLearner)
```

Ensure your data is in the correct format, with treatment, outcome, and covariate variables clearly defined. Using tmle with the SuperLearner library allows a flexible approach to estimating the ATE while adjusting for covariates in a high-dimensional setting:

```
# ADVANCED DOUBLY ROBUST ESTIMATION USING TMLE
# Define the list of candidate learners
SL.library <- c("SL.glm", "SL.randomForest")
# Applying TMLE for causal effect estimation
tmle_result <- tmle(
  Y = data$sales,
  A = data$packaging,
  W = data[, c("weather", "festival_proximity",
               "store_capacity", "competitors_pricing")],
  Q.SL.library = SL.library,
  cvQinit = 2,  # Reduce cross-validation folds
  g.SL.library = SL.library)
# Print TMLE results
print(tmle_result)
```

```
# Extracting and printing the estimated Average Treatment Effect (ATE)
print(paste("TMLE Estimated ATE:", tmle_result$estimates$ATE$psi))
print(paste("TMLE ATE 95% CI:",
            tmle_result$estimates$ATE$CI[1], "to",
            tmle_result$estimates$ATE$CI[2]))
```

The `tmle` package's output provides an estimate of the ATE and confidence intervals, allowing for inference about the treatment effect. The use of `SuperLearner` ensures that the estimation process is informed by the best-performing model or combination of models from the specified library, according to prediction performance. This methodology is particularly useful in settings where the true data-generating process is complex, and only some models may capture all relevant aspects of the data.

This advanced DR technique, facilitated by the `tmle` and `SuperLearner` packages, offers a robust and flexible approach to causal inference that's particularly suitable for observational studies with complex relationships among variables. By leveraging machine learning models within a theoretically grounded framework, it aims to provide accurate and reliable estimates of causal effects, even in the presence of high-dimensional covariates. We'll touch on a few interesting discussion points regarding DR estimation next.

Balancing flexibility and reliability with DR estimation

In causality, graphical intuition meets practical estimation challenges. This means that while visual or graphical methods that are used to understand causality may seem straightforward and intuitive, they often face significant challenges when it comes to practically estimating the effects and addressing errors in real-world data. Though deconfounding strategies might seem interchangeable graphically, their estimation errors differ. Consider two estimators: one based on covariates, X, and another on propensity scores, $e(X)$. While they aim for the same causal effect, their errors vary. One might perform well, while the other might suffer from significant bias.

DR methods minimize bias by leveraging the individual errors of outcome and treatment models, becoming asymptotically unbiased if at least one model is accurate. This robustness is part of the broader meta-learner framework, which allows us to use various estimators for the outcome and treatment models. DR methods are consistent with large samples, though in smaller samples, their robustness might be less pronounced.

Another aspect is that, even if DR methods emphasize flexibility and reliability, they're not invulnerable. If the outcome model is wrong, substantial bias can remain, even with an accurate treatment model. If both models are moderately inaccurate, high bias and variance can result.

Despite various challenges in causal inference, the true merit of DR estimators becomes evident under specific conditions. When both the outcome and treatment models are specified correctly, DR estimators often exhibit lower variance compared to IPW methods, particularly in large sample scenarios. While they can be compared to meta-learners – advanced techniques that combine multiple base learners to estimate causal effects – it's important to note that DR estimators aren't simply enhanced versions of SuperLearners, which use a single model to predict outcomes with treatment included as a feature. Instead, DR estimators uniquely integrate outcome regression and propensity score models, offering double robustness that helps reduce bias and improve the accuracy of causal effect estimates. Understanding these distinctions is crucial for applying these methodologies effectively in causal analysis.

In conclusion, while DR estimation isn't infallible, its strategic integration of two models furnishes it with a significant edge in mitigating bias and reducing variance, provided the models are specified accurately. This dual-model approach not only enhances the estimator's reliability but also positions it as a valuable tool in the arsenal of causal inference methodologies, particularly when we're handling the complexities of observational studies with high-dimensional data.

The concept of DR estimation, while featuring prominently in contemporary statistical discourse, has roots extending back several decades. The term itself emerged from a pivotal 1994 study by Robins et al. [1], which explored regression coefficient estimation under conditions of partially observed regressors, framing the discussion within the realm of missing data challenges.

Yet, the foundational principles of DR estimation can be traced even further back to a 1976 publication by Cassel et al. [2]. This earlier work laid down the groundwork for what would later be recognized as the DR estimator. The notable distinction between the two eras of development lies in Cassel et al.'s assumption of known propensity scores, as opposed to the subsequent realization that these scores often require estimation.

From this, we can see that the evolution of DR estimators underscores a broader narrative of statistical innovation where initial ideas matured into complex tools, reflecting both an expansion of application contexts and a deeper understanding of underlying statistical properties.

Summary

In this chapter, you were introduced to DR estimation, a method that cleverly balances two models – the exposure/treatment model and the outcome model – to ensure our causal inference is on solid ground, even if one of the models decides to go off track. This chapter also looked through R code so that we weren't just learning but also on an electrifying journey to uncover the truth behind the data. It was all about getting our hands dirty with R, making sense of complex datasets, and ensuring our analysis remained robust, regardless of our challenges.

We walked through the mathematics and practical applications of the DR estimation method, employing R to bring theoretical concepts to life. This discussion transitioned from the basics to more complex techniques, emphasizing the method's flexibility and resilience. At this point, we're not just acquainted with DR estimation; we're empowered to apply it confidently in our analyses. This method stands out for its ability to provide reliable estimates, even when we're faced with potential model misspecifications. It's an invaluable tool for any data scientist or researcher diving into the depths of causal inference. Now that we've learned about DR estimation, we are well-positioned to discuss a very interesting topic in the next chapter: instrumental variables.

References

1.   Robins, J., Rotnitzky, A., and Zhao, L. *Estimation of regression coefficients when some regressors are not always observed.* Journal of American Statistical Association, 89(427):846–866, 1994.

2.   Cassel, C., Sarndal, C., and Wretman, J. *Some results on generalized difference estimation and generalized regression estimation for finite populations.* Biometrika, 63:615–620, 1976.

Part 3:
Advanced Topics and
Cutting-Edge Methods

The final part discusses more complex and innovative aspects of causal inference. It explores instrumental variables, mediation analysis, sensitivity analysis, and heterogeneity in causal effects. This part also introduces cutting-edge approaches such as causal forests and causal discovery algorithms, showing how to apply these advanced techniques using R to tackle complex causal questions and data analysis challenges.

This part has the following chapters:

- *Chapter 10, Analyzing Instrumental Variables*
- *Chapter 11, Investigating Mediation Analysis*
- *Chapter 12, Exploring Sensitivity Analysis*
- *Chapter 13, Scrutinizing Heterogeneity in Causal Inference*
- *Chapter 14, Harnessing Causal Forests and Machine Learning Methods*
- *Chapter 15, Implementing Causal Discovery in R*

10

Analyzing Instrumental Variables

In this chapter, we'll delve deeper into a new technique called **instrumental variables** that will significantly enhance our understanding of causal inference. Instrumental variables address the complexities of confounding and endogeneity (this occurs when an explanatory variable in the model is correlated with the error term). Instrumental variables provide precise analyses to reveal causal relationships to discover hidden truths. This chapter is designed to equip you with the skills to identify potential instrumental variables, understand the criteria for an effective instrumental variable, and execute instrumental variable analysis using R. With a combination of theoretical insights, practical R applications, and in-depth explanations, we aim to explain the utility of this methodology in causal science. By the chapter's conclusion, you'll possess the expertise to employ instrumental variable analysis confidently, approaching your data with a new level of statistical acumen.

In this chapter, we will cover the following topics:

- Introduction to instrumental variables
- Criteria for instrumental variables
- Strategies for identifying valid instrumental variables
- Instrumental variable analysis in R
- Challenges and limitations

Technical requirements

You can find the code examples for this chapter in this book's GitHub repository: `https://github.com/PacktPublishing/Causal-Inference-in-R/tree/main/chap_10`.

Introduction to instrumental variables

First, let's lay the foundation by defining instrumental variables and illustrating their importance in tackling endogeneity and omitted variable biases. Wait – we will learn about these new terms in depth in a minute. Through engaging examples and clear explanations, you will be acquainted with such new concepts and their critical role in identifying true causal effects. This section sets the stage for a deeper exploration of how instrumental variables can illuminate the path to causality in complex datasets, providing a prelude to the practical applications and methodologies that follow.

The concept of instrumental variables

Instrumental variables offer a sophisticated approach to infer causal relationships in situations laden with confounding or when a direct causal path cannot be feasibly manipulated. To qualify as an instrumental variable, a variable must satisfy three core criteria:

- They should influence the treatment
- They should have an impact on the outcome only through this treatment
- The effect on the outcome should be devoid of any confounding

This framework allows you to extract the pure causal effect of a treatment on an outcome by leveraging variations in the treatment, induced by the instrument.

The mathematical essence of instrumental variables can be illustrated through a simple causal model where

$$Y = \alpha + \delta T + \varepsilon \qquad (1)$$

with Y representing the outcome, T the treatment, α the intercept, δ the causal effect of the treatment on the outcome, and ε encapsulating all other causes of Y. In this setting, the treatment T might be correlated with ε, thereby confounding the estimation of δ. An instrumental variable Z, which influences T but is uncorrelated with ε, can be used to obtain a consistent estimate of δ, the causal effect of interest.

Let's learn about this through an example. A relevant scenario involving new renters and real estate companies could be examining the impact of online reviews T on the demand for rental properties Y . In this context, an instrumental variable might be the sudden appearance of a highly influential real estate blogger Z or a viral social media post Z featuring specific properties or companies. This event, external to both the renters' decision-making processes and the companies' operational strategies, could significantly influence the volume of online reviews, thereby affecting potential renters' perceptions and, consequently, demand. This instrumental variable leverages the exogenous variation introduced by the blogger's or social media influence, which is independent of any inherent qualities of the properties or the companies' actions. Such an approach allows you to isolate the causal effect of online reviews on rental demand, providing insights into the dynamics of consumer behavior in the real estate market.

This scenario underscores the instrument's ability to exploit external variations in influencing factors, enabling a clearer understanding of causal relationships within the real estate domain.

You should know that the utility of instrumental variables is widespread across various disciplines, where they help address endogeneity issues, especially when randomized control trials are not feasible.

The importance of instrumental variables in causal inference

Instrumental variables are crucial for addressing endogeneity, providing a strong method for causal inference when traditional methods struggle due to confounding or simultaneity. For example, in evaluating new parenthood drugs or the impact of gaming systems on the elderly, instrumental variables help reveal causal relationships more clearly. We have discussed endogeneity a couple of times. What is it then?

Endogeneity arises in causal inference when a predictor variable in a regression model correlates with the model's error term due to omitted variable bias, measurement error, or simultaneity. This correlation contravenes the essential assumption of classical linear regression that predictors (regressors) should be exogenous, that is, uncorrelated with the error term. It's interesting to know that there is also a phenomenon called simultaneity, which is a form of endogeneity that occurs when variables exhibit bidirectional causality, meaning a predictor not only influences but is also influenced by the outcome variable, blurring the lines between cause and effect.

Endogeneity introduces bias and inconsistency in the estimation of regression coefficients, thereby obscuring authentic causal relationships. This happens because the variable intended to explain the outcome is also influenced by unseen factors affecting the outcome, resulting in a confounding effect.

Instrumental variables offer a solution to endogeneity. *An instrumental variable is associated with the endogenous predictor but not with the outcome variable, except indirectly through the endogenous predictor.* Utilizing an instrumental variable allows for isolating the endogenous predictor's effect from its correlation with the error term, enabling a more accurate estimation of causal effects.

Consider evaluating a new drug designed to combat postpartum depression. The challenge here is that variables such as socioeconomic status or pre-existing health conditions can both influence who gets the drug and the health outcomes observed. An instrumental variable could be the geographic distribution of the drug's availability, which we assume is random and only influences the outcome (mental health of the parents) through the mechanism of drug intake. This setup provides a mathematically clean framework for extracting the drug's effect from the confounding variables.

In a different context, say, assessing cognitive enhancement games for the elderly, we face self-selection bias: those who opt to use these games might inherently differ from non-users in ways that also impact cognitive outcomes. Here, an instrumental variable might be the availability of community programs that provide free access to these games. Assuming this availability is unrelated to individual traits directly linked to cognitive outcomes, this instrumental variable allows us to separate the true effect of gaming on cognitive health from external influences.

These scenarios highlight the mathematical and technical implications of instrumental variables in clarifying causal relationships in complex situations. Through careful selection and application of instrumental variables, you can handle complex layers of endogeneity, leading to more precise and meaningful insights into causal dynamics.

Let's address the key assumptions underlying the use of instrumental variables:

- **Relevance condition**: The instrumental variable Z must be correlated with the endogenous explanatory variable T. If Z does not have a significant effect on T, the instrumental variable approach will not be informative. This condition ensures that the instrumental variable is capable of influencing the treatment variable.

- **Exclusion restriction**: The instrumental variable Z should affect the outcome variable Y only through its effect on T. This means that there should be no direct path from Z to Y other than through T. This condition is crucial to avoid confounding and ensure that any observed relationship between Z and Y is mediated solely by T.

- **Instrument validity**: Z must not be correlated with the error term in the outcome equation. This means that Z should not be related to any omitted variables that directly affect Y. Ensuring instrument validity prevents bias that could arise from Z having any unobserved effects on Y.

Thus, to effectively use an instrumental variable, it is essential to ensure that the instrument satisfies these conditions. By doing so, you can derive robust and credible estimates of causal effects. Next, we will explore the criteria that qualify a variable to serve as an instrumental variable.

Criteria for instrumental variables

In this section, we will explore the foundational principles underpinning the criteria to be an instrumental variable. We will be emphasizing the critical assumptions of relevance, exogeneity, and exclusion restriction. Let's look at these in more detail.

Relevance of the instrumental variable

Using an instrumental variable helps to identify causal relationships without bias. The instrumental variable, Z, influences the treatment variable T and highlights its effects on the outcome Y without directly affecting Y.

The key conditions are as follows:

- $Cov(Z, T) \neq 0$: Z must influence T

- **Exclusion restriction**: Z should not directly affect Y except through T

- **Independence condition**: Z should not be related to any unobserved factors that affect Y

These conditions help to avoid confounding variables and ensure a clear causal effect between T and Y. This strategy is useful when dealing with non-experimental data to reveal cause and effect accurately.

Exogeneity of the instrumental variable

In instrumental variable estimation, an important rule is the instrumental variable's **exogeneity**. This means your instrumental variable, Z, cannot be correlated with the error term in the outcome equation. It ensures we get a clear understanding of causality.

Consider T as your variable of interest and Y as the outcome. The instrumental variable estimator relies on how Z and T move together and how Z influences Y, without being affected by the error term ϵ.

For example, imagine Z is the amount of rainfall and T is the crop yield, while Y is the income of farmers. The exogeneity condition requires that rainfall Z affects farmers' income Y only through its impact on crop yield T. If rainfall directly impacts farmers' income by causing floods or other disruptions not related to crop yield, it violates the exogeneity condition. We need Z to influence Y only through T to ensure a clear cause-and-effect relationship.

In instrumental variable analysis, the instrument's neutrality is key. Essentially, your instrumental variable (Z) needs a clean slate with the outcome's error term ϵ, ensuring it only affects the outcome Y through your variable of interest T. The math behind it boils down to this: Z must shake hands with T without any secret dealings with ϵ. This clean relationship allows the instrumental variable estimator to accurately gauge T's effect on Y, keeping the causal story straight and free from bias. So, Z and T must link up $Cov_N(t_i, z_i) \neq 0$, but Z can't influence or affect ϵ, ensuring our causal estimates aren't swayed by anything unseen. Next, let's discuss exclusion.

Exclusion restriction

It states that an instrumental variable used to determine causality must affect the outcome only through its effect on the variable being studied. This means that the instrumental variable should only influence the outcome (such as health or income) through the main variable of interest (such as a specific policy or treatment) and not through any other hidden pathways.

For example, if we're using an instrumental variable to study the effect of education on earnings, the exclusion restriction requires that the instrumental variable only affects earnings by changing education levels. It should not directly influence earnings or have any other indirect effects.

This rule is crucial because if the instrumental variable affects the outcome through other pathways, we can't be sure that the observed changes are truly due to the main variable of interest, such as education. The challenge lies in ensuring that the instrumental variable strictly follows this rule, which can be difficult because it requires confidence that there are no hidden pathways. If the instrumental variable violates this rule, our conclusions about causality could be incorrect. Thus, choosing an appropriate instrumental variable and ensuring it adheres to the exclusion restriction is vital in these studies.

Now that we know what makes a variable worthy of an instrumental variable, let's focus on how to find them.

Strategies for identifying valid instrumental variables

Finding the right instrumental variables is essential for uncovering and learning about true cause-and-effect relationships, especially when dealing with tricky data that hides those relationships. *For an instrumental variable to be up to the task, it must pass three tests: it has to be relevant, meaning it actually influences the treatment; it must be exogenous, not tangled up with other factors that could bias the outcome; and it must adhere to the exclusion restriction, affecting the outcome only through the treatment, not by any backdoor means.*

In simple terms, relevance ensures our instrumental variable and the treatment are connected, $Cov(Z, D) \neq 0$; exogeneity confirms the instrumental variable is not secretly affecting the outcome through unknown paths, $Cov(Z, \epsilon) = 0$; and the exclusion restriction guarantees the instrumental variable's influence is strictly through its effect on the treatment.

Practically, finding such instrumental variables often involves spotting natural experiments—situations where external changes mimic random assignment to treatments. This could be policy shifts, geographic differences, or other sudden changes that impact the treatment but are not expected to directly sway the outcome.

Experts in the field are key to picking out these instrumental variables, using their understanding to identify variables likely to influence the treatment but not directly related to other outcome factors.

In the math behind validating instrumental variables, we look for strong instrumental variables with

tests such as the F-test in **two-stage least squares** (**2SLS**) regression, which flags weak instrumental variables, and use overidentification tests (such as the Sargan or Hansen J-test) to indirectly check for exogeneity and the exclusion restriction by examining whether extra instrumental variables hold up under scrutiny. Feel free to learn more about these tests here [1]. Next, let's learn more about the relevance condition.

Relevance condition

The relevance of an instrumental variable is typically *assessed through the first-stage F-statistic in a 2SLS regression framework.* Now, what is an F-statistic? An F-statistic [2] is a ratio of two variances that measures the overall significance of a statistical model by comparing the variance explained by the model to the unexplained variance.

The first-stage F-statistic measures the strength of the relationship between the instrumental variables and the endogenous explanatory variable. A higher F-statistic indicates a stronger relationship, suggesting that the instrumental variables are more relevant. A low F-statistic (generally below 10) suggests that the instrumental variable is weak, potentially leading to biased and inconsistent estimates of the causal effect. Therefore, demonstrating that the F-statistic is sufficiently large is crucial for confirming the instrumental variable's relevance.

The code here shows how you may apply first stage F-statistic in R:

```
install.packages("AER")
library(AER)
set.seed(42) # Ensure reproducibility

n <- 100 # Number of observations
Z <- rnorm(n, mean=0, sd=1) # Instrumental variable: Z
T <- 0.5 * Z + rnorm(n, mean=0, sd=1) # Treatment variable: T,
influenced by Z
Y <- 2 * T + rnorm(n, mean=0, sd=1) # Outcome variable: Y, influenced
by T
# Combine into a data frame
data <- data.frame(Z, T, Y)
```

Next, you perform the first-stage regression to evaluate the instrumental variable's relevance, focusing on the relationship between Z (instrumental variable) and T (treatment):

```
# First stage regression using lm
first_stage_lm <- lm(T ~ Z, data=data)
summary(first_stage_lm)$fstatistic
```

The F-statistic from the first stage regression is approximately 15.99, with 1 numerator degree of freedom and 98 denominator degrees of freedom. Here, degrees of freedom indicate independent information for estimation. The numerator (1) represents instruments minus 1, while the denominator (98) suggests a sample size of about 100, minus the estimated parameters. This indicates that the instrumental variable Z is significantly correlated with the treatment variable T, suggesting that Z is a relevant instrument in this context. Generally, an F-statistic greater than 10 is considered strong evidence of instrument relevance in instrumental variable analysis.

Exogeneity condition

Evaluating exogeneity is trickier than checking relevance because it involves the relationship between the instrument and unseen factors that influence the outcome. The Sargan test, or the test for overidentifying restrictions, is a technique to evaluate instrument exogeneity in scenarios with multiple instrumental variables. This test examines whether the instrumental variables are correlated with the outcome's error term, suitable for overidentified models—where there are more instrumental variables than endogenous variables. It calculates whether the additional instrumental variables are statistically unrelated to the error term. A non-significant outcome implies the collective instrumental variables might be exogenous, indicating they don't share a correlation with the error term. However, a key point is that the Sargan test validates the group of instrumental variables, not each one individually.

For technical implementation, such as in R, the Sargan test can be performed using econometric packages such as AER. After estimating a model via 2SLS using, for example, the `ivreg` function in AER can facilitate the Sargan test.

Next, we'll create a synthetic dataset with several instrumental variables, apply a 2SLS regression, and proceed with the Sargan test using the AER package to see this in action:

```
set.seed(42) # Ensure reproducibility
n <- 100 # Number of observations
Z1 <- rnorm(n) # First instrumental variable
Z2 <- rnorm(n) # Second instrumental variable
X <- 0.5 * Z1 + 0.3 * Z2 + rnorm(n) # Endogenous regressor, influenced
by Z1 and Z2
Y <- 2 * X + rnorm(n) # Outcome variable, influenced by X


# Combine into a data frame
data <- data.frame(Y, X, Z1, Z2)
install.packages("AER")
library(AER)


# Perform 2SLS regression
model <- ivreg(Y ~ X | Z1 + Z2, data=data)


# Conduct the Sargan test
sargan_test = summary(model, diagnostics = TRUE)
# Print the test results
print(sargan_test)
```

Here, $Y \sim X \mid Z1 + Z2$ specifies the regression model with Y as the outcome, X as the endogenous regressor, and $Z1$ and $Z2$ as instrumental variables. *summary*(*model, diagnostics = TRUE*) performs the Sargan test on the specified model, testing whether the instrumental variables are uncorrelated with the error term in the outcome equation. The estimated coefficient for X is 2.06334 (standard error = 0.17824, p-value < 2e-16), indicating a strong relationship with Y. The instruments $Z1$ and $Z2$ are strong and relevant (weak instruments test statistic = 14.300, p-value = 3.61e-06) with no strong evidence of endogeneity in X (Wu-Hausman test statistic = 1.123, p-value = 0.292). The Sargan test (statistic = 0.007, p-value = 0.932) supports the validity of the instruments, and the model explains 83.49% of the variability in Y. However, remember, this does not individually validate each instrumental variable but rather the collective set of instrumental variables used in the model.

Assessing the validity of instrumental variables through diagnostic tests such as the Sargan test and verifying the relevance condition are pivotal steps in ensuring the reliability of instrumental variable estimates. These tests provide empirical checks against theoretical assumptions, thereby enhancing the credibility of causal inference analyses using instrumental variables.

With a solid understanding of the theory, we are ready to test our knowledge of instrumental variables in R in the next section.

Demonstrating instrumental variable analysis in R

To illustrate more instrumental variable analysis in R, focusing on the context of assessing rental unit prices tested by a real estate company, we'll go through setting up the R environment, *utilizing the* ivreg *package for 2SLS regression again. We'll be implementing the* **Generalized Method of Moments (GMM)** *using the* gmm *package, and conducting diagnostics tests.* First, this example will involve creating synthetic data to simulate a realistic scenario involving multiple variables that might influence rental prices.

To begin, let's install several R packages essential for instrumental variable analysis:

```
install.packages("AER") # For 2SLS regression
install.packages("gmm") # For GMM estimators
install.packages("lmtest") # For diagnostic tests
# Load the packages
library(AER)
library(gmm)
library(lmtest)
```

For this example, we'll generate synthetic data representing various factors affecting rental prices, including a potential endogeneity problem to be addressed by instrumental variable analysis.

The variables might include square footage (sqft), location quality (loc_quality), age of the building (age), and a binary variable for recent renovation (renovated). We hypothesize that renovation status might be endogenous due to omitted variables such as the owner's investment capacity or unobserved property characteristics influencing both the decision to renovate and rental prices:

```
set.seed(123) # For reproducibility
n <- 1000 # Number of observations
sqft <- runif(n, 500, 1500) # Square footage
loc_quality <- sample(1:10, n, replace = TRUE) #Location quality
age <- sample(1:100, n, replace = TRUE) # Building age
renovated <- rbinom(
  n, 1, prob = 0.3) # Whether the unit is recently renovated
rent_price <- 500 + 0.5*sqft + 150*loc_quality -
  2*age + 300*renovated + rnorm(n, 0, 250) # Rental price

# Endogenous variable
investor_interest <- 0.3*loc_quality + 0.02*sqft + rnorm(n, 0, 1)
data <- data.frame(sqft, loc_quality, age, renovated,
                   rent_price, investor_interest)
```

To address the potential endogeneity of `renovated`, we use `investor_interest` as an instrumental variable. The `ivreg` function from the AER package allows for straightforward 2SLS regression:

```
# 2SLS regression with ivreg
iv_model <- ivreg(
  rent_price ~ sqft + loc_quality + age + renovated |
    sqft + loc_quality + age + investor_interest,
  data = data
)
summary(iv_model)
```

In this code, `rent_price` is regressed on `sqft`, `loc_quality`, `age`, and the endogenous variable `renovated`, with `investor_interest` serving as the instrumental variable. The `summary` function provides estimates of the coefficients, from which we can interpret the causal effect of renovation on rental prices, adjusting for other factors.

Using gmm for generalized method of moments

For more complex models or when dealing with multiple instrumental variables, the gmm package can be used. The generalized method of moments is a more flexible approach to instrumental variable estimation:

```
gmm_model <- gmm(
  rent_price ~ sqft + loc_quality + age + renovated,
  x = cbind(sqft, loc_quality, age, investor_interest),
  data = data
)
summary(gmm_model)
```

The gmm package in R efficiently handles multiple instrumental variables, surpassing traditional instrumental variable methods in flexibility. It adeptly manages over-identified models by effectively using multiple instrumental variables for robust causal estimation. GMM tackles endogeneity, enhancing causal analysis robustness, and offers tools for testing the validity of instrumental variables through overidentifying restriction tests. These tests verify the exogeneity of instrumental variables.

GMM broadens the scope of instrumental variable techniques for complex or multiple-instrument models, addressing endogeneity and assumption deviations in linear regression. GMM's foundation in moment conditions allows for comprehensive econometric and complex variable relationship analysis. Its key advantages include the following:

- Versatility in application, including over-identified models and error term issues.
- Efficiency improvement over 2SLS estimators by utilizing all available information.
- Robustness against specification errors and model specification testing capability.

Diagnostics and tests in instrumental variable analysis

For diagnostic tests, the code first conducts Hansen's J test to check for overidentifying restrictions. This test helps verify whether the instrumental variables used are valid and not correlated with the error term. The results of Hansen's J test are extracted and printed.

Next, the Wald test is performed to check the joint significance of the coefficients for sqft and age. This test helps determine whether these variables significantly contribute to the model. The test is executed using the waldtest function:

```
## Diagnostics and tests
# Test for overidentifying restrictions using Hansen's J test
hansen_j_test <- summary(iv_model, diagnostics = TRUE)$diagnostics
hansen_j_test
# Wald test to check joint significance of coefficients
# Example: Test if 'sqft' and 'age' coefficients are jointly zero
waldtest(iv_model, . ~ . - sqft - age)
```

To test for weak instruments, a first-stage regression is performed where the endogenous variable (sqft) is regressed on the instruments (investor_interest, loc_quality, age, renovated). The summary of this regression is printed, and the F-statistic is extracted and printed. The F-statistic is used to assess the strength of the instruments, with a higher value indicating stronger instruments:

```
# Test for weak instruments
# First stage regression:
# endogenous variable on instruments and exogenous variables
first_stage <- lm(
  sqft ~ investor_interest + loc_quality + age + renovated,
  data = data
)
# F-test for weak instruments
summary(first_stage)
# Calculate the F-statistic
F_statistic <- summary(first_stage)$fstatistic[1]
F_statistic
```

Interpretation of results

The result from the code shared includes both exogenous variables (sqft, loc_quality, age) and an endogenous variable (renovated), with investor_interest serving as an instrumental variable for the endogenous variable. Let's interpret the results in detail:

The model estimates the effect of square footage (sqft), location quality (loc_quality), age of the unit (age), and renovation status (renovated) on rental prices (rent_price), correcting for potential endogeneity of the renovation status by using investor_interest as an instrumental variable .

The summary output from the code shows the following numbers:

- The residuals, which are the differences between the observed and predicted rental prices, range from -1308.5 to 894.5.

- The median residual is 113.5, indicating that the middle value of the residual distribution is slightly positive, suggesting a slight underestimation of rental prices by the model for half of the data.

- `Intercept`: The estimated intercept is 252.78473, with a standard error of 1125.56007. The t-value is 0.225, with a p-value of 0.822, indicating that the intercept is not statistically significant. This means the baseline rental price, when all other variables are zero, is not significantly different from zero according to this model.

- `sqft`: For every additional square foot, the rental price is estimated to increase by 0.53541, with a highly significant p-value (5.43e-14). This suggests a strong positive relationship between square footage and rental price.

- `loc_quality`: Each unit increase in location quality is associated with a 154.51848 increase in rental price, also highly significant (p-value: 7.44e-12). This underscores the importance of location quality in determining rental prices.

- `age`: The coefficient for age is -3.09387, with a p-value of 0.279, indicating that the effect of age on rental price is negative, though not statistically significant at conventional levels. This implies that older units might be cheaper, but the evidence is not strong enough to confirm this effect.

- `renovated`: The coefficient for renovated units is 1152.44340, with a p-value of 0.763, suggesting that, according to this analysis, renovated units do not have a statistically significant premium on rental price when controlling for other factors and using `investor_interest` as an instrumental variable .

The residual standard error is 443.6, showing how much the actual observations vary from the model's predictions. The model explains about 31.27% of the variation in rental prices, as shown by the Multiple R-Squared value, while the Adjusted R-squared value is a bit lower at 0.3099, hinting that the model does a decent job considering it's dealing with complex real-life data. The extremely low p-value (less than 2.2e-16) from the Wald test reveals that the variables in the model significantly affect rental prices. This result strongly counters the idea that the predictors have no impact, affirming that the chosen factors indeed play a substantial role in determining rental prices.

This analysis demonstrates the importance of square footage and location quality in determining rental prices. The lack of statistical significance for the `renovated` variable suggests that, after accounting for endogeneity via instrumental variable analysis, renovation status alone does not significantly impact rental prices in the presence of other factors. This outcome highlights the nuanced relationships between property characteristics and rental prices, as well as the critical role of proper instrumentation in identifying causal relationships in real estate economics.

It's important to note that the statistical significance and magnitude of the coefficients are sensitive to the choice of instrumental variable and the assumption that the instrumental variable is valid (i.e., it is strongly correlated with the endogenous variable but not with the error term of the outcome equation). The selection of `investor_interest` as an instrumental variable for `renovated` presupposes that investor interest influences renovation decisions without directly affecting rental prices, except through renovations. This assumption, like all instrumental variable assumptions, requires careful consideration and justification.

Challenges and limitations of instrumental variable analysis

Instrumental variables analysis is a powerful tool in econometrics and statistics for estimating causal effects when dealing with endogeneity issues. Despite its usefulness, instrumental variable analysis comes with several challenges and limitations that you must carefully address. This discussion delves into common pitfalls such as weak instrumental variables, measurement errors in instrumental variables, and the nuances of interpreting instrumental variable estimates.

Weak instrumental variables

An instrumental variable is considered weak if its correlation with the endogenous explanatory variable it aims to instrument is minimal. Such weak instrumental variables compromise the instrumental variable estimation, yielding biased results akin to those of **ordinary least squares** (**OLS**) in endogeneity scenarios. Furthermore, they elevate the instrumental variable estimator's variance, diminishing its precision.

The instrumental variable 's adequacy is typically evaluated using the first-stage regression F-statistic, where the endogenous variable is regressed on the instrumental variables. A commonly accepted benchmark is an F-statistic exceeding 10, indicating a sufficiently strong instrumental variable . However, the literature specifies more nuanced thresholds that vary with sample size and endogeneity levels.

From a mathematical standpoint, given an instrument Z and an endogenous variable X, a weak instrumental variable implies a near-zero correlation, $Corr(Z, X)$. This low correlation directly challenges the instrumental variable criterion that necessitates a strong predictive relationship between the instrumental variable and the endogenous variable.

Measurement errors in instrumental variables

Instrumental variable measurement errors introduce attenuation bias, compromising the validity of instrumental variable estimates. This bias aligns the instrumental variable estimates closer to OLS estimates due to the errors-in-variables issue, where measurement errors in the instrumental variable skew the coefficient estimates.

For an instrument Z observed with measurement error u, denoted $Z^* = Z + u$, this error u adds noise, obscuring the true Z and endogenous variable X relationship. This weakens the instrumental variable's effectiveness, risking biased instrumental variable estimates.

Interpretation of instrumental variable estimates

Interpreting instrumental variable estimates demands precision, as these estimates reflect **local average treatment effects** **(LATE)**, pinpointing the treatment's impact on compliers—individuals swayed by the instrumental variable. This contrasts with the **average treatment effect** **(ATE)** across all subjects. Instrumental variable estimates' relevance is grounded in the monotonicity assumption, asserting a unidirectional influence of the instrumental variable on treatment uptake.

Instrumental variable estimate validity critically depends on two assumptions: relevance, indicating a correlation between the instrumental variable and the endogenous variable, and exogeneity, ensuring the instrumental variable's independence from the outcome equation's error term. Breaching these assumptions compromises the reliability of conclusions.

Efficiently handling instrumental variable analysis intricacies necessitates a comprehensive grasp of its foundational assumptions and potential complications. It's essential for researchers to rigorously validate instrumental variable strength, account for measurement errors, and contextualize instrumental variable estimates through the lens of LATE. Employing multiple instrumental variables, leveraging the GMM in overidentified setups, and conducting sensitivity analyses to evaluate the instrumental variable estimates' resilience against assumption violations are pivotal strategies for overcoming these analytical hurdles.

Summary

In this chapter, we explored instrumental variables regression, focusing on addressing unobserved confounders in estimating causal relationships between a treatment (D) and an outcome (Y) It highlighted instrumental variable regression as a technique for mitigating confounder bias by using an exogenous instrument (Z) that affects the outcome exclusively through the treatment. This method isolates the causal effect of the treatment by identifying variations in the outcome resulting from instrumental variable-driven changes in the treatment. The chapter addressed the fulfillment of instrumental variable assumptions, such as the exclusion restriction, and introduced two-stage least squares estimation for practical implementation. By combining theoretical discussions with economic history examples, the chapter elucidated instrumental variable regression's utility and constraints, emphasizing its effectiveness in resolving endogeneity when conventional approaches are inadequate. In the next chapter, let's discuss mediation analysis in detail.

References

1. Extended instrumental variables/2SLS, GMM and AC/HAC, LIML, and k-class regression: `https://search.r-project.org/CRAN/refmans/plm/html/sargan.html`

2. F Statistic / F Value: Simple Definition and Interpretation: `https://www.statisticshowto.com/probability-and-statistics/f-statistic-value-test/`

11

Investigating Mediation Analysis

Alright readers, get ready to dig deep to learn about a new topic called **mediation analysis**, which captures the hidden pathways between variables. We'll start by defining causal mediation analysis and exploring its mathematical implications. Next, we'll suit up with essential packages such as `mediation` and `lavaan`, amazingly useful tools of mediation in the R world.

First, we'll generate a synthetic dataset to practice on, complete with real-world complexities. Then, we'll lay the foundational knowledge for identifying those hard-to-find mediation effects.

With that base, we'll get our hands dirty testing mediation in R step by step. But that's not all. We'll level up to multiple **mediators**, explore mediation's complicated relationship with moderation, and even tackle the unique challenges of longitudinal data. By the end, you'll be a mediation master. Let's dig in and enjoy learning!

In this chapter, we'll cover the following topics:

- What is mediation analysis?
- Identifying mediation effects
- Mediation analysis in R

Technical requirements

You can find the code examples for this chapter in this book's GitHub repository: `https://github.com/PacktPublishing/Causal-Inference-in-R/tree/main/chap_11`.

What is mediation analysis?

Mediation analysis enables you to rigorously test hypotheses about indirect effects and extricate the complex causal pathways underlying outcomes of interest across diverse fields such as psychology, sociology, medicine, and public health. The following section further introduces the fundamental concepts of mediation analysis within the context of causal inference, distinguishing it from the related but distinct concept of moderation.

Definition and overview

Mediation analysis elucidates how an independent variable influences a dependent variable through intermediary variables known as mediators [1]. Specifically, this technique seeks to identify and quantify the indirect effects that mediate the relationship between an **independent variable** and a **dependent variable** through one or more intervening variable mediators . It decomposes the total effect of the independent variable on the dependent variable into direct and indirect paths, providing insight into the underlying causal mechanisms.

The traditional approach of mediation analysis, proposed by Baron and Kenny [2], involves four steps:

1. Assessing the total independent variable-dependent variable association.
2. Evaluating the independent variable-mediator relationship.
3. Examining the mediator-dependent variable association adjusting for the independent variable.
4. Determining the extent of mediation based on the attenuation of the total effect after accounting for the mediator.

This approach quantifies the indirect effect via the product of the coefficients method (the independent variable-mediator and mediator-dependent variable effects) or the difference between the total and direct effects [1].

However, in this approach, a few fundamental limitations arise:

- First, valid decomposition requires linear regression models for the mediator, and dependent variable without independent variable-mediator interaction; for instance, if the mediator is lung inflammation and an interaction exists whereby smoking amplifies inflammation more for genetically predisposed individuals, this assumption is violated [3].

- Second, unmeasured confounding between the mediator and the dependent variable, a common issue in observational studies and potentially in randomized trials with non-randomized mediators, can lead to biased estimates [4].

In summary, the traditional approach provides a framework for mediation analysis but has restrictive assumptions that may be violated in practice, motivating more flexible causal mediation methods to handle such complexities while maintaining robust causal interpretations. Next, let's learn how traditional mediation analysis can be applied in R, and why R is a powerful tool suitable for this kind of analysis.

Why R for this technique?

R is a great tool for mediation analysis in causal inference because it has a vast repository of open source packages tailored for advanced statistical analysis, making it cost-effective and highly versatile. For instance, the `mediation` package provides functions for conducting causal mediation analysis with both parametric and non-parametric approaches, making it easier to estimate indirect effects with bootstrapping methods. Then the `lavaan` package allows for structural equation modeling, enabling you to specify complex models that include multiple mediators and moderators.

Building on these strengths, R offers various estimation approaches for mediation analysis. While regression-based methods such as SAS macros [7] are available, the R mediation package [8] stands out with its simulation-based methods. This package not only conducts sensitivity analyses to assess potential violations of the no unmeasured mediator-outcome confounding assumption but also offers remarkable flexibility. It supports various statistical models, handles multiple causally dependent mediators, and accommodates treatment noncompliance for robust analysis. Furthermore, the package includes comprehensive sensitivity analysis to evaluate the robustness of findings and provides detailed graphical and numerical summaries. These features facilitate easy interpretation and communication of results. Perhaps most importantly, its user-friendly design requires minimal programming knowledge, making it accessible to researchers and practitioners with varying levels of expertise in statistical programming.

Moreover, R's extensive visualization capabilities help in illustrating mediation models and interpreting results effectively. For example, you can create detailed path diagrams using the `semPlot` package, which enhances the understanding of the relationships between variables in your mediation model. The flexibility of R also allows for custom script development, enabling you to tailor your analysis to specific research questions and data structures. Later in this chapter, we will go through a hands-on mediation analysis exercise in R, but first, let's learn the math behind this technique.

Math behind mediation

Mathematically, mediation can be conceptualized through a series of equations that represent the causal pathways involved. Consider a simple mediation model where X represents the independent variable, Y the dependent variable, and M the mediator. The total effect of X on Y can be decomposed into two components: the direct effect (the effect of X on Y not mediated by M) and the indirect effect (the effect of X on Y that is mediated by M).

The total effect c of X on Y can be expressed as the following:

$$c = c' + ab \qquad (1)$$

Let's break this down:

- c' is the direct effect of X on Y
- a is the effect of X on M

- b is the effect of M on Y after controlling for X

- ab represents the indirect effect of X on Y through M

These relationships are typically modeled through **regression analyses**, where a and b are estimated from the data, allowing for the computation of the indirect effect. The significance of the indirect effect can be tested using statistical methods, such as bootstrap confidence interval, which provide a robust assessment of mediation.

Bootstrap confidence intervals in mediation analysis involve repeatedly resampling the original data with replacement, calculating the indirect effect for each resample, and constructing a confidence interval from the resulting distribution of indirect effects. This method doesn't assume the normality of the sampling distribution, making it robust for assessing the significance of indirect effects, especially in complex models or with non-normal data. If the resulting confidence interval doesn't include zero, the indirect effect is considered statistically significant.

This is also often termed **causal mediation analysis**. Based on the potential outcomes framework, a causal mediation analysis addresses the limitations of traditional mediation analysis [5]. It defines natural direct and indirect effects that allow for effect decomposition even in the presence of $X - M$ interaction. For example, the natural indirect effect is the difference in Y when setting M to the level it would naturally take in the exposed ($M1$) versus unexposed ($M0$), integrating over the $X - M$ interaction.

Causal mediation also clearly states the key assumptions required:

- Consistency

- No unmeasured independent variable-dependent variable confounding

- No unmeasured mediator-dependent variable confounding

- No unmeasured independent variable-mediator confounding [6]

Consistency means that the observed outcomes (dependent variable) for a given treatment and mediator level should match the expected outcomes defined by the causal model. In other words, if we observe a particular treatment and mediator level, the resulting outcome should be what we would predict based on our model. The consistency assumption essentially states that the way we observe outcomes in reality matches how we've defined potential outcomes in our causal model. It ensures that our observed data can be used to estimate causal effects defined in terms of potential outcomes.

Furthermore, the first two assumptions are needed for controlled direct effects, while all of them are required for natural direct and indirect effects.

Thus, causal mediation provides a rigorous framework for effect decomposition and understanding mechanisms, explicitly stating key assumptions. When assumptions are met, it reduces to the traditional approach in simple settings but allows for handling complexities like dependent variable-mediator interaction. Sensitivity analyses probe the implications of violating assumptions [9]. Now, in the next phase of the chapter, let's highlight the distinction between mediation and moderation.

Distinction between mediation and moderation

Mediation is all about finding the middleman – how one thing affects another through a third party. It's like trying to figure out how a new training program boosted employee productivity. Did it make people more satisfied with their jobs, which in turn made them work harder? If so, job satisfaction is the mediator, explaining the pathway the training took to improve productivity.

To illustrate, consider the effect of this training program, X, on employee productivity, Y. If we are interested in whether the impact of the training program on productivity is through increased job satisfaction, M, we are asking a mediation question. However, if we are interested in whether the impact of the training program on productivity varies depending on the level of support from supervisors, Z, we are asking a moderation question.

Moderation is more about figuring out the situation or conditions where an effect is stronger or weaker. Using that same training example, a moderation analysis might look at whether the training was more effective for employees who had supportive bosses compared to those with unsupportive ones. The boss' supportiveness is moderating the impact of the training.

While mediation analysis focuses on explaining the pathway through which an effect occurs, moderation analysis examines how the strength or direction of the relationship between two variables changes as a function of a third variable (the moderator). In other words, mediation addresses the question of *how* an effect occurs, while moderation addresses the question of *when* or *under what conditions* the effect is observed. They're two different analytical lenses for getting a clearer picture of what's really going on.

Let's consider the act of mastering a new skill, say, juggling. In this scenario, you're determined to become the life of your college party with your juggling prowess, Y, so you start with practice sessions, X, and, to spice things up, you decide to wear a pair of sporty gloves, Z, that are rumored to improve coordination. The magical formula that might predict your success could look something like this:

$$Y = \beta_0 + \beta_1 X + \beta_2 Z + \beta_3 X \times Z + \epsilon \qquad (2)$$

In this, we can see the following:

- β_0 is your natural talent for juggling (the starting point before practice and gloves).
- β_1 reveals how each practice session improves your skill.
- β_2 assesses the gloves' power on their own – do they really make a difference?
- β_3 uncovers the synergy between practice and the gloves, showing if they truly enchant each other's effects.
- ϵ is the error term capturing noise, because some days, the balls just keep dropping, no matter what.

The interaction term, $\beta_3 X \times Z$, is where it gets interesting. If β_3 is positive, it's as though the gloves amplify the effect of your practice sessions, making each hour spent juggling even more valuable. A negative β_3, on the other hand, suggests otherwise: perhaps the gloves are too bulky or distracting, making the practice less effective.

This scenario isn't just about juggling or sporty gloves; it's a metaphor for how moderation analysis, β_3, helps us understand under which circumstances (wearing gloves) our efforts (practice) yield better results. It's a neat mathematical approach, pointing us toward the most efficient path to mastering new tricks.

By embedding this analysis within our broader exploration (which could include mediation analysis to understand *how* practice makes us better), we get a more nuanced view of our causal analysis. Next, let's further drill down on mediation analysis and its relevance.

The importance of mediation analysis

Mediation's importance spans various disciplines, offering insights into complex causal relationships that direct effects alone cannot fully explain. This section explores the significance of mediation analysis across different application areas and presents real-world examples where its application is pivotal.

Real-world examples where mediation analysis is crucial include the following:

- **Mental health interventions**: In mental health research, mediation analysis is used to figure out how mindfulness-based stress reduction programs help improve mental well-being. Things such as better self-awareness and less rumination often play a key role in these studies, providing targets to make therapies even more effective. Take psychology, for example. Mediation analysis helps us understand how psychological treatments impact mental health. It can show us how **cognitive behavioral therapy** (**CBT**) reduces depression symptoms by changing negative thought patterns. In this case, CBT is the treatment, changes in thought patterns are the middle step, and feeling less depressed is the end result.

- **Educational interventions**: Sociologists use mediation analysis to investigate the societal structures and processes that influence individual and group behaviors. For instance, they might study how socioeconomic status affects educational attainment through the mediator of parental involvement. This kind of analysis helps us understand the indirect effects of broader social factors on educational outcomes. Mediation analysis has also shown how educational interventions can boost academic performance by enhancing motivational factors or learning strategies. By identifying specific mediators, educators can design interventions that more effectively target these underlying mechanisms. A relevant study on this topic is by Reardon and Portilla (2016) [10], who examined how socioeconomic status affects educational outcomes through various mediating factors. They found that parental involvement and access to educational resources partially mediated the relationship between family income and children's academic achievement.

- **Behavioral change in public health**: In public health, mediation analysis is often used to understand how interventions change health-related behaviors. For example, studying how a campaign to increase physical activity reduces the risk of heart disease by improving physical fitness levels can help refine campaign strategies. Public health researchers use this method to see how interventions work through changes in behavior, policy adherence, or environmental modifications. A relevant example is examining how smoking cessation programs reduce lung cancer rates by decreasing cigarette consumption. In medical sciences, mediation analysis is key to identifying the biological pathways through which treatments affect health outcomes. For instance, analyzing how a new drug lowers blood pressure by affecting kidney function provides insights into the drug's mechanism of action, guiding further drug development and clinical practice.

- **Economic policy analysis**: Mediation analysis is essential for assessing the effectiveness of economic policies. For example, evaluating how tax incentives drive economic growth by increasing business investments provides insights into the mechanisms of the policy and areas for improvement. Economists use mediation analysis to explore how economic policies impact outcomes. For instance, understanding how changes in interest rates affect consumer spending through consumer confidence can inform better monetary policy decisions.

- **Environmental policy**: In environmental research, mediation analysis can elucidate how policies aimed at reducing carbon emissions impact climate change through changes in industrial practices or consumer behavior. For instance, a policy that incentivizes the use of renewable energy sources may reduce emissions by encouraging industries to adopt cleaner technologies. Similarly, consumer behavior, such as increased use of public transportation, can significantly mediate the relationship between emission-reduction policies and climate outcomes. Understanding these pathways helps in designing more effective environmental policies and interventions.

In conclusion, mediation analysis offers a powerful tool for uncovering the indirect pathways through which causal effects are transmitted across a wide range of disciplines. By identifying and quantifying these mediatory processes, researchers and practitioners can design more effective interventions, policies, and treatments that directly target the underlying mechanisms of change.

Identifying mediation effects

Identifying mediation effects is a critical step in understanding the complex causal pathways that link the instrumental variables to their outcomes through one or more intervening variables. This section dives into the methodologies and statistical frameworks essential for detecting and quantifying these indirect effects. By outlining criteria for mediation, introducing statistical tests for assessing mediation significance, and discussing the assumptions underlying these analyses, you will have the tools to explore the nuanced mechanisms that drive observable relationships in your data. Through a comprehensive and technical exploration, this segment illuminates the processes that mediate causal relationships, enhancing the interpretability of empirical findings and contributing to more informed decision-making across various fields of study.

Criteria for mediation

Let's first outline the criteria necessary to establish a mediation effect, and then provide a step-by-step approach for testing mediation, grounded in both technical explanations and mathematical formulations.

Criteria for establishing a mediation effect

The identification of mediation effects hinges on satisfying several criteria that collectively demonstrate the presence of an indirect pathway through which the independent variable affects the dependent variable. These criteria, rooted in both theory and empirical testing, are crucial for substantiating mediation claims:

- **Significant relationship between the independent variable and the mediator (Path a):** For a variable to act as a mediator, it must be influenced by the independent variable. Mathematically, this relationship is tested by estimating the coefficient a in the regression equation $M = aX + e_M$, where M represents the mediator, X the independent variable, and e_M the error term. A statistically significant a coefficient indicates that changes in X are associated with changes in M.

- **Significant relationship between the mediator and the dependent variable. Controlling for the independent variable (Path b):** The mediator must have a significant effect on the dependent variable when the independent variable's effect is controlled for. This is tested through the regression equation $Y = c'X + bM + e_Y$, where Y is the dependent variable, c' is the direct effect of X on Y (controlling for M, b represents the effect of M on Y, and e_Y is the error term). A significant b coefficient suggests that M contributes to changes in Y.

- **Significant indirect effect (Path ab):** The product of paths a and $b \rightarrow (ab)$ represents the indirect effect of X on Y through M. The significance of this indirect effect is crucial for establishing mediation. It is assessed using bootstrapping techniques to obtain confidence intervals for ab, with a confidence interval that does not include zero indicating a significant indirect effect.

- **Direct effect (Path c') is not a prerequisite:** Early mediation models posited that a significant total effect of X on Y (before M) is introduced was necessary. However, contemporary approaches acknowledge that a significant indirect effect can exist even if the total direct effect c' is not significant, allowing for the possibility of full mediation (where $c' = 0$) or in cases where direct and indirect effects counterbalance each other (see *Figure 11.1*).

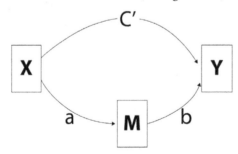

Figure 11.1 – The a and b paths and the unknown c' path highlighting the impact of mediation

Next, let's look at how to establish a mediation effect in practice.

Establishing a mediation effect – a step-by-step approach

The examination of mediation encompasses a meticulously structured series of analyses aimed at uncovering and quantifying the indirect effects within a causal pathway. This exploration integrates rigorous mathematical scrutiny with an underlying human curiosity about how one variable may influence another through an intermediary. Here is a nuanced, stepwise guide that blends technical rigor with an appreciation for the subtleties of causal inquiry:

1. **Foundational review**: Begin by evaluating the overarching impact of X on Y, setting aside the mediator for the moment, via the regression equation $Y = cX + e$. Here, c symbolizes the total effect, offering an initial glance at the direct relationship between X and Y. While this step doesn't directly test for mediation, it lays the groundwork for deeper analysis.

2. **Exploring the X to M pathway**: Advance to a regression analysis with M cast as the outcome and X as the predictor. A statistically significant coefficient for X here signals that the precursor to mediation – the influence of X on M – holds.

3. **Investigating the M to Y connection**: Further analysis entails a regression with Y as the outcome and both X and M as inputs. The significance of $M's$ coefficient b in this model, while adjusting for X, underpins the second critical aspect of mediation.

4. **Quantifying the indirect effect**: The indirect effect, represented as ab, is then calculated, with bootstrapping techniques employed to derive its confidence interval. A confidence interval that excludes zero attests to the mediation effect's significance.

5. **Delving into the direct effect (optional)**: Although the essence of mediation analysis lies in unveiling the indirect route, examining the direct influence of X on Y post-mediation c' can shed light on whether the mediation is partial or complete.

6. **Robustness checks via sensitivity analysis**: To solidify the findings, sensitivity analyses are pivotal. They evaluate how the identified mediation effect withstands the presence of unaccounted confounding variables, thereby bolstering the mediation model's integrity.

This structured approach to mediation analysis provides a clear and precise method to uncover and understand the underlying causal mechanisms at work. Now, we are up for testing mediation.

Testing for mediation

The process of testing for mediation effects involves statistical methods that quantify and evaluate the indirect effects of an independent variable on a dependent variable through one or more M. Now, what are the technical aspects of these methods, alongside the assumptions underpinning them and their limitations? Let's learn these.

Statistical methods for testing mediation effects

This section covers the foundational approach proposed by Baron and Kenny, the Sobel test for directly assessing the significance of the indirect effect, the more robust bootstrapping method, and the comprehensive structural equation modeling framework. Each method has its strengths and limitations, offering different insights into mediation analysis.

- One of the earliest approaches proposed for testing mediation was **Baron and Kenny's steps** [2], as mentioned previously. This method involves a series of regression analyses to test for the following:

  - The effect of the independent variable on the dependent variable.

  - The effect of the independent variable on the mediator.

  - The effect of mediator on the dependent variable, controlling for the independent variable.

  A significant reduction in the independent variable's effect on the dependent variable when mediator is included in the model indicates mediation. However, this approach has been criticized for its reliance on significance testing and for not directly quantifying the indirect effect.

- The **Sobel test** offers a more direct approach by testing the significance of the indirect effect (the product of paths a and b). The test statistic is calculated as:

$$\frac{ab}{\sqrt{a^2 s b^2 + b^2 s a^2}}$$

  Where, $s a^2$ and $s b^2$ are the variances of the a and b path coefficients, respectively. Despite its widespread use, the Sobel test is limited by its assumption of normal distribution for the indirect effect, which is often violated in practice.

- As we mentioned before, modern mediation analysis frequently employs **bootstrapping**, a non-parametric resampling technique, to estimate the distribution of the indirect effect. By repeatedly resampling the data and recalculating the indirect effect, bootstrapping generates an empirical distribution of the indirect effect from which confidence intervals can be derived. This method does not assume the normality of the indirect effect distribution and is thus considered more robust than the Sobel test.

- **Structural equation modeling** (SEM) allows for the simultaneous estimation of multiple regression equations and is particularly useful for complex mediation models involving multiple mediators or moderated mediation. SEM frameworks, such as path analysis, can directly estimate and test the indirect effects, offering a comprehensive tool for mediation analysis.

Assumptions and limitations underlying mediation analysis

Mediation analysis is underpinned by a few essential assumptions and a set of limitations. Let's cover the assumptions first:

- **Linearity**: It's generally assumed that the paths connecting the independent variable, any mediator, and the dependent variable are linear. If these relationships bend or twist in unexpected ways (nonlinear), our understanding of how mediation works might get muddled.

- **No hidden complications**: The analysis assumes we're not missing any critical variables that might throw a wrench in both our mediator and outcome. Missing these could skew the results.

- **Order matters**: The cause (independent variable) needs to come before the mediator, which needs to come before the effect (dependent variable). Mixing up this order can make it hard to claim one thing leads to another.

- **Trustworthy measurements**: Especially for the mediator and dependent variable, we need to measure things accurately. Errors here can dilute the connections we're trying to observe, potentially hiding mediation effects that are really there.

In addition, there are a few assumptions for mediation packages in R (e.g., `lavaan` and `mediation`), which are listed here:

- **Sequential ignorability**: This assumption assumes no unmeasured confounders affect both the mediator and dependent variable given the treatment and observed covariates

- **Correct model specification**: This requires that the specified statistical models accurately represent the relationships between variables

- **Measurement without error**: This error assumes that the treatment, mediator, and outcome variables are measured accurately without error

- **No unmeasured mediator -outcome confounders**: This assumption assumes no unmeasured confounders exist that affect both the mediator and outcome, critical for unbiased estimates

- **Sensitivity to sample size and model assumptions**: Accurate estimation in causality requires a sufficiently large sample size and the results can be highly sensitive to the assumptions about the data-generating process

However, mediation analysis isn't a catch-all solution; the plethora of limitations is discussed next:

- **Causal conundrums**: Just because we find a path through the mediation analysis doesn't mean we've proved cause and effect. The analysis hints at possible causal links, but these need extra evidence, often from experiments or detailed studies over time.

- **A tangle of mediators**: When you've got several mediators in the mix, figuring out how they all interact gets tricky. We sometimes have to make educated guesses about how these mediators relate to each other.

- **Fine-tuning the model**: The insights we gain can really depend on the specifics of how we set up our analysis – what mediators or control variables we include. Leave something important out, or throw in a red herring, and you might end up misleading yourself.

- **Sticking to assumptions**: If we stray from the assumptions of linearity, avoiding hidden variables, and keeping things in the right order, we're in danger of drawing the wrong conclusions.

To wrap up, diving into mediation analysis means walking a tightrope with statistical methods, chosen and applied with the study's goals and data quirks in mind. You need to keep a sharp eye on the foundational assumptions and the method's boundaries, and always be ready to validate the findings through various approaches or concrete evidence.

Mediation analysis in R

Now, let's see how we can use R to apply mediation analysis. Let's look at it through a problem-solving frame of mind with notable R packages.

In this case study, we aim to dissect the factors leading to an increase in policy termination within an insurance company, focusing on how these factors interact with gender and age groups.

First, we'll construct a synthetic dataset to mimic real-world complexities, then apply mediation analysis in R to uncover potential mediators of the observed trends.

Setting up the R environment

Before we begin our analysis, ensure our R environment is equipped with the necessary packages. For mediation analysis, we will utilize the `mediation` and `lavaan` packages. You can install them using the following commands:

```
install.packages("mediation")
install.packages("lavaan")
library(mediation)
library(lavaan)
```

Preparing data for mediation analysis

To simulate a dataset that reflects insurance policy terminations segmented by gender and age, we first set the random seed to 123 for reproducibility and generate synthetic data for 1,000 observations.

Next, we introduce several variables: `Gender` (`Male`, `Female`), `AgeGroup` (`Under 30`, `30-60`, `Over 60`), `IncomeLevel`, `PolicyDuration`, `NumberofClaims`, `CustomerSatisfaction`, and `Termination` (policy terminated: `Yes`, `No`).

Here, gender and age group are categorical variables, while income level, policy duration, number of claims, and customer satisfaction are numeric. Termination status is a binary factor variable, with a higher probability assigned to No than Yes:

```
set.seed(123) # For reproducibility
n <- 1000 # Number of observations

# Creating synthetic data
data <- data.frame(
  Gender = factor(sample(
    c("Male", "Female"), n, replace = TRUE)),
  AgeGroup = factor(sample(
    c("Under 30", "30-60", "Over 60"), n, replace = TRUE,
    prob = c(0.2, 0.5, 0.3))),
  IncomeLevel = sample(20000:100000, n, replace = TRUE),
  PolicyDuration = sample(1:30, n, replace = TRUE),
  NumberofClaims = rpois(n, lambda = 2),
  CustomerSatisfaction = sample(1:10, n, replace = TRUE),
  Termination = factor(sample(
    c("Yes", "No"), n, replace = TRUE, prob = c(0.3, 0.7)))
)
```

Let's then inspect the structure of our dataset and perform some initial data explorations:

```
str(data)
summary(data)
table(data$Gender, data$Termination)
```

The `str()` function provides an overview of the data types, while `summary()` offers descriptive statistics for each variable. The `table()` function cross-tabulates policy terminations by gender, giving us a glimpse into potential gender differences in policy termination rates.

Conducting mediation analysis

For this case study, let's hypothesize that `CustomerSatisfaction` mediates the relationship between `NumberofClaims` and `Termination`, and we're interested in how this mediation effect differs across `Gender` and `AgeGroup`.

First, we recode `Termination` into a binary variable called `TerminationBinary` (Yes = 1, No = 0) for logistic mediation analysis:

```
data$TerminationBinary <- ifelse(data$Termination == "Yes", 1, 0)
```

Using the `mediation` package, we specify our mediator and outcome models. Given the categorical nature of our outcome, we'll use logistic regression for the outcome model:

```
# Mediator model
med.model <- lm(
  CustomerSatisfaction ~ NumberofClaims + GenderNumeric +
    AgeGroupNumeric + IncomeLevel + PolicyDuration,
  data = data
)
# Outcome model
out.model <- glm(
  TerminationBinary ~ CustomerSatisfaction + NumberofClaims +
    GenderNumeric + AgeGroupNumeric + IncomeLevel + PolicyDuration,
  family = "binomial", data = data
)
```

The independent variable in this analysis is `NumberofClaims`, and the dependent variable is `TerminationBinary`, with `CustomerSatisfaction` serving as the mediator.

Next, we use the `mediate()` function from the `mediation` package to analyze the mediation effect. Since our focus includes examining how gender and age groups might influence the mediation effect, we'll investigate interactions in subsequent analyses.

```
# Conducting mediation analysis
med.out <- mediate(
  med.model, out.model, treat = "NumberofClaims",
  mediator = "CustomerSatisfaction", robustSE = TRUE, sims = 500
)
summary(med.out)
```

In *Figure 11.2*, we see the output from `summary(med.out)`.

```
Causal Mediation Analysis

Quasi-Bayesian Confidence Intervals

                            Estimate 95% CI Lower 95% CI Upper p-value
ACME (control)              6.03e-05    -6.53e-04         0.00    0.86
ACME (treated)              6.01e-05    -6.68e-04         0.00    0.86
ADE (control)              2.12e-03    -1.87e-02         0.02    0.81
ADE (treated)              2.12e-03    -1.87e-02         0.02    0.81
Total Effect               2.18e-03    -1.84e-02         0.02    0.81
Prop. Mediated (control) -1.17e-06    -4.50e-01         0.60    1.00
Prop. Mediated (treated) -1.00e-06    -4.51e-01         0.60    1.00
ACME (average)             6.02e-05    -6.60e-04         0.00    0.86
ADE (average)              2.12e-03    -1.87e-02         0.02    0.81
Prop. Mediated (average) -1.08e-06    -4.51e-01         0.60    1.00

Sample Size Used: 1000
```

Figure 11.2 – The results of the mediation analysis

The output further provides insights into the indirect effect of `NumberofClaims` on `Termination` through `CustomerSatisfaction`. The `robustSE` = `TRUE` option requests robust standard errors, and `sims` = `500` specifies the number of simulations for bootstrapping, enhancing the reliability of our inference. A significant **average causal mediation effect** (ACME) would suggest that customer satisfaction is a mediator in the relationship between the number of claims and policy termination. Let's continue to understand how to interpret the results.

Interpretation and further steps

Here's a breakdown of the terms and the interpretation of the results:

- **ACME**: ACME measures the effect of the independent variable on the dependent variable that is mediated through the mediator variable. For both control and treated groups, the ACME is around 6.02×10^{-5}, indicating a very small effect of the number of claims on termination probability through customer satisfaction. The negative lower bounds and the high p-value ($p = 0.86$) suggest this mediation effect is not statistically significant.

- **Average direct effect (ADE)**: ADE quantifies the direct effect of the independent variable on the dependent variable, bypassing the mediator. The ADE is approximately (2.12×10^{-3}) for both control and treated groups, which, although slightly larger than the ACME, is still relatively small and, given the high p-value ($p = 0.81$), not statistically significant.

- **Total effect**: This represents the combined effect of the ACME and ADE, essentially the overall effect of the independent variable on the dependent variable. The total effect is around (2.18×10^{-3}), and with a p-value of (0.81), it's also not statistically significant.

- **Proportion mediated (Prop. Mediated)**: This term refers to the proportion of the total effect that is mediated through the mediator. The negative values and high p-value (($p = 1.00$)) suggest that the proportion mediated is negligible and statistically indistinguishable from zero.

The interpretation is summarized here:

- The mediation effect of customer satisfaction between the number of claims and termination probability is negligible, as evidenced by the very small ACME values and the non-significant p-values.

- The direct effect of the number of claims on termination probability, bypassing customer satisfaction, is also small and not statistically significant.

- Overall, the number of claims has a negligible total effect on termination probability, with both the mediated and direct paths proving to be statistically insignificant.

- The proportion of the total effect mediated by customer satisfaction is effectively zero, indicating that customer satisfaction does not play a significant role in the relationship between the number of claims and the likelihood of termination in this analysis.

The results suggest that, in this dataset, customer satisfaction does not mediate the relationship between the number of claims and termination probability in a statistically meaningful way. Additionally, the direct relationship between the number of claims and termination is also not significant. This indicates that other factors not included in the analysis might influence the probability of termination, or that the relationship between these variables is complex and not linear.

For a deeper dive, one could extend this analysis to explore moderated mediation effects by `Gender` and `AgeGroup`, possibly utilizing the `lavaan` package for SEM-based approaches.

This case study illustrates how to set up, prepare, and conduct mediation analysis in R, leveraging synthetic data to explore real-world business questions. Each step, from data preparation to detailed mediation analysis, is crucial for uncovering underlying mechanisms and guiding strategic decisions. Next, let's get our hands dirty with more advanced analysis and models.

Advanced mediation models

Given the scenario involving an insurance company seeking to understand the cause of the rise in policy terminations segmented by gender and age groups, advanced mediation models can provide deeper insights. These models can explore the effects of **multiple mediators**, the interaction between mediation and moderation (**moderated mediation**), and changes over time (**longitudinal mediation analysis**). We will continue with the synthetic dataset created in the previous section, and test some of these approaches.

Multiple mediators and their effects

Let's assume `CustomerSatisfaction` and `IncomeLevel` act as parallel mediators in the relationship between `NumberofClaims` and `TerminationBinary`. We aim to understand how these variables collectively mediate the effect of claims on policy termination:

```
# ADVANCED MEDIATION MODELS
# Setting up the model with multiple mediators
model <- '
# Mediation paths
CustomerSatisfaction ~ b1*NumberofClaims
IncomeLevel ~ b2*NumberofClaims
TerminationBinary ~ c1*CustomerSatisfaction + c2*IncomeLevel +
c3*NumberofClaims

# Indirect effects
CustomerSatisfactionMediation := b1 * c1
IncomeLevelMediation := b2 * c2
# Total effect
TotalEffect := b1 * c1 + b2 * c2 + c3
'

fit <- sem(model, data = data, missing = "ML",
```

```
          estimator = "MLR", fixed.x = FALSE)
summary(fit, standardized = TRUE, fit.measures = TRUE)
```

In this setup, b1 and b2 are telling us how NumberofClaims influences CustomerSatisfaction and IncomeLevel. Then, c1 and c2 step in to reveal the impact these middle players have on the outcome, TerminationBinary. Then there's c3, the straightforward path, showing the direct influence NumberofClaims has on TerminationBinary. This model cleverly juggles all these connections at once, letting us see the distinct role each mediator plays in the story.

Mediation in the presence of moderation

To explore how gender may moderate the mediation effect, we can extend the model to include interactions between gender and the number of claims on both mediators. This examines if the mediation effect varies by gender:

```
# MEDIATION IN THE PRESENCE OF MODERATION
model_modmed <- '
# Moderated mediation paths
CustomerSatisfaction ~ b1*NumberofClaims + b3*GenderNumeric +
b4*NumberofClaims:GenderNumeric
IncomeLevel ~ b2*NumberofClaims + b5*GenderNumeric +
b6*NumberofClaims:GenderNumeric
TerminationBinary ~ c1*CustomerSatisfaction + c2*IncomeLevel +
c3*NumberofClaims + c4*GenderNumeric
'
fit_modmed <- sem(model_modmed, data = data, missing = ""ML",
                  estimator = "MLR", fixed.x = FALSE)
summary(fit_modmed, standardized = TRUE, fit.measures = TRUE)
```

In this moderated mediation model, b3, b4, b5, and b6 capture the moderating effects of gender on the paths from NumberofClaims to the mediators. c4 represents the moderation of gender on the direct effect of NumberofClaims on TerminationBinary. Analyzing these coefficients helps identify whether gender influences the mediation process.

Longitudinal mediation analysis

Longitudinal mediation analysis considers changes over time. Suppose we extend our dataset to include yearly observations (Year) for policy terminations. This allows us to explore how the mediation effect evolves:

```
# LONGITUDINAL MEDIATION ANALYSIS
# Extending the dataset for a longitudinal perspective
data$Year <- rep(1:5, each = n/5)
# Building a simple longitudinal mediation model for demonstration
# Assuming CustomerSatisfaction as the mediator for Year 1
```

```
# impact on TerminationBinary in Year 5
model_long <- '
TerminationBinary ~ a*CustomerSatisfaction + b*NumberofClaims + c*Year
CustomerSatisfaction ~ d*NumberofClaims + e*Year
'
fit_long <- growth(model_long, data = data, estimator = "MLR")
summary(fit_long, standardized = TRUE, fit.measures = TRUE)
```

This simplified longitudinal model explores how the mediation by `CustomerSatisfaction` of the impact of `NumberofClaims` on `TerminationBinary` is influenced by time (`Year`). A more sophisticated approach could involve growth modeling or time-varying effects to fully capture the dynamics over time.

Advanced mediation analysis, especially when dealing with multiple mediators, moderated mediation, and longitudinal data, offers nuanced insights into the complex processes underlying observable phenomena. Through careful specification of models in R, researchers can disentangle these effects, providing a richer understanding of causal mechanisms.

Each type of advanced analysis enriches our understanding of the data, guiding more informed decisions and strategies in various domains, including insurance policy management. With the knowledge provided here, you can handle more complex use case scenarios and data samples.

Summary

In this chapter, our focus was on investigating the mechanisms by which variables influence each other within complex systems, utilizing R for mediation analysis. We learned how to set up the R environment, including the installation of essential packages such as `mediation` and `lavaan`. We also explored creating a synthetic dataset to simulate real-world complexities, helping you gain a comprehensive understanding of mediation analysis. Foundational aspects were covered in detail, including establishing mediation effects and a step-by-step approach to testing these effects in R.

As the chapter progressed, we looked deep into sophisticated models, examining multiple mediators, the interplay between mediation and moderation, and the challenges posed by longitudinal data. This exploration aims to equip you with a profound understanding of mediation analysis, enabling you to conduct investigations with confidence and clarity.

In the next chapter, we will investigate the concept of sensitivity analysis in causality.

References

1. MacKinnon, D.P., Fairchild, A.J. and Fritz, M.S. (2007). *Mediation analysis.* Annual Review of Psychology, 58, 593-614.

2. Baron, R.M. and Kenny, D.A. (1986). *The moderator-mediator variable distinction in social psychological research: Conceptual, strategic, and statistical considerations.* Journal of Personality and Social Psychology, 51(6), 1173-1182.

3. Valeri, L. and VanderWeele, T.J. (2013). *Mediation analysis allowing for exposure-mediator interactions and causal interpretation: Theoretical assumptions and implementation with SAS and SPSS macros.* Psychological Methods, 18(2), 137-150.

4. VanderWeele, T.J. and Vansteelandt, S. (2009). *Conceptual issues concerning mediation, interventions and composition.* Statistics and Its Interface, 2, 457-468.

5. VanderWeele, T.J. (2015). *Explanation in Causal Inference: Methods for Mediation and Interaction.* Oxford University Press.

6. Imai, K., Keele, L. and Yamamoto, T. (2010). *Identification, inference and sensitivity analysis for causal mediation effects.* Statistical Science, 25(1), 51-71.

7. Valeri, L. and VanderWeele, T.J. (2013). *Mediation analysis allowing for exposure-mediator interactions and causal interpretation: Theoretical assumptions and implementation with SAS and SPSS macros.* Psychological Methods, 18(2), 137-150.

8. Tingley, D., Yamamoto, T., Hirose, K., Keele, L. and Imai, K. (2014). *Mediation: R package for causal mediation analysis.* Journal of Statistical Software, 59(5), 1-38.

9. VanderWeele, T.J. (2015). *Explanation in Causal Inference: Methods for Mediation and Interaction.* Oxford University Press.

10. Reardon, S. F., and Portilla, X. A. (2016). *Recent trends in income, racial, and ethnic school readiness gaps at kindergarten entry.* AERA Open, 2(3), 2332858416657343.

12

Exploring Sensitivity Analysis

In this chapter, we'll thoroughly examine sensitivity analysis within the framework of causal inference. The real power of this approach is in its ability to scrutinize the robustness of causal conclusions against the backdrop of underlying assumptions. Drawing upon the foundational contributions of pioneers such as Cornfield et al. [1], we aim to discover the intricacies of sensitivity analysis, exploring its significance, historical evolution, and practical applications across various domains. This exploration is motivated by a desire to verify the credibility of causal claims while addressing the notable gap in its application within current research practices. Throughout this chapter, you will not only deepen your understanding of sensitivity analysis but also learn how to advocate for its broader adoption so that you can construct causal narratives.

In this chapter, we'll cover the following topics:

- Introduction to sensitivity analysis

- Sensitivity analysis for causal inference

- Implementing sensitivity analysis in R

- Practical guidelines for conducting sensitivity analysis

- Advanced topics in sensitivity analysis

Technical requirements

You can find the code examples for this chapter in this book's GitHub repository: https://github.com/PacktPublishing/Causal-Inference-in-R/tree/main/chap_12.

Introduction to sensitivity analysis

Sensitivity analysis stands as a methodological approach within the realm of causal inference studies that focuses on evaluating how variations in the underpinning assumptions of a model may influence the conclusions drawn about causal relationships. This analytical technique plays a pivotal role in appraising the stability of conclusions that are inherently predicated on assumptions that might not be directly verifiable through the data at hand. By identifying and quantifying the extent to which the outcomes are affected by changes in these foundational assumptions, sensitivity analysis offers a more nuanced understanding of the level of confidence that researchers can place in their causal inferences.

Why do we need sensitivity analysis?

That's a good question. Why do we need it? Well, sensitivity analysis primarily aims to evaluate the reliability of outcomes or conclusions across various scenarios or assumptions, particularly in causal studies. This involves scrutinizing how our understanding of causal links shifts if our assumptions about the data, model, or unaccounted confounders change. Since observational studies lack the controlled conditions of randomized trials and may include unmeasured confounding, this type of analysis is vital. It enables you to probe the effects of hidden biases and determine the circumstances under which your results remain consistent. This enhances your credibility in reporting causal assertions and directing future research by pinpointing influential factors on causal robustness.

Sensitivity analysis is indispensable in fields such as prevention science, where ethical or practical constraints prevent randomized studies. It helps researchers measure the stability of their findings, acknowledging that unobserved variables might distort observed associations. Through this method, we can deepen our investigation into whether observed patterns genuinely reflect the relationships under study or are influenced by overlooked factors. So, our goal is to seek a more comprehensive understanding of our findings.

Next, let's learn a bit about the history of sensitivity analysis.

Historical context

The formalization of sensitivity analysis has paralleled advancements in statistical methods and causal inference, evolving from an era where assessing model assumptions was often an informal, expert-driven process. With the rise of sophisticated models and enhanced computational power, the statistical and epidemiological communities now underscore the necessity of explicitly evaluating the resilience of causal inferences against underlying assumptions. This shift toward a more rigorous and systematic approach marks a significant development in how research robustness is conceived and scrutinized.

The seminal work by Cornfield et al. [1] in 1959 marked a pivotal moment for sensitivity analysis, addressing the influence of unobserved variables, such as genetic factors, on the relationship between smoking and lung cancer. They posited that an unseen confounder would need to increase the likelihood of smoking by ninefold to undermine the causal link, a threshold they found implausible. This argument not only reinforced the causal role of smoking in lung cancer but also set a precedent for accounting for unseen influences in scientific investigations.

Cornfield's method provided a robust tool against the tobacco industry's denials of the smoking-lung cancer link, challenging assertions that overlooked genetic factors could be responsible. This contribution significantly impacted epidemiology and the social sciences, equipping researchers with a method to affirm the reliability of their results amidst the uncertainties of unmeasured variables. Beyond its immediate rebuttal to industry claims, this innovation has reshaped research validation across various disciplines.

Sensitivity analysis has transitioned from a qualitative, expert-driven process to a quantitative, structured evaluation of uncertainties in research findings. This shift highlights the advancement in statistical methodologies and the importance placed on drawing reliable causal inferences from data. The emphasis on transparency, rigor, and replicability reflects a broader movement within scientific research toward openly addressing and scrutinizing the assumptions that underpin causal conclusions, recognizing this as vital for enhancing the credibility and dependability of research outcomes.

The formal adoption and refinement of this methodology during the latter half of the 20th century underscores the scientific community's growing awareness of the challenges inherent in observational studies, including the risk of bias. Influential work by statisticians and epidemiologists introduced techniques to measure the impact of unmeasured, confounding errors in model specification, as well as other biases regarding research conclusions. These methods have since become fundamental in causal analysis across fields such as health sciences, economics, and social sciences, where **randomized controlled trials (RCTs)** may not be feasible or ethical, necessitating reliance on observational data for causal inferences.

Therefore, sensitivity analysis is an invaluable tool, albeit underutilized, in contemporary studies across sociology, criminology, psychology, and prevention science. Its application to scrutinize the effects of unobserved confounding on outcomes – from gang involvement and school dropout rates to criminal offending – shines a light on its potential to solidify causal inferences in non-experimental research. Despite its proven utility in revealing deeper insights into complex social phenomena, the relatively sparse use of sensitivity analysis highlights a critical area for methodological expansion. The limited application of sensitivity analysis in research stems from several factors. Many researchers lack familiarity with these techniques as they are often absent from standard training. The complexity of implementation can also deter use, particularly for those without advanced statistical skills. Furthermore, traditional reliance on established statistical methods contributes to inertia against adopting newer techniques, limiting the exploration of valuable insights that sensitivity analysis can provide. Embracing this technique more broadly could markedly enhance the reliability and depth of our understanding when conducting social science research.

Sensitivity analysis is crucial because it helps you understand how factors you haven't observed might affect your conclusions. Non-experimental studies can't control all possible confounders like randomized controlled trials can, so there's always a risk that observed associations could be due to unobserved variables. Sensitivity analysis allows you to see how much the causal findings might be influenced by these unaccounted-for confounders, giving a clearer picture of the results. Now, let's explore how this method is used in studying causality.

Sensitivity analysis for causal inference

Understanding how to conduct sensitivity analysis properly is essential for those aiming to strengthen their research findings. Let's discuss this further.

How do we use sensitivity analysis?

Sensitivity analysis is pivotal for evaluating how findings from randomized trials hold up when the assumptions of the primary analysis are challenged [5]. However, there's been a noticeable gap in guidance on selecting appropriate sensitivity analyses, resulting in their sporadic use despite their critical role in establishing causal inferences.

Morris, Kahan, and White (2014) [2] tackle this issue by outlining a principled framework for identifying meaningful sensitivity analyses through three essential inquiries:

1. Does the sensitivity analysis explore the same question as the primary analysis?
2. Could the sensitivity analysis potentially lead to a conclusion that diverges from that of the primary analysis?
3. If differing conclusions are reached, will there be real uncertainty about which conclusion is more credible?

The trio contends that a genuine sensitivity analysis aimed at testing the durability of the primary analysis's underpinnings must yield affirmative answers to all three questions. Labeling an analysis as a sensitivity check without satisfying these criteria might lead to confusion and the study's outcomes being misinterpreted.

Furthermore, *Morris, Kahan, and White illustrate how these guidelines can be practically applied in scenarios such as addressing missing data, defining study outcomes, and considering clustering effects.* They encourage people to engage with their work thoroughly so that they can refine their ability to judiciously select and report sensitivity analyses. This, in turn, sharpens the evaluation of how resilient the conclusions of randomized trials are against the bending or breaking of initial assumptions.

By sidestepping analyses that don't hold water and focusing on those that genuinely test the study's assumptions, this methodology streamlines the process of identifying sensitivity analyses that truly matter. It prevents unnecessary worry over the robustness of findings and stops false comfort from being drawn from sensitivity analyses that don't provide us with much information.

Types of sensitivity analysis

In causal inference research, sensitivity analysis is divided into parametric and non-parametric methods, each offering distinct advantages and fitting different research scenarios.

Parametric methods

These techniques require setting a range for parameters that represent the assumptions being tested so that we can explore how the calculated causal effect shifts with these parameter values. A typical use case is to evaluate the potential bias introduced by unmeasured confounders. Researchers hypothesize the existence of these confounders and assign them specific relationships with both the treatment and the outcome, aiming to quantify how much these unmeasured variables could skew the causal effect estimate.

In statistical models, such as regression, we often ascertain the causal impact of a treatment or intervention, adjusting it for known confounders. So, parametric sensitivity analysis examines the resilience of the causal effect estimate against breaches in the model's assumptions.

The following process occurs when parametric sensitivity analysis is conducted regarding causal inference:

1. First, a parameter is defined that represents a breach in model assumptions, such as deviations from expected treatment assignment mechanisms.

2. Then, the sensitivity parameter is adjusted across a realistic spectrum to simulate varying degrees of assumption breach.

3. Finally, with each adjustment, the estimated causal effect is recalculated.

This process sheds light on how the causal effect estimate fluctuates as assumptions are systematically relaxed or tightened. A causal conclusion that varies significantly with slight assumption adjustments indicates a high sensitivity to those assumptions, questioning the conclusion's robustness.

Here are some examples of parametric sensitivity analysis applications that are worth exploring:

- Unmeasured confounding

- Misclassification in treatment assignment

- Errors or corruption within a dataset

Adjusting sensitivity parameters allows us to investigate how assumption breaches might modify key causal conclusions, offering insights into the conclusions' stability against potential assumption violations.

Non-parametric methods

These methods stand apart by not depending on predefined assumptions about data distribution or the exact nature of relationships between variables. Techniques such as bootstrapping and permutation tests are employed to evaluate how changes in data or model specifics might influence causal effect estimates. This approach is particularly advantageous when the data's distribution or the relationships between variables don't conform well to standard assumptions or are unknown.

In the context of causal inference, non-parametric sensitivity analysis evaluates the dependability of causal effect estimates without leaning on parametric models. This contrasts with parametric methods, which necessitate detailed statistical models and parameters, as they rely less on assumptions about the underlying relationships and distributions.

The following are some essential features of non-parametric sensitivity analysis in causal inference:

- The omission of a detailed parametric model for mapping treatment/exposure or outcome mechanisms.

- The adoption of strategies such as bounding, weighting, or imputation, which hold minimal assumptions (we'll take a closer look at this later in this section).

- A spectrum of causal effect estimates is provided that reflect different scenarios of assumption violations, rather than singular point estimates.

- The range or bounds width signifies the extent of sensitivity to unseen confounding or other assumption breaches.

Methods such as partial identification, non-parametric **inverse probability weighting** (IPW), and flexible multiple imputation are used to implement the aforementioned features [7, 8]. Specifically, they are tools or approaches that are used in conducting non-parametric sensitivity analysis.

An example of this is Rosenbaum's sensitivity analysis [6], which delivers a span of causal effect estimates under various degrees of unmeasured confounding. A tighter range implies a stronger robustness to assumption violations.

Non-parametric methods excel in offering insights into the sensitivity of conclusions without being bound by potentially inaccurate parametric assumptions. Nonetheless, they might yield less precise estimates compared to parametric methods when the original assumptions hold true.

Let's discuss a few strategies that are used in non-parametric sensitivity analysis for causal inference, specifically bounding, weighting, and imputation. These approaches aim to reduce reliance on strong modeling assumptions and produce more robust causal estimates:

- **Bounding**: This involves calculating upper and lower limits for causal effects without making strong assumptions about missing data mechanisms. This aligns with the general goal of non-parametric methods to minimize assumptions.

- **Weighting**: IPW aims to create a pseudo-population that mimics a randomized experiment by assigning weights to observed data points. It's used to adjust for confounding and selection bias in observational studies. IPW can be applied in point treatment scenarios and with time-varying exposures and confounders. It involves estimating propensity scores (probability of treatment) and using their inverse as weights.

- **Imputation**: Multiple imputation is used as an alternative or complementary approach to IPW. Specifically, it's used to fill in missing data while accounting for uncertainty. Multiple imputation can be combined with IPW in analyses, though care must be taken in how the methods are integrated.

For more details on these, please refer to [10].

From this, we understand that both parametric and non-parametric approaches are critical tools for scrutinizing the stability of causal inferences against assumption violations in observational research, each with its unique strengths and applications.

Key concepts and measures

In the process of dissecting the complexity of sensitivity analysis in causal inference, it becomes imperative to learn how the potential biases, which spring from assumption violations, can distort the accuracy of causal effect estimates. This understanding is illuminated through several key methodologies and metrics. Let's take a closer look.

Bias formulas

At the core of sensitivity analysis lies the formulation of **bias formulas**. These mathematical constructs delineate the extent to which an estimate might veer from the true causal effect, attributing such deviations to specific sources of bias such as unmeasured confounding or inaccuracies in model specifications. *These formulas serve a dual purpose: not only do they offer a quantitative assessment of the bias introduced by assumption violations, but they also provide a means to adjust the estimated causal effects by incorporating alternative hypotheses about the bias sources.*

These are instrumental in quantifying the potential bias in causal effect estimates. This arises from violations of the assumptions necessary for point identification – that is, where a singular estimate of the causal effect can be achieved in the absence of any assumption breaches. The essence of employing bias formulas is encapsulated in their ability to do the following:

- Quantify the estimated causal effect's deviation from the true effect while considering the following:

  - The magnitude of assumption violation, such as the level of unmeasured confounding.

  - The interplay between unmeasured confounders, the treatment, and the outcome.

- Cater to various sources of assumption violations, including unmeasured confounding, differential measurement error, sample selection bias, and treatment misclassification, through specific bias formulas tailored for each scenario.

- Utilize inputs, potentially bounded or specified from external data or knowledge, such as the association between treatment and outcome due to confounding, the prevalence of unmeasured confounders, and error rates in treatment or outcome measurement.

- Allow researchers to explore a range of plausible input values within the bias formula, leading to the calculation of bias-adjusted causal effect estimates.

- Present a spectrum of estimates that reflect the potential impact of the violated assumption on the original estimate, thus enriching the sensitivity analysis by highlighting how assumption violations could bias the causal effect estimate.

For instance, when considering an unmeasured binary confounder, the bias formula illuminates how the original estimate might be biased – either overstated or understated – based on the dynamics between the unmeasured confounder, treatment, and outcome. Such analytical insights from bias formulas not only augment traditional computational techniques, such as simulations, but also fortify the foundation of sensitivity analysis by embedding a principled approach to examining the robustness of causal effect estimates against assumption violations. This mathematical sophistication ensures that causal inference is not only grounded in rigorous statistical analysis but also remains adaptable to the complex realities of empirical research.

Bias-adjusted estimators

These are refined tools in the domain of causal inference that are designed to refine initial causal effect estimates by counteracting potential biases identified through meticulous sensitivity analysis. This adjustment process integrates corrective measures based on insights derived from bias formulas or equivalent sensitivity analysis outcomes, aiming to enhance the precision of causal effect estimations and diminish their dependency on initial assumptions.

The methodology unfolds as follows:

1. Begin with a foundational causal effect estimator, such as a regression coefficient or a propensity score weighting estimator.

2. Execute a thorough sensitivity analysis by employing bias formulas to evaluate possible biases under various scenarios of assumption violation.

3. Calculate an adjustment factor that's been analytically derived from the bias formula to neutralize or minimize the bias.

4. Apply this adjustment factor to the original estimator, thus yielding a bias-adjusted estimator.

The following are some illustrative examples:

- Regression estimators refined for unmeasured confounding bias through techniques such as residual regression.

- Inverse probability weighting estimators rectified for biases due to treatment misclassification.

- Doubly robust estimators that synergize regression and weighting adjustments to address multiple bias sources.

The primary benefits of bias-adjusted estimators are manifold:

- They offer a singular, coherent point estimate of the causal effect that incorporates considerations for potential biases.

- They embed the analytical insights from sensitivity analysis directly into the estimator.

- They circumvent the necessity to present a spectrum of estimates under diverse scenarios, simplifying their interpretation.

Nonetheless, their efficacy is contingent upon the precise formulation and applicability of the bias formula used and its assumptions regarding the nature of the bias.

Bias-adjusted estimators are pivotal for delivering a purified, unified causal estimate that withstands the scrutiny of assumption violations, enhancing both the clarity and robustness of the causal analysis. Their success hinges on the thoroughness of the underlying sensitivity analysis.

Addressing unmeasured confounders

In observational studies, unmeasured confounding stands as a critical challenge and a pivotal focus of sensitivity analysis. Techniques for sensitivity analysis typically involve evaluating the potential influence of hypothetical unmeasured confounders on the estimated causal effect. This critical scrutiny aids in assessing the degree to which conclusions might sway due to unaccounted confounding, thus bolstering the credibility and robustness of causal conclusions.

By adeptly applying concepts such as bias formulas and bias-adjusted estimators, researchers are better equipped to comprehend and alleviate the effects of unmeasured confounding and other bias sources, significantly reinforcing the integrity and dependability of their causal findings.

Implementing sensitivity analysis in R

R is an impactful tool if you're aiming to assess the robustness of your causal conclusions under various assumptions. This section will provide a step-by-step overview of using R to perform sensitivity analysis. Here, we'll understand the impact that reading comics at a young age has in terms of their engagement in higher studies later on.

Using R for sensitivity analysis

In this section, we'll consider a study that's designed to explore how reading comics at a young age influences students' likelihood of pursuing higher studies. In this hypothetical study, we'll consider variables such as the frequency of comic reading (`comic_freq`), parents' level of education (`parents_edu`), time spent on homework (`homework_time`), and an indicator of enrolling in higher education (`higher_studies`).

First, let's generate some synthetic data for this study:

```
# Set seed for reproducibility
set.seed(123)
# Number of observations
n <- 500
# Generate synthetic data
data <- data.frame(
  comic_freq = sample(
    1:5, n, replace = TRUE), # 1: Rarely, 5: Very Frequently
  parents_edu = sample(
    c('Low', 'Medium', 'High'), n, replace = TRUE,
    prob = c(0.3, 0.4, 0.3)),
  homework_time = rnorm(n, mean = 2, sd = 0.5), # Hours per day
  higher_studies = sample(
    c(0, 1), n, replace = TRUE) # 0: No, 1: Yes
)
```

Now, let's perform sensitivity analysis to assess how an unmeasured confounder could affect the observed association between comic reading frequency and higher education engagement.

We'll start by fitting a logistic regression model, which helps us understand the relationships between various factors and the likelihood of higher studies. We're particularly interested in the impact of comic reading frequency, so we'll extract its odds ratio and confidence interval. The odds ratio tells us how the chances of pursuing higher studies change as comic reading frequency increases, while the confidence interval gives us a measure of how certain we are about this relationship:

```
# Fit a logistic regression model
model <- glm(
  higher_studies ~ comic_freq + parents_edu + homework_time,
  data = data, family = binomial
)
# Print model summary
summary(model)
# Extract odds ratio and confidence interval for comic_freq
or <- exp(coef(model)["comic_freq"])
ci <- exp(confint(model)["comic_freq",])
# Print odds ratio and confidence interval
cat("Odds Ratio for comic_freq:", or, "\n")
cat("95% Confidence Interval:", ci[1], "-", ci[2], "\n")
```

Next, we'll calculate the prevalence of students pursuing higher studies in our dataset. Then, we'll create a special function to calculate E-values, which are like sensitivity meters for our results. E-values are a measure used in sensitivity analysis for observational studies to assess how robust an observed association is to a potential unmeasured confounding. We'll use this function to compute E-values for both our main estimate and the lower bound of our confidence interval. These E-values tell us how strong an unmeasured factor would need to be to potentially explain away our observed relationship between comic reading and higher studies. With E-values, we're giving ourselves a clearer picture of how confident we can be in our results. It's like we're stress-testing our findings, asking, "How strong would an unknown factor need to be to shake our confidence in this relationship?" In other words, E-values provide insights into the strength of unmeasured confounding that would be necessary to explain away the observed association. This approach helps us be more cautious and thorough in our interpretation, ensuring we don't overstate our conclusions:

```
# Calculate the prevalence of the outcome
outcome_prevalence <- mean(data$higher_studies)

# Function to calculate E-value
calculate_evalue <- function(estimate) {
  if (estimate < 1) estimate <- 1 / estimate
  return(estimate + sqrt(estimate * (estimate - 1)))
}

# Calculate E-values
evalue_point <- calculate_evalue(or)
evalue_lower <- calculate_evalue(ci[1])

# Print E-values
cat("E-value (point estimate):", evalue_point, "\n")
cat("E-value (lower confidence limit):", evalue_lower, "\n")
```

Visualizing our findings

Visualization can greatly aid in interpreting the results of sensitivity analysis.

In this final stage of our analysis, we must create a visual representation of our sensitivity analysis. Here, we're crafting a plot that will help us understand and communicate our findings more effectively. First, we must generate a range of potential confounder strengths and calculate their corresponding bias factors.

Then, we must create a line plot showing how these bias factors relate to confounder strengths. We'll add horizontal lines to represent our observed odds ratio and E-values, each in a different color for clarity. By labeling these lines, we're making the plot easy to interpret. This visualization is a map of our study's sensitivity – it shows us, at a glance, how strong an unmeasured factor would need to be to challenge our conclusions. As we look at this plot together, we can see the strength of our findings and discuss potential limitations (*Figure 12.1*):

```
# Create data for plotting
conf_strengths <- seq(
  1, max(evalue_point, evalue_lower) + 0.5, by = 0.1)
bias_factors <- sapply(
  conf_strengths, function(x) x + sqrt(x * (x-1)))

# Plot
plot(conf_strengths, bias_factors, type = "l",
     xlab = "Confounder-Outcome Relative Risk",
     ylab = "E-value",
     main = "Sensitivity Analysis Plot")

# Calculate a suitable x-position for text
x_pos <- min(conf_strengths) + (max(conf_strengths) -
                                min(conf_strengths)) * 0.02

# Calculate y-axis offset (adjust this value as needed)
y_offset <- (max(bias_factors) - min(bias_factors)) * 0.02

abline(h = or, col = "red", lty = 2)
text(x_pos, or + y_offset, "Observed OR", pos = 4,
     col = "red", adj = 0)

abline(h = evalue_point, col = "blue", lty = 2)
text(x_pos, evalue_point + y_offset, "E-value (point estimate)",
     pos = 4, col = "blue", adj = 0)

abline(h = evalue_lower, col = "green", lty = 2)
text(x_pos, evalue_lower + y_offset, "E-value (lower CI)",
     pos = 4, col = "green", adj = 0)
```

Here's the output:

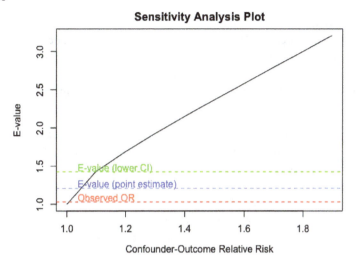

Figure 12.1 – The output of the sensitivity analysis conducted on the E-value package

Next, we'll go through a case study.

Case study

While the preceding example serves as a synthetic illustration, the real-world applicability of sensitivity analysis spans a broad and diverse array of contexts. A notable instance is the in-depth exploration of sensitivity analyses within clinical trials, as detailed by Thabane et al. (2013) [3] in their contribution to BMC Medical Research Methodology. The imperative for conducting sensitivity analysis emerges from its critical function in affirming the reliability of clinical trial results. It looks into the potential alterations in outcomes precipitated by shifts in underlying assumptions or the adoption of alternate analytical strategies, thereby evaluating the repercussions of such modifications on the conclusions that are drawn.

The tutorial in their paper underscores several compelling reasons delineating the significance of sensitivity analyses. *Paramount among these is the reality that clinical trials frequently rest upon assumptions that may not universally apply across distinct contexts or demographic groups. By executing sensitivity analyses, you're empowered to ascertain the extent to which these presumptions influence the study's conclusions.*

Following a strict and disciplined methodology not only makes the research findings more reliable but also ensures that the interpretations you draw from the results are grounded in a deep understanding of their consistency and accuracy.

Next, let's learn about the key takeaways.

Key takeaway

Our exploration of sensitivity analysis in R has revealed its crucial role in validating research findings across various fields, from clinical trials to environmental studies and education research. This powerful tool allows you to examine how factors such as outliers, non-compliance, missing data, and outcome definitions can significantly impact study conclusions. Despite its importance, a review of major medical and health economics journals found that only about ~27% of studies reported using sensitivity analysis, highlighting a concerning gap in research practices [3].

Thabane et al. emphasize the urgent need for the wider adoption of sensitivity analysis to ensure the robustness and reliability of research outcomes. Their work serves as a valuable guide for implementing these techniques in clinical trials and beyond. By incorporating sensitivity analysis into both the planning and reporting stages of research, we can enhance the credibility and trustworthiness of our findings. This approach is not limited to medical research; it proves equally valuable in assessing the impact of unmeasured confounders in other fields such as environmental studies or examining how socio-economic factors might influence educational program effectiveness.

Simulation in R

Let's illustrate the tutorial set forth by Thabane et al. [3] by creating a simple R example. This example will focus on a hypothetical scenario involving a clinical trial that compares the effect of two treatments on a continuous outcome (that is, blood pressure reduction). We'll perform a primary analysis and then conduct sensitivity analysis by changing the method of handling missing data, which is a common issue in clinical trials.

First, let's create synthetic data for our example. We'll assume that we have data for 100 patients, with 50 receiving treatment A and 50 receiving treatment B:

```
set.seed(123) # For reproducibility

# Generating synthetic data
n <- 100 # Number of patients
treatment <- rep(c("A", "B"), each = n/2)
outcome <- c(rnorm(n/2, mean = 20, sd = 5),
             rnorm(n/2, mean = 18, sd = 5))
               # Treatment A has a slightly higher effect
patient_id <- 1:n

data <- data.frame(patient_id, treatment, outcome)
# Introduce some missing data in the outcome
set.seed(234)
missing_indices <- sample(1:n, 20) # Randomly choose 20 outcomes to be
missing
data$outcome[missing_indices] <- NA
```

We must remove any missing data (if any) to ensure that our analysis is effective:

```
complete_cases_data <- na.omit(data) # Remove missing data
# Perform t-test to compare treatments
primary_result <- t.test(outcome ~ treatment,
                         data = complete_cases_data)
print(primary_result)
```

In the preceding code, the **Welch two-sample t-test** we conducted compares the effects of two treatments (A and B) on a continuous outcome:

- The t-statistic of 2.4866, with 77.685 degrees of freedom, resulted in a p-value of 0.01504. This indicates a statistically significant difference between the treatments, meaning it's quite unlikely to observe such a difference by chance if there are no actual effects.

- The confidence interval ranging from 0.4668 to 4.2173, which doesn't include 0, further supports the significance of this difference. The confidence interval not including 0 is a crucial indicator of statistical significance, complementing the p-value interpretation. When the interval excludes 0, it suggests that even at the most conservative estimate, there's still a meaningful effect, ruling out the possibility of no difference between treatments. This aligns with the p-value being less than 0.05. For a two-tailed test at the 5% significance level, a 95% confidence interval not containing 0 is mathematically equivalent to a p-value < 0.05, providing a consistent and robust interpretation of the results.

- Essentially, the means of the outcomes for treatments A and B were 20.41514 and 18.07307, respectively, suggesting that treatment A is more effective on average than treatment B.

Next, let's impute missing data using a simple approach (mean imputation) and then re-analyze it for the purpose of sensitivity analysis:

```
# Simple imputation with the mean of available outcomes
mean_outcome <- mean(data$outcome, na.rm = TRUE)
data$outcome[is.na(data$outcome)] <- mean_outcome

# Perform t-test after imputation
sensitivity_result <- t.test(outcome ~ treatment, data = data)
print(sensitivity_result)
```

In the end, you would compare the results of the primary analysis (complete case analysis) with the sensitivity analysis (after imputation). This comparison helps us understand how handling missing data impacts the study's conclusions. In real-world scenarios, additional sensitivity analyses would be conducted, varying from not just handling missing data but also other assumptions and definitions used in the analysis.

After imputing missing data and re-running the Welch two-sample t-test, we discover the following:

- A t-statistic of 2.4734 with 97.989 degrees of freedom, and a p-value of 0.0151. This result, similar to the primary analysis, confirms a statistically significant difference between treatments A and B.

- The confidence interval, stretching from 0.3701 to 3.3748, excludes 0, affirming the difference's significance.

- The calculated means – 20.20962 for treatment A and 18.33714 for treatment B – demonstrate that treatment A remains more effective on average, even after handling missing data.

This consistency validates the robustness of our initial findings against the chosen method for managing missing data.

Remember, this example has been simplified to illustrate the process. Actual clinical trials and sensitivity analyses can be much more complex and involve various statistical models, multiple imputations, and consider different types of missing data mechanisms. Next, we'll learn about practical guidelines for conducting sensitivity analysis.

Practical guidelines for conducting sensitivity analysis

As you've seen, sensitivity analysis is a powerful tool in causal inference for assessing the robustness of research findings under various assumptions. However, its effectiveness depends on thoughtfully selecting parameters for analysis, accurately interpreting the results, and understanding its limitations. In this section, we'll cover some practical guidelines for conducting sensitivity analysis that can help you tackle these aspects effectively. First, we'll learn how to choose parameters.

Choosing parameters for sensitivity analysis

Begin by identifying the key assumptions underpinning your causal model. This includes assumptions about the absence of unmeasured confounders, the type relationship between variables (for example, linearity), and the homogeneity of effects across subpopulations.

Parameter deep dive

Once you've done this, you must prioritize parameters for which there is substantial uncertainty or debate within your field. These might include parameters related to unmeasured confounding, model specifications, or extrapolation beyond the observed data.

For each parameter that's selected, determine a plausible range of values based on prior studies, theoretical considerations, or expert judgment. It's crucial to explore a wide enough range to capture potential variability but also to remain within bounds that are considered realistic and relevant to the context of your study.

Also, reflect on how changes in each parameter might impact your causal conclusions. Parameters that have a substantial effect on the outcomes of interest should be prioritized for sensitivity analysis.

To delve deeper into how varying parameters for sensitivity analysis impact causality mathematically, let's formalize a scenario of loan defaulters based on age, gender, or family structure. We'll construct a basic regression model to predict loan default rates and then illustrate how sensitivity analysis can be represented and interpreted mathematically. Let's see it in action by using a regression model to understand these factors more.

Basic regression model for loan defaults

Suppose our initial regression model to predict the probability of loan default, $P_{default}$, based on age A, gender G, and family structure F, is as follows:

$$P_{default} = \beta_0 + \beta_{1A} + \beta_{2G} + \beta_{3F} + \epsilon \qquad (1)$$

Here, we have the following:

- $P_{default}$ is the probability of loan default
- A represents the age of the loan borrower
- G is a categorical variable representing gender
- F is a categorical variable representing family structure
- β_0 is the intercept of the model
- $\beta_{1A}, \beta_{2G}, \beta_{3F}$ are the coefficients for age, gender, and family structure, respectively
- ϵ is the error term

Diving into our model's sensitivity analysis, we're not just poking at parameters $\beta_0, \beta_1, \beta_2$ for the fun of it – though a little academic mischief never hurt. We're here to see how these tweaks affect $P_{default}$, because who doesn't like to play "What if?" with their data?

1. **Playing with age (β_{1A}):** Let's suppose we're feeling adventurous and decide to give β_{1A} ±10% a whirl based on past data or intuition based on domain expertise. Our model now reads as follows:

$$P_{default} = \beta_0 + \beta_{1A} \pm 10\% + \beta_{2G} + \beta_{3F} + \epsilon \qquad (2)$$

   This little experiment lets us peek into how tweaking the age factor by just a tad influences our loan default scenario, keeping everything else unchanged.

2. **Gender bender (β_{2G}):** With gender coded as 0 for males, 1 for females, and 2 for those who defy binary norms, what if we speculate that the gender impact on loan defaults got a bit lost in translation? If β_{2G} is shrouded in mystery for females and non-binary folks, a strategic nudge might just unveil some intriguing causal insights.

3. **Family structure shuffle** (β_{3F}): Envision family structure as our narrative backdrop – singles, married sans kids, and the married folks with kids. Pondering over the potential effect of kids on loan defaults, we take a gamble:

$$P_{default} = \beta_0 + \beta_{1A} + \beta_{2G} + \beta_{3F} \pm 15\% + \epsilon \qquad (3)$$

By playing around with the parameters, especially for the "married with kids" crew, we dive into how such a life choice sways the likelihood of greeting a loan default.

Mathematical interpretation

The equation provided (3) explores the effect of increasing or decreasing the impact of being married with children on the default rate by 15%, which might reflect the financial burden associated with children. The sensitivity analysis, through these equations, mathematically quantifies the robustness of our model's predictions to changes in our assumptions about the relationship between borrower characteristics and loan default rates. By adjusting the coefficients within plausible ranges and observing the variations in $P_{default}$, we can identify which parameters our model is most sensitive to. This approach not only informs us about the stability of our conclusions but also highlights areas where additional data or research may be necessary to refine our understanding.

In general, the guideline is to use the results of your sensitivity analysis to quantify the robustness of your causal claims. *This can be done by reporting how much your estimates change across the range of parameter values being considered. E-values or bias-adjusted estimates can provide a concise summary of this robustness,* which we will learn next.

What are E-values?

E-values and bias-adjusted estimates are effective tools in sensitivity analysis, particularly for quantifying the robustness of causal claims in the presence of potential unmeasured confounding or biases. These methods allow you to assess how resilient their findings are to the assumptions made during the analysis. Let's dive into each concept and provide a mathematical explanation for their application.

An E-value is a measure that quantifies the minimum strength of association that an unmeasured confounder would need to have with both the treatment and the outcome to fully explain away a specific treatment-outcome association, assuming no other confounding. The concept of an E-value provides a straightforward way to communicate the potential impact of unmeasured confounding on causal claims.

Mathematically, if R_{obs} is the observed risk ratio for the association between the treatment and outcome, the E-value can be calculated as follows:

$$E = RR_{obs} + \sqrt{RR_{obs} \times (RR_{obs} - 1)} \qquad (4)$$

For an observed risk ratio of less than 1, the formula uses the inverse of RR_{obs}. The larger the E-value, the more robust the causal claim is to potential unmeasured confounding. An E-value close to 1 suggests that the observed association could be easily explained by unmeasured confounding.

Bias-adjusted estimates come into play when we tweak our initial findings to account for possible skewing factors such as confounding, selection bias, or measurement errors. This process usually involves figuring out which way and how much these biases might be tilting our results, and then adjusting our original estimates to correct for these biases. To clarify, E-values quantify the minimum strength of unmeasured confounding needed to nullify an observed association, providing a measure of robustness of the result. Bias-adjusted estimates, on the other hand, directly modify initial findings to account for potential biases, offering corrected results based on assumed magnitudes and directions of these biases.

Imagine that you've got a result, E_{obs}, but suspect it's off-kilter because of some confounding factors you couldn't measure. Let's call the influence of this unseen confounder on your treatment and outcome U. If you have a guess for U, you can tweak E_{obs} to get a cleaner, bias-adjusted figure, E_{adj}:

$$E_{adj} = E_{obs} - U \qquad (5)$$

This approach assumes we're dealing with straightforward, additive biases, though real-world scenarios might demand wrestling with more complex, perhaps multiplicative, biases.

In sensitivity analysis, once you've pinned down a plausible range for these unseen confounders or biases, you can churn out E-values for each possibility. This gives you a spectrum of E-values that shed light on how sturdy your findings are under various hypotheticals. You can also get bias-adjusted figures for each scenario, offering a deeper dive into how your original conclusions hold up under scrutiny.

Sharing E-values and bias-adjusted figures is essentially telling the world how much sway unmeasured confounders or other biases would need to have to topple your observed relationship. It's a transparency boost, giving everyone a clearer picture of how solid your causal connections are, and making your interpretations all the richer.

In essence, E-values and bias-adjusted estimates are your best friends in sensitivity analysis, helping you quantitatively probe the susceptibility of your causal stories to unseen confounders and biases. They lay down a mathematical basis for weighing and communicating the sturdiness of your findings, letting you and your peers gauge just how much confidence to place in the causal links you've drawn from the data.

When it comes to presenting your work, it's vital to be upfront about the uncertainties surrounding your causal assertions, based on your sensitivity checks. Point out the assumptions your arguments hinge on the most and discuss whether they're on solid ground. If it turns out your conclusions are based on shaky assumptions, it might be a cue to either add some caveats to your claims or dive back into the data for more clarity. Next, we'll talk about the limitations and challenges in this area.

Limitations and challenges

The application of sensitivity analysis presents its own set of challenges, notably in defining the plausible range of values for the parameters under scrutiny. *Opting for ranges that are too wide may render the analysis less meaningful, while ranges that are too narrow might not fully encapsulate the actual scope of uncertainty.*

A critical aspect to acknowledge is that sensitivity analysis, despite its utility in gauging the potential effects of unmeasured confounding, stops short of proving its non-existence. This underscores an essential truth: sensitivity analysis is better seen as a lens to examine, rather than dispel, the clouds of uncertainty.

Diving deeper, we may encounter additional hurdles and constraints:

- **Data quality and availability**: The integrity and completeness of the data at the heart of a sensitivity analysis shape the trustworthiness of its outcomes. Gaps or biases in data can seed uncertainties that defy precise quantification.

- **Model specification**: The task of choosing a suitable model framework for sensitivity analysis is pivotal. Missteps in model specification or foundational assumptions can skew the results, leading to misinterpreted sensitivity measures.

- **Interaction effects**: Adequately accounting for the interplay between variables poses its own set of challenges. Overlooking these dynamics may leave the analysis short of offering a full picture of the study's robustness.

- **Assumption sensitivity**: The insights drawn from a sensitivity analysis are inextricably linked to the assumptions about variable relationships and the character of confounding factors. Shifts in these assumptions have the power to sway the analysis's narrative.

Dealing with these challenges requires paying careful attention and having a clear approach to explaining findings, which helps make sensitivity analyses more trustworthy in research. Sensitivity analysis is a key part of studying cause and effect, helping us understand how strong our conclusions are. By choosing the right factors to look at, understanding what the results mean, and being open about their limits, you can use sensitivity analysis to make your insights more reliable and useful. Now that we've covered the basics of sensitivity analysis, we're ready to explore more complex ideas in this area.

Advanced topics in sensitivity analysis

Sensitivity analysis in causal inference has traditionally zeroed in on binary treatment effects within straightforward models. Yet, as we march forward in time, there's an emerging consensus on the necessity to push these methods into more intricate arenas. This includes dealing with continuous treatments, interventions that span multiple levels, and even weaving in **machine learning** (**ML**) techniques. Let's dive into these advanced areas.

Venturing beyond binary treatment

First, we have **continuous and multi-level treatments**. The classic picture of sensitivity analysis often paints treatments in stark black and white - you either get the treatment or you don't. But the real world isn't always so binary. Treatments can vary in dosage or come in different forms (think varying educational programs). To tackle these realities, sensitivity analysis is stretching its muscles to understand how changes in the intensity or the nature of treatments can sway our causal insights. Methods such as dose-response modeling and approaches for dissecting multi-level treatments are stepping into the spotlight, enabling us to grapple with these more intricate treatment effects.

You can refer this pioneering paper that unveils a **continuous treatment-effect marginal sensitivity model (CMSM)** [4]. This model is a game-changer for sizing up the effects of continuous interventions, especially when we're tangled up in the typical assumptions that often trip us up in areas ranging from climate science to healthcare. With its scalable algorithm and foray into deep learning, it opens up new avenues for estimating bounds on treatment effects, even when potential confounders lurk in the shadows. The paper presents a compelling case study on climate, leveraging 15 years of satellite data to make sense of human impacts on climate.

We also have **complex models**. Modern causal inference involves complex models with multiple pathways and interactions. Sensitivity analysis must consider a broader range of assumptions, including the structure of the causal model itself. Advanced tools such as graphical models and **structural equation models (SEMs)** help with navigating these complex scenarios. **Directed acyclic graphs (DAGs)** are particularly useful for mapping out assumed causal relationships. They allow us to visualize and analyze the causal structure of our models. In this context, sensitivity analysis examines how robust our causal conclusions are when key assumptions are challenged, such as when potential confounders are overlooked or causal sequences are misspecified.

SEMs are powerful tools for investigating causality in complex scenarios. By combining path analysis and factor analysis, SEMs allow us to model both observable and latent variables, along with their relationships. When using SEMs for sensitivity analysis, we can examine how different model specifications – such as factor structures, error correlations, or causal pathways – affect our causal estimates. We can use tools such as model fit indices and parameter perturbations to assess the stability of causal estimates under various assumptions.

These developments enhance our ability to uncover and validate causal relationships more accurately. As we explore these complex areas, the core of rigorous research remains: expanding our knowledge and understanding its limitations. Next, let's consider how ML might contribute to sensitivity analysis.

ML approaches

ML techniques offer advanced tools for identifying potential biases and confounders in sensitivity analysis. Algorithms such as the **Least Absolute Shrinkage and Selection Operator (LASSO)**, elastic net, and random forests can unearth variables that may act as unmeasured confounders by analyzing their associations with both treatment and outcomes. This capability extends to uncovering non-linear relationships and interactions that traditional methods might miss, offering a data-driven complement to expert-driven variable selection.

Here are some other applications and techniques that involve using ML:

- **Variable selection algorithms**: Tools such as LASSO and random forests identify unmeasured confounders by examining associations with treatment and outcomes, enhancing the accuracy of causal estimates.

- **Detecting non-linear relationships**: ML models, including neural networks and gradient-boosting machines, can reveal complex patterns and interactions overlooked by linear analyses.

- **Counterfactual ML methods**: Approaches such as causal forests and targeted maximum likelihood estimation estimate counterfactual outcomes, offering a robust framework for causal inference that accommodates flexible ML models.

- **Automated sensitivity analysis**: The development of automated sensitivity analysis tools powered by ML can systematically assess the robustness of causal conclusions across a wide range of models, parameters, and assumptions. This automation broadens the scope of sensitivity analysis, improving comprehensiveness and reducing the likelihood of oversight.

- **Addressing missing data and measurement error**: ML techniques also address challenges such as missing data or measurement error through imputation methods, such as **multiple imputation with chained equations** (MICE) or denoising autoencoders. These approaches allow for analyses that consider various assumptions about missing data or measurement errors.

While integrating ML into sensitivity analysis offers significant advantages, challenges such as interpretability, model complexity, and overfitting necessitate careful consideration. The causal assumptions of ML models, often derived from observational data, require critical evaluation to ensure they accurately reflect the causal dynamics at play. The integration of ML with traditional sensitivity analysis methods presents a promising avenue for addressing potential biases and confounders in causal inference, especially in settings with complex, nonlinear data-generating processes. By combining the strengths of both approaches, you can achieve a more nuanced and reliable understanding of their causal findings. You can find a new approach to sensitivity analysis using ML by taking a look at the *References* section at the end of this chapter [9].

Future directions

A promising avenue in sensitivity analysis is its synergy with causal discovery algorithms (covered later in this book), which seek to unearth causal connections directly from datasets. Sensitivity analysis can augment these algorithms by evaluating the stability of identified causal relationships against different underlying assumptions. This dual approach enriches our understanding of causal dynamics, offering a more comprehensive view of causality derived from data.

Advancing personalized sensitivity analysis

With the rise of personalized medicine and custom-tailored interventions for a business goal or a set objective, the demand for sensitivity analysis capable of evaluating causal effects at an individual level is increasing. This necessitates the development of innovative methods that can navigate the variability in treatment responses and potential biases unique to different subgroups or individual profiles. Such personalized sensitivity analysis will be crucial in tailoring treatments to achieve optimal outcomes for individual patients.

Prioritizing transparency and reproducibility

As sensitivity analysis methods grow in complexity, maintaining transparency and ensuring reproducibility are becoming critical challenges. Future efforts in the field are expected to focus on establishing uniform reporting standards and developing open source tools that support the widespread sharing and replication of sensitivity analysis findings. These initiatives aim to bolster the integrity and accessibility of sensitivity analysis research, enabling a broader community of researchers to contribute to and benefit from advancements in the field.

The landscape of sensitivity analysis within causal inference is evolving rapidly, with ongoing innovations addressing the nuances of continuous treatments, the complexities of advanced models, and the incorporation of ML. These developments are set to significantly improve our capacity to verify the reliability of causal conclusions amidst the intricacies of the modern world. By pushing the boundaries of sensitivity analysis, researchers are paving the way for more precise and trustworthy causal inferences across diverse fields, enhancing our ability to make informed decisions based on robust causal evidence.

Summary

In this chapter, we explored sensitivity analysis, a crucial tool in causal inference that helps us evaluate how stable our findings are under different assumptions. We saw how this technique, pioneered by researchers such as Cornfield, is especially valuable in observational studies where uncertainties are common. Sensitivity analysis not only makes our causal claims stronger but also shows us where to focus future research by revealing the impact of factors we can't observe directly. While it's incredibly useful, we noticed it's not used as widely as it should be across different fields of study. This gap highlights how important it is for more researchers to adopt this method. By using sensitivity analysis more often, we can make our understanding of cause and effect more robust and detailed, leading to better-informed policies and scientific investigations. As you continue your research journey, consider how incorporating sensitivity analysis could enhance the depth and reliability of your findings.

In the next chapter, we'll learn about heterogeneity in causality.

References

1. Cornfield J, Haenszel W, Hammon E, Lilienfeld A, Shimkin M, Wynder E. *Smoking and lung cancer: recent evidence and a discussion of some questions.* Journal of National Cancer Institute. 1959;22:173–203. Available at http://ije.oxfordjournals.org/content/38/5/1175.full.

2. Morris, T.P., Kahan, B.C. & White, I.R. *Choosing sensitivity analyses for randomized trials: principles.* BMC Med Res Methodol 14, 11 (2014). Available at https://doi.org/10.1186/1471-2288-14-11.

3. Thabane, L., Mbuagbaw, L., Zhang, S. et al. *A tutorial on sensitivity analyses in clinical trials: the what, why, when and how.* BMC Med Res Methodol 13, 92 (2013). Available at https://doi.org/10.1186/1471-2288-13-92.

4. *Scalable Sensitivity and Uncertainty Analysis for Causal-Effect Estimates of Continuous-Valued Interventions,* by Andrew Jesson, Alyson Douglas, Peter Manshausen, Maëlys Solal, Nicolai Meinshausen, Philip Stier, Yarin Gal, and Uri Shalit. Available at https://arxiv.org/abs/2204.10022.

5. Sullivan, Adam J. 2015. *Sensitivity Analysis for Linear Structural Equation Models, Longitudinal Mediation With Latent Growth Models and Blended Learning in Biostatistics Education.* Doctoral dissertation, Harvard University, Graduate School of Arts & Sciences. Available at https://dash.harvard.edu/handle/1/17467398.

6. Rosenbaum, P. R. (2002). *Observational Studies.* Springer Series in Statistics. Springer-Verlag.

7. Manski, C.F. (2010). *Partial Identification in Econometrics.* In: Durlauf, S.N., Blume, L.E. (eds) Microeconometrics. The New Palgrave Economics Collection. Palgrave Macmillan, London. Available at https://doi.org/10.1057/9780230280816_21.

8. Robins, J. M. (1999). *Association, Causation, and Marginal Structural Models.* Synthese, 121(1-2), 151-179.

9. *A Neural Framework for Generalized Causal Sensitivity Analysis,* by Dennis Frauen, Fergus Imrie, Alicia Curth, Valentyn Melnychuk, Stefan Feuerriegel, and Mihaela van der Schaar. Available at https://arxiv.org/html/2311.16026v2.

10. Hernán, M. A., & Robins, J. M. (2020). *Causal Inference: What If.* Chapman & Hall/CRC.

13

Scrutinizing Heterogeneity in Causal Inference

In this chapter, we explore what heterogeneity means. It's a fancy way of saying treatments don't affect everyone equally. Some folks benefit more, some less. This matters because a one-size-fits-all approach to causality from every unit (e.g., a person or event) of the data may lead to sub-optimal results.

We'll explore different types of heterogeneity, how to spot them in data, and how to use this knowledge to design treatments that target specific needs. Think of it like giving the right medicine to the right person.

By using powerful statistics, we can see beyond averages and understand who benefits most from treatments. This is key for creating interventions that are both effective and fair, reaching those who need them the most. Buckle up, as we're about to make better causal decisions that reflect the real world!

In this chapter, we'll cover the following topics:

- What is heterogeneity?
- Understanding the types of heterogeneity
- Estimation methods for identifying heterogeneous causal effects
- Case study – Heterogeneity in R
- Tailoring interventions to different groups

Technical requirements

You can find the code examples for this chapter in this book's GitHub repository: `https://github.com/PacktPublishing/Causal-Inference-in-R/tree/main/chap_13`.

What is heterogeneity?

Let's start with the basics—that is, let's start with heterogeneity!

Heterogeneity in causal effects reveals the diversity in how individuals or groups respond to the same intervention—a principle that can be humorously likened to the age-old mystery of why some people seem to thrive on a diet that consists largely of pizza and energy drinks, while others must adhere to a regime of kale and quinoa to maintain the same level of health. This variation underscores the fallacy of the one-size-fits-all approach, such as assuming everyone will enjoy the same Netflix series. Just as one person's trash is another's treasure, an intervention's success can vary widely across different populations, influenced by a myriad of factors, including genetic makeup, environmental conditions, and even personal preferences.

To dissect these intricate variations, researchers investigate, employing advanced statistical methods similar to how a chef explores different culinary techniques to perfect a complex recipe. Techniques such as subgroup analysis and **machine learning** (**ML**) models are used extensively in the research world, used to slice through data and uncover the nuanced ways in which different ingredients (or factors) affect the final outcome. The goal is to understand who benefits from an intervention and under what specific conditions—essentially, decoding the secret sauce of causal effects.

Playing around with heterogeneity not only challenges the academic taste buds but also brings to light profound implications for crafting policies and personalized interventions. However, much like experimenting with a new recipe, it presents its own set of challenges, including the risk of ending up with a kitchen disaster (or spurious findings) due to the complexity of the ingredients involved. Despite these culinary mishaps, embracing the diversity captured by heterogeneity in causal effects allows for the concoction of more effective, equitable, and palatable interventions tailor-made to suit the varied tastes and needs of the global population. Next, let's dive deeper into this.

Definition of heterogeneity in causality

In layman's terms, it's the idea that interventions – whether a medicine, a program, or anything else – can have wildly different impacts on different people.

In causality, it is a crucial concept. Let me explain it further mathematically.

Imagine a group of people (we'll call them units) – let's say unit number 1, unit number 2, and so on. Each person could receive a treatment, such as taking a new drug (represented by a 1) or not (represented by a 0). We can even imagine what might happen if they didn't get the treatment – that's called a potential outcome. So, for each person, we have two potential outcomes: one if they get the treatment and one if they don't. The true effect of the treatment for that person is simply the difference between those two outcomes. We can write this mathematically as follows:

$$\tau_i = Y_{i(1)} - Y_{i(0)} \qquad (1)$$

Here, τ_i represents the causal effect of the treatment on unit i, $Y_{i(1)}$ represents the potential outcome for unit i if they receive the treatment, and $Y_{i(0)}$ represents the potential outcome for unit i if they don't receive the treatment.

Here's where things get interesting. What if the effect of the treatment isn't the same for everyone? That's where heterogeneity comes in. Maybe the new drug works wonders for some people (big difference between their potential outcomes) but has little effect on others (smaller difference). Mathematically, this translates to $\tau_i \neq \tau_j$ for some units i and j where $i \neq j$. In other words, the causal effect τ can vary across units.

To understand this variability better, we can look at things that might influence how a person responds to treatment – things such as age, health background, or even lifestyle choices. These are called covariates, denoted by X_i. By factoring these in, we can build a model that helps us see how these different characteristics might modify the treatment effect for each person. One way to represent this is as follows:

$$\tau_i = f(X_i) \tag{2}$$

Here, f is a function that captures how the covariates X_i modulate the treatment effect for unit i. This model allows us to explore how different characteristics interact with the treatment to produce varying outcomes.

But how do we measure this variation across a whole group of people? Here, statistics come to our rescue. We can look at the spread of the treatment effects for everyone or use fancy models that account for heterogeneity directly. These models can tell us, on average, how much the treatment helps (the **average treatment effect**, or **ATE**), but also how much that effect can vary from person to person (*the magic term here is variance*). The bigger the variance, the more spread out the effects are, highlighting significant heterogeneity. One common approach uses a normal distribution to represent the spread of treatment effects:

$$\tau_i \sim N(\mu, \sigma^2\tau) \tag{3}$$

Here, μ represents the ATE, providing a general picture of the impact. The crucial element is $\sigma^2\,\tau$. This term captures the variance of the treatment effects across units, essentially telling us how much heterogeneity exists. A larger $\sigma^2\,\tau$ implies a greater spread of effects, highlighting significant heterogeneity.

So, why do we get bogged down in math? By carefully considering heterogeneity, we can draw much more accurate and insightful conclusions about how well interventions work and for whom. We can avoid oversimplifying things and ensure that the unique responses of different people are considered. Even better, understanding the mathematical underpinnings of heterogeneity allows us to design treatments that are targeted and personalized – a win-win for everyone involved. Now, let's take a deep dive into a real-world example and see how these concepts play out in action!

Case studies and discussion

Causal understanding often begins with the ATE. This singular metric, though informative, paints a rather broad picture. It assumes a uniform response to an intervention across an entire population. However, as you may recognize already, real-world complexities challenge this notion. This is where the fascinating concept of heterogeneity in causal effects steps in.

Heterogeneity acknowledges the fundamental truth that individuals are not interchangeable cogs in a machine. We come with a rich tapestry of experiences, genetic predispositions, and environmental contexts that influence how we respond to interventions. A medication, for example, might be a life-saver for some patients with a specific genetic profile, while proving ineffective or even detrimental for others. Similarly, an educational program, meticulously crafted, might flourish in a well-resourced school with experienced educators but falter in a setting with limited resources and high student-teacher ratios.

To uncover and quantify this heterogeneity, we must move beyond simple average effects and employ a sophisticated array of statistical methods. Subgroup analysis allows us to segment the population into more homogeneous groups, examining treatment effects within each. Interaction tests help identify how the impact of an intervention may depend on other factors. This deeper understanding enables us to pinpoint who benefits most from a given treatment, in which specific circumstances, and to what extent. Such nuanced insights are crucial for moving beyond the limitations of ATEs and toward a more personalized and effective approach to causal inference in real-world scenarios.

Next, let's go one by one through a few different scenarios where **heterogeneous treatment effects (HTEs)** may play a crucial role.

Examples (more of them)

The following examples, though fictional, are grounded in plausible scenarios often encountered in public health, education, and economic policy research, reflecting the types of studies that could be conducted:

- **Smoking cessation**: Genetic variations can influence how people respond to programs. A study might show a modest overall decline in smoking rates, but subgroup analysis could reveal a significant drop for those with a specific genetic marker, favoring personalized medicine approaches.

- **Technology-enhanced learning**: Socioeconomic background matters. A study comparing traditional and digital learning might show students from high-income families thrive with digital platforms due to better home technology access. Conversely, students from lower-income families might see limited benefits or even setbacks due to the digital divide.

- **Universal basic income (UBI)**: Location is key. UBI's impact on employment might be negligible in high-cost urban areas where the income might not cover living expenses. However, in lower-cost rural areas, UBI could boost economic stability and even encourage entrepreneurship, highlighting the context-dependent effects of economic policies.

- **Urban green spaces**: Age and mobility play a role. Green spaces might significantly improve mental health and physical activity for younger, mobile residents. However, for older adults or those with limited mobility, the benefits might be less pronounced due to accessibility challenges. This underlines the need for inclusive urban planning that considers the diverse needs of city dwellers.

These scenarios emphasize the importance of considering heterogeneity in causal effects when designing, implementing, and evaluating interventions. By acknowledging and understanding these variations, you can tailor your causal approaches to achieve more equitable and impactful outcomes. Now, we should consider learning about different types of heterogeneity in causal inference.

Understanding the types of heterogeneity

Heterogeneity can be categorized into pre-treatment heterogeneity, post-treatment heterogeneity, and contextual heterogeneity. Understanding these distinctions is crucial for designing studies, interpreting results, and formulating policies or interventions. Next, we explore each type in detail, incorporating mathematical formulations and examples.

Pre-treatment heterogeneity

Pre-treatment heterogeneity *refers to variability in baseline characteristics among units before receiving treatment*. Now, what does that mean? This means that the baseline characteristics may include demographics, health status, prior knowledge, and any others that can influence the treatment's effect.

Let X_i represent the vector of baseline covariates for unit i, and let $Y_{i(1)}$ and $Y_{i(0)}$ denote potential outcomes under treatment and control, respectively. The causal effect for unit i, $\tau_i = Y_{i(1)} - Y_{i(0)}$ may depend on X_i, indicating pre-treatment heterogeneity. Refer to [2] for more details on learning pre-treatment heterogeneity. In an educational intervention aiming to improve math skills, pre-treatment heterogeneity might arise from students' initial math proficiency levels. Students with higher baseline skills might experience smaller gains compared to those with lower baseline skills due to ceiling effects.

To identify pre-treatment heterogeneity, you can employ various methods to calculate it. These range from descriptive analysis and statistical tests to more advanced techniques such as propensity score analysis and ML approaches. To expand on this, you need to examine distributions of baseline characteristics, conduct statistical comparisons between groups, estimate propensity scores, perform subgroup analyses, and utilize ML algorithms to identify important predictors and estimate HTEs. Additionally, sensitivity analyses are used to assess the robustness of findings to unmeasured confounding.

When identifying pre-treatment heterogeneity, you should focus on theoretically relevant covariates, consider both statistical significance and practical importance of differences, adjust for multiple comparisons, and interpret results cautiously. By systematically examining baseline characteristics through these methods, you can effectively identify potential sources of treatment effect heterogeneity, leading to more targeted and informative analyses. This comprehensive approach allows for a deeper understanding of how pre-existing differences among study units may influence treatment outcomes.

Post-treatment heterogeneity

Post-treatment heterogeneity refers to *differences in how units respond to a treatment, which are not solely attributable to their baseline characteristics.* This form of heterogeneity is often linked to how individuals interact with the treatment or intervention itself.

If we let Z_i denote the post-treatment response or behavior that affects the outcome, the causal effect might then be written as follows:

$$\tau_i(Z_i) \; = \; Y_i(1, Z_i) - Y_i(0, Z_i) \qquad (4)$$

This highlights the dependency on post-treatment variables.

Consider a medication for hypertension. Post-treatment heterogeneity might manifest in side effects experienced by patients, influencing medication adherence. Those who experience severe side effects (high Z_i) may have less favorable outcomes due to lower adherence.

Contextual heterogeneity

Contextual heterogeneity [3] *accounts for how external factors or settings influence the treatment's effectiveness. This includes environmental variables, cultural norms, policy landscapes, and other contextual elements.*

Let C represent the context in which the treatment is administered. The causal effect, considering contextual heterogeneity, can be expressed as follows:

$$\tau(C) \; = \; \mathbb{E}[Y(1, C) - Y(0, C)] \qquad (5)$$

Here, the expectation is over the distribution of potential outcomes in context C.

Consider a job training program that might have different effects in regions with varying unemployment rates. In areas with high unemployment (C high), even effective training may not lead to job placement, unlike in areas with lower unemployment (C low).

These types of heterogeneity interact in complex ways to shape the overall causal effect of an intervention. For instance, pre-treatment characteristics X_i might interact with contextual factors C, leading to nuanced patterns of post-treatment responses Z_i. Mathematically, this interaction can be represented as $\tau_i(X_i, C, Z_i)$, acknowledging the multifaceted nature of treatment effects.

Heterogeneity in treatment effects is not just a minor consideration for the causal inference scholar, it's essential to uncovering the true impact of interventions. By exploring how treatment effects differ across individuals and contexts, we gain the following valuable insights:

- **Targeted interventions**: We can identify subgroups that reap the most (or least) benefit, allowing for targeted interventions that maximize impact.

- **Contextualization**: We can tailor interventions to specific situations, ensuring effectiveness across diverse settings.

- **Realistic expectations**: We can anticipate variations in treatment efficacy, avoiding overly optimistic (or pessimistic) assessments.

Ignoring heterogeneity may mean that we might miss crucial details that influence how well an intervention works. A comprehensive analysis goes beyond average effects, diving deep to explore the distribution of causal effects. This unveils the nuanced picture – who benefits most, in what contexts – informing policy and practice with greater precision.

In essence, embracing heterogeneity isn't just good research practice; it's the key to designing effective and equitable interventions that truly make a difference. Let's dive into more details next.

Heterogeneous causal effects deep dive

Understanding and identifying heterogeneity in causal effects is central to tailoring interventions and policies more effectively. Several strategies are employed in statistical analysis to detect and estimate heterogeneous effects, each with its mathematical framework and application.

Interaction terms in regression models

Interaction terms are a powerful tool for causal inference scholars to explore how an intervention's impact changes based on another variable.

Consider a linear regression model examining the effect of a treatment T on an outcome Y. We suspect this effect might vary by a covariate X. To capture this, we extend the model with an interaction term:

$$Y = \beta_0 + \beta_1 T + \beta_2 X + \beta_3 (T \times X) + \varepsilon \qquad (6)$$

Here, β_0 is the intercept, β_1 is the ATE, β_2 captures the effect of X on Y, and the key term β_3 is the interaction effect. It reveals how the treatment effect (β_1) varies with changes in the covariate X. Finally, ε is the error term.

For instance, imagine studying a math tutoring program's effect on student performance (outcome: Y, treatment: participation T, coded 1/0). We suspect a student's initial proficiency (covariate: X) might influence program effectiveness. Including the interaction term $T + X$ allows us to test this formally.

A significant and positive β_3 suggests the program is more effective for students with higher initial proficiency (higher X). Conversely, a negative β_3 implies the program benefits struggling students more. *This interaction term allows us to move beyond the ATE and explore how program impact differs across subgroups defined by the covariate X.*

That is, **interaction terms** highlight **heterogeneity**. Knowing this is critical for designing targeted interventions, tailoring programs to specific populations, and ultimately, maximizing their real-world impact. Instead of a uniform approach, we should embrace a nuanced understanding of how treatment effects vary according to individual characteristics and specific contexts.

Subgroup analysis

Subgroup analysis is a cornerstone technique [4] in causal inference, that helps peel back the layers and expose heterogeneity in treatment effects. This approach goes beyond the ATE and delves into how the impact of an intervention might differ across distinct subgroups within a population.

Here's the essence of subgroup analysis:

- **Divide and conquer**: The population is strategically divided into subgroups based on relevant characteristics, such as gender or age or other relevant factors.

- **Model within**: For each subgroup (denoted by i), a separate model is estimated to capture the treatment effect specifically within that group. This model can be written as:

$$Y_i = \alpha_0^i + \alpha_1^i T + \epsilon_i. \qquad (7)$$

Here, Y_i represents the outcome variable for subgroup i, α_0^i is the intercept specific to subgroup i, α_1^i captures the estimated treatment effect for subgroup i (think of it as the effect within that subgroup), and ε_i denotes the error term.

- **Unearthing differences**: By analyzing the estimated treatment effects (α_1^i) across different subgroups, we can identify variations that might be masked by the ATE.

Imagine the tutoring program example again. Subgroup analysis allows us to assess the program's effectiveness separately for girls and boys. The analysis might reveal a significant and positive $\alpha_1^{(female)}$ (indicating improved math scores for girls), while $\alpha_1^{(male)}$ might be statistically insignificant (suggesting no effect for boys). This shows a critical hidden difference – the program might be particularly beneficial for female students.

ML techniques

ML techniques offer advanced tools for identifying heterogeneity, especially in high-dimensional data. These methods can capture complex, nonlinear relationships and interactions without explicitly specifying them. Refer to more details here [5, 6]:

- **Random Forests**: This ensemble learning method can be used to estimate **conditional ATEs (CATEs)** by training separate trees on treated and control groups and comparing predictions.

- **Gradient Boosting Machines (GBMs)**: GBMs can be adapted to estimate HTEs by optimizing loss functions that directly target estimations of CATE.

These methods leverage the strengths of ML algorithms to handle high-dimensional data and capture complex interactions while addressing causal inference issues such as confounding and selection bias.

Random Forests have been extended into Causal Forests, which estimate CATE by growing trees that split on covariates to maximize treatment effect differences between subgroups. Double ML uses Random Forests in a two-stage process to predict outcomes and treatment assignments, then estimate treatment effects using residuals. Similarly, GBM has been adapted for causal inference through methods such as Causal Boosting, which modifies the loss function to directly target CATE estimation, and the R-learner, which uses GBM to estimate nuisance parameters and optimize a CATE-focused loss function. These approaches provide flexible, data-driven alternatives to traditional econometric methods in causal inference, offering new tools for estimating HTEs.

Identifying and estimating **heterogeneous causal effects** (HCEs) are critical for understanding the nuanced impacts of interventions. By employing interaction terms in regression models, conducting subgroup analyses, and leveraging ML techniques, you can uncover how and for whom treatments work, leading to more effective and personalized interventions. These methods illuminate the path toward evidence-based, data-driven decision-making in various fields. Now, we'll look into various estimation methods in HTE.

Estimation methods for identifying HCEs

In this section, we dive into sophisticated causal inference methodologies that are crucial for revealing the variable impacts of treatments across different subgroups within a population. Techniques such as **regression discontinuity design (RDD)**, **instrumental variables analysis**, and **propensity score matching (PSM)** stand at the forefront. RDD capitalizes on a pre-set cutoff within an assignment variable to estimate causal effects near this threshold, simulating a randomized experiment environment. Instrumental variables analysis, on the other hand, addresses endogeneity and unobserved confounding by leveraging external instruments to uncover treatment effect heterogeneity. Meanwhile, PSM aims to reduce selection bias in observational studies, enabling a comparative analysis of treatment effects across varied strata or covariates.

These methods collectively enhance our understanding of how and why treatment effects differ among individuals or settings. By employing these advanced techniques, you can tailor interventions more effectively, ensuring a deeper comprehension of causal relationships.

Regression Discontinuity Designs

Regression Discontinuity Designs (RDD) offers a robust framework for estimating causal effects by exploiting a predetermined cutoff point within an assignment variable that dictates treatment allocation. *This design is particularly valuable in revealing HTEs around the cutoff, which serves as a quasi-randomization threshold.* This threshold separates those who receive the treatment from those who do not. This cutoff mimics randomization near the threshold, allowing for the estimation of causal effects and the exploration of differences in treatment effects around this point.

Consider a treatment T assigned to observations based on a running variable X, such that $T = 1$ if $X \geq c$ and $T = 0$ if $X < c$, where c is the cutoff point. The causal effect of T on an outcome Y can be modeled as follows:

$$Y_i = \alpha + \tau T_i + \beta X_i + \epsilon_i \qquad (8)$$

Here, α is a constant term, τ represents the treatment effect, and ϵ_i is an error term. The RDD focuses on estimating τ by comparing observations just above and below the cutoff c.

To explore heterogeneity, in addition to the RDD framework, you can include interaction terms between the treatment indicator and subgroups of interest or covariates:

$$Y_i = \alpha + \tau T_i + \gamma(T_i \times Z_i) + \beta X_i + \epsilon_i \qquad (9)$$

Here, Z_i represents a covariate or subgroup indicator, and γ captures the differential treatment effect due to Z. This model allows for the identification of variations in the treatment effect across different values of Z. In evaluating the impact of a scholarship program on student test scores, where scholarships are awarded based on a score threshold, RDD can reveal how effects differ across demographic groups (for example, gender or socioeconomic status) by including interactions between the treatment indicator and demographic indicators.

Instrumental variables

Instrumental variables (as you have read in *Chapter 10*) are used in *causal inference to address endogeneity issues; they also offer a path to uncover hidden heterogeneities in treatment effects when there's an unobserved confounding or when the treatment is not randomly assigned.* Consider the following model:

$$Y_i = \alpha + \tau T_i + \epsilon_i \qquad (10)$$

Here, Y_i is the outcome, T_i is the endogenous treatment variable, and ϵ_i is an error term. An instrument Z_i is introduced to satisfy two conditions: (1) Z_i is correlated with T_i (relevance), and (2) Z_i is not correlated with ϵ_i (exogeneity). The instrumental variable estimate of τ can be obtained through **two-stage least squares (2SLS)**:

$$T_i = \pi_0 + \pi_1 Z_i + v_i \qquad (11)$$

$$Y_i = \alpha + \tau \hat{T}_i + \epsilon_i \qquad (12)$$

Here, $\hat{T}_i$ is the predicted value of T_i from the first stage.

Heterogeneity can be explored by examining how the estimated treatment effect τ varies with changes in Z, or by interacting Z with other covariates.

In assessing the effect of a new medication on blood pressure, where prescription bias exists, *an instrumental variable such as distance to the nearest clinic dispensing the medication can reveal heterogeneities in medication effectiveness across different patient groups.*

Propensity Score Matching

PSM, as you have learned already in preceding chapters (for example, *Chapter 6*) is an invaluable tool for you to untangle heterogeneity in treatment effects within observational studies. If you recall precisely, *PSM tackles the challenge of confounding variables by statistically adjusting for them, allowing for a more accurate estimation of the true treatment effect. This technique shines when we want to reveal how treatment effects vary across different subgroups or based on specific covariates.*

At the heart of PSM lies the propensity score, denoted by $e(X_i)$. This score represents the probability of an individual receiving the treatment, given their specific set of covariates, X_i. Mathematically, we can express it as follows:

$$e(X_i) = P(T_i = 1 \mid X_i) \qquad (13)$$

Here, T_i indicates treatment assignment (1 for receiving treatment, 0 for not).

PSM leverages these propensity scores to match or stratify units in the treatment and control groups. By doing so, we can compare outcomes, Y_i, as if the treatment had been randomly assigned. This effectively reduces the influence of confounding variables and allows us to isolate the true effect of the treatment.

Following matching, we can delve deeper into differential effects across subgroups. This can be achieved by either stratifying the sample based on propensity scores or by incorporating interaction terms into the outcome model. Here's an example of such a model:

$$Y_i = \alpha + \tau T_i + \gamma (T_i \times Z_i) + \beta X_i + \delta e(X_i) + \epsilon_i \qquad (14)$$

This model allows us to estimate the ATE represented by τ, while also examining how the treatment effect might vary depending on a subgroup indicator, Z_i (for example, demographic group). The interaction term, $T_i \times Z_i$, captures this potential variation.

Imagine we're studying the impact of a job training program on employment outcomes. PSM can account for pre-existing differences between participants (for example, education level or work experience) and their propensity to receive the training. This allows us to assess how the program's effectiveness changes across different demographic or socioeconomic groups, revealing valuable insights into heterogeneity.

In conclusion, PSM, alongside other techniques such as RDD and instrumental variables, empowers you to uncover and analyze the intricacies of HTEs. This knowledge tailors interventions for specific groups and maximizes their impact. It also deepens our understanding of treatment effects across diverse groups.

Case study – Heterogeneity in R

This case study aims to illustrate the application of various R programming techniques and methodologies in analyzing heterogeneity in causal inference, particularly in the context of selling bicycles to a diverse group of customers. The scenario encompasses multiple factors affecting bicycle sales, including purposes of biking (sports, commuting, carrying heavy items, occasional biking, city biking, rural biking), as well as demographics (age), and environmental conditions (price, weather, road conditions). The goal is to generate synthetic data that mimics this complex scenario and apply advanced statistical methods to understand the causal impact of different factors on bicycle sales. You can learn more about a similar study in R shown here [7]:

```r
packages <- c("tidyverse", "caret", "MatchIt", "panelMatch",
"ggplot2", "synthpop", "fixest", "dplyr", "lubridate")
install.packages(setdiff(packages, rownames(installed.packages())))
```

Generating synthetic data

First, let's create a synthetic dataset that reflects the scenario. This dataset will include a mix of continuous and categorical variables, representing different factors affecting bicycle usage and preferences:

```r
set.seed(123)
n <- 1000
time_periods <- 5

data <- tibble(
  id = rep(1:(n/time_periods), each = time_periods),
  year = rep(2019:2023, n/time_periods),
  age = sample(18:65, n, replace = TRUE),
  biking_purpose = sample(c(
    "sports", "commute", "heavy_carrying", "occasional",
    "city", "rural"), n, replace = TRUE, prob = c(
      0.2, 0.3, 0.1, 0.2, 0.1, 0.1)),
  weather_condition = sample(c(
    "sunny", "rainy", "windy"), n, replace = TRUE),
  road_condition = sample(c(
    "good", "moderate", "poor"), n, replace = TRUE),
  price_sensitivity = sample(1:5, n, replace = TRUE),
  treatment = sample(0:1, n, replace = TRUE)
)

data$sales = with(
  data, 200 + 20 * treatment - 5 * price_sensitivity + ifelse(
    biking_purpose == "commute", 30, 0) + ifelse(
      weather_condition == "sunny", 15, -10) + rnorm(n, 0, 50))
```

Exploratory data analysis

Let's perform some basic **exploratory data analysis** (**EDA**) to understand our synthetic data better. This bar plot (*Figure 13.1*) visualizes the distribution of biking purposes across treatment groups in our study. The *x* axis shows biking purposes, the *y* axis shows participant counts, and different colors represent treatment groups. This allows us to quickly compare the prevalence of various biking purposes between groups, helping us identify potential patterns or effects of the treatment on biking behaviors. This exploratory analysis guides our further statistical investigations and helps form initial hypotheses about the study outcomes:

```
ggplot(data, aes(x = biking_purpose, fill = as.factor(treatment))) +
  geom_bar(position = "dodge") +
  labs(title = "Biking Purpose by Treatment",
       x = "Biking Purpose", y = "Count") +
  theme_minimal() +
  theme(
    axis.text.x = element_text(angle = 45, hjust = 1, vjust = 1),
    plot.margin = unit(c(1, 1, 1, 1), "cm")
  )
```

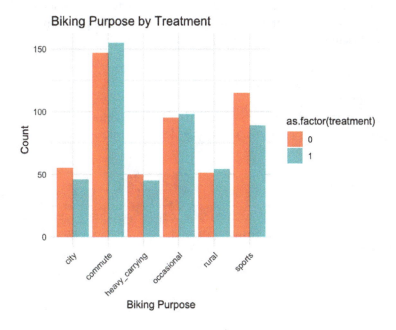

Figure 13.1 – The count of various biking purposes colored by treatment assignment

Matching for causal inference

Before estimating the causal impact of promotions on bicycle sales, we use propensity score matching to create comparable groups of treated and untreated observations:

```
library(MatchIt)
matchit_model <- matchit(
  treatment ~ age + biking_purpose + weather_condition +
    road_condition + price_sensitivity,
  data = data, method = "nearest")
matched_data <- match.data(matchit_model)
```

Estimating the ATE

Next, we estimate the ATE using a regression model on the matched data:

```
library(fixest)
ate_model <- feols(sales ~ treatment, data = matched_data)
summary(ate_model)
```

To account for the complexity and heterogeneity in treatment effects, we can employ techniques such as interactive fixed effects models or panel data methods. These techniques help capture the subtle impacts of different factors and their interactions on bicycle sales:

```
# Using 'fixest' for two-way fixed effects model
# (individual and time fixed effects)
fe_model <- feols(
  sales ~ treatment + age + biking_purpose + weather_condition +
    road_condition + price_sensitivity | id + year, data = data)
summary(fe_model)

# fixed effects model
ife_model <- feols(
  sales ~ treatment + age + biking_purpose + weather_condition +
    road_condition + price_sensitivity | id, data = data)
summary(ife_model)
```

The output for the fixed effects model is shown here:

```
OLS estimation, Dep. Var.: sales
Observations: 1,000
Fixed-effects: id: 200
Standard-errors: Clustered (id)
                               Estimate Std. Error   t value    Pr(>|t|)
treatment                      22.283842  3.683881  6.049013 7.1037e-09 ***
age                             0.001585  0.127797  0.012405 9.9011e-01
biking_purposecommute          39.612185  6.201285  6.387738 1.1665e-09 ***
biking_purposeheavy_carrying    5.162328  7.664915  0.673501 5.0141e-01
biking_purposeoccasional       -0.489823  6.335249 -0.077317 9.3845e-01
biking_purposerural            -4.800061  7.704544 -0.623017 5.3399e-01
biking_purposesports            0.970362  6.055177  0.160253 8.7284e-01
weather_conditionsunny         22.893025  4.095986  5.589136 7.4486e-08 ***
weather_conditionwindy         -0.455779  3.966076 -0.114919 9.0862e-01
road_conditionmoderate         -2.419622  4.389007 -0.551291 5.8205e-01
road_conditionpoor             -0.677905  4.276341 -0.158525 8.7420e-01
price_sensitivity              -6.810576  1.310445 -5.197146 4.9969e-07 ***
---
Signif. codes:  0 '***' 0.001 '**' 0.01 '*' 0.05 '.' 0.1 ' ' 1
RMSE: 44.8     Adj. R2: 0.178803
               Within R2: 0.209378
```

Figure 13.2 – Summary output of the fixed effects model

The results reveal significant heterogeneity in the treatment effects on bicycle sales. Being in the treatment group increased sales by 22.28 units ($p < 0.001$), indicating an overall meaningful effect. However, this effect varied across different contexts. The largest increase in sales was seen for commuters (39.61 units, $p < 0.001$), showing that biking purpose influences the responsiveness to the treatment, with commuters benefiting more than others. Sunny weather also boosted sales (22.89 units, $p < 0.001$), whereas windy or poor road conditions had no effect, suggesting that environmental factors like the weather can impact the effectiveness of the treatment. Price sensitivity was another important factor: for each unit increase in price sensitivity, sales decreased by 6.81 units ($p < 0.001$), indicating that more price-sensitive customers respond less to the treatment. Age did not significantly affect sales.

Overall, these findings emphasize that the impact of the treatment is not uniform; rather, it varies across customer characteristics and contexts. This underscores the need to tailor interventions to specific segments and conditions, as factors like biking purpose, weather, and price sensitivity shape the effectiveness of the treatment. The model explained 17.88% of the total variation in sales, with 20.94% of within-individual variance captured, highlighting treatment, commuting, sunny weather, and price sensitivity as the key influences on sales.

This case study demonstrated a simplified process of generating synthetic data, performing EDA, and applying causal inference techniques to understand the impact of promotions on bicycle sales. The approach highlighted the importance of considering heterogeneity in treatment effects and the usefulness of advanced statistical methods in uncovering causal relationships in complex scenarios. Next, let's discuss how to tailor our interventions to various groups.

Tailoring interventions to different groups

To design tailored interventions that account for heterogeneous effects, a conceptual framework incorporating both statistical and substantive considerations is essential. This framework involves identifying the sources of heterogeneity, quantifying these effects, and then designing interventions that specifically address the varied needs of different groups. Let's discuss this more next.

Conceptual framework

Heterogeneity in treatment effects may sound simple or complex, depending on how you see it, but nevertheless, in applications, it is a challenge every causal inference scholar must address. Here's the plan:

- **Unmasking the culprits**: We begin with detective work – EDA. Our goal? To identify potential suspects – demographic factors (age, gender), socioeconomic background, location, or pre-existing conditions – that might be influencing the treatment effect differently across subgroups. Advanced statistical models such as interaction models or **hierarchical linear models** (**HLMs**) become our helper tools, pinpointing these interacting factors with statistical significance.

  Take a regression model, for example: $Y = \alpha + \beta T + \gamma X + \delta (T \times X) + \epsilon$. Here, Y is the outcome, T is the treatment, X is the covariate of interest, α is the intercept, β captures the ATE, γ reflects the effect of X on Y, and the critical term δ represents the interaction effect. A significant δ tells us the effect of T on Y changes depending on the level of X.

- **Quantifying the variations**: Once we've identified the suspects, we need to quantify their influence. We turn to causal inference methods such as RDD or instrumental variable analyses. They can handle complex interactions and provide unbiased estimates of causal effects for different subgroups.

- **Tailoring the intervention**: Armed with the knowledge of heterogeneity sources and their quantified effects, we can finally craft a tailored intervention strategy. This might involve differentiated content, delivery methods, or intensity levels, all designed to maximize impact for specific subgroups.

By following these steps, we, as causal inference scholars, can move beyond the limitations of ATEs to tackle heterogeneity. This allows us to design interventions with laser focus, ensuring they hit the right target and deliver the greatest possible impact. We discuss this concept further in the following case studies.

Case study 1 – Educational interventions and their varied effects on different student demographics

Consider an educational intervention aimed at improving math skills among high school students. Analysis reveals significant interaction effects between the intervention and variables such as prior math achievement, **socioeconomic status** (**SES**), and whether the student attends an urban or a rural school.

For students with low prior achievement and from a lower SES, a version of the intervention incorporating foundational math skills reinforcement and provided in smaller, more intensive sessions could be developed. For students in rural schools, where access to advanced courses might be limited, supplementing the intervention with online resources could address this gap.

A hierarchical linear model might reveal that the intervention effect is β_1 for urban students with high prior achievement but $\beta_1 + \delta$ for rural students, where δ captures the additional effect of being in a rural setting. The tailored intervention can then adjust resources accordingly.

Case study 2 – Public health campaigns and their differential impacts on various population segments

A public health campaign aimed at increasing vaccination rates discovers heterogeneous responses influenced by factors such as age, health literacy levels, and trust in medical institutions.

For older adults, particularly those with lower health literacy, the campaign could include straightforward, easy-to-understand messages delivered through traditional media channels. For younger populations, leveraging social media platforms and involving trusted influencers could be more effective. Additionally, addressing trust in medical institutions may require community-based interventions where healthcare professionals engage directly with the community.

Suppose a logistic regression model for the likelihood of getting vaccinated Y as a function of exposure to the campaign T, trust in medical institutions X, and an interaction between the two $T \times X$ yields an estimate for δ that is positive and significant. This suggests that increasing trust in medical institutions amplifies the campaign's effectiveness, leading to the design of a tailored component of the campaign that focuses on building trust.

Through these comprehensive examples, the conceptual framework for designing tailored interventions demonstrates the importance of a customized approach to intervention design that accounts for the varied effects across different groups. This approach not only maximizes the effectiveness of interventions but also ensures equity by addressing the specific needs and circumstances of diverse populations.

Summary

In this chapter, we looked into how responses to interventions can differ significantly across various subgroups within a population. This concept, known as heterogeneity in causal effects, challenges the simplistic notion of ATEs and highlights the need for tailored interventions. We explored these differences by employing statistical methods such as subgroup analysis and ML to design more effective policies. Additionally, we addressed methodological challenges in studying heterogeneity, such as requiring larger sample sizes [1] and sophisticated analytics, which are crucial for accurately understanding and predicting the varied impacts of interventions. This nuanced approach allows us to create interventions that are not only more effective but also equitable, meeting the diverse needs of different groups.

In the next chapter, we'll dive into the implementation of causal forests in R.

References

1. Williams DN, Williams KA. Sample Size Considerations: *Basics for Preparing Clinical or Basic Research*. Ann Nucl Cardiol. 2020;6(1):81-85. doi: 10.17996/anc.20-00122. Epub 2020 Aug 31. PMID: 37123495; PMCID: PMC10133938.

2. Xie Y. *Causal Inference and Heterogeneity Bias in Social Science*. Inf Knowl Syst Manage. 2011 Jan 1;10(1):279-289. doi: 10.3233/IKS-2012-0197. PMID: 23970824; PMCID: PMC3747843.

3. Hecht, C. A., Dweck, C. S., Murphy, M. C., Kroeper, K. M., & Yeager, D. S. (2023). *Efficiently exploring the causal role of contextual moderators in behavioral science*. PNAS Proceedings of the National Academy of Sciences of the United States of America, 120(1), 1–11.

4. Farrokhyar, F., Skorzewski, P., Phillips, M.R. et al. *When to believe a subgroup analysis: revisiting the 11 criteria*. Eye 36, 2075–2077 (2022). https://doi.org/10.1038/s41433-022-01948-0

5. Brand, J. E., Zhou, X., & Xie, Y. (2023). *Recent developments in causal inference and machine learning*. Annual Review of Sociology, 49, 81–110. https://doi.org/10.1146/annurev-soc-030420-015345

6. Knittel, Christopher R., and Samuel Stolper. 2021. *Machine Learning about Treatment Effect Heterogeneity: The Case of Household Energy Use* AEA Papers and Proceedings, 111: 440-44.

7. *Treatment for causal panel analysis*: https://yiqingxu.org/tutorials/panel.html#Example_1:_wo_Treatment_Reserveal

14

Harnessing Causal Forests and Machine Learning Methods

In this chapter, we will guide you through the theoretical foundations and practical implementation of causal forests using R. You will learn why causal forests are essential for causal inference, their ability to estimate **heterogeneous treatment effects** (**HTE**), and how they differ from traditional random forests. We will dig deep into the mathematical principles and mechanisms that underpin causal forests, including their use of a splitting criterion that maximizes differences in treatment effects and the concept of honest estimation. By the end of this chapter, you will understand the importance of causal forests in providing fine-tuned insights into how different subpopulations respond to interventions, setting up the necessary environment, and building and tuning these models in R. This knowledge will empower you to make informed, data-driven decisions based on reliable estimates of treatment effects.

So, let's get started.

In this chapter, we will cover the following topics:

- Introduction to causal forests
- Historical development and key researchers
- Theoretical foundations of causal forests
- Math behind causal forests
- Using R to understand causal forests
- Machine learning approaches to heterogeneous causal inference
- Impact of social media using causal forests in R

Technical requirements

You can find the code examples for this chapter in this book's GitHub repository: `https://github.com/PacktPublishing/Causal-Inference-in-R/tree/main/chap_14`.

Introduction to causal forests for causal inference

Causal forests are an extension of traditional random forests, which are machine learning methods used for classification and regression tasks. A random forest builds many decision trees, each trained on different random subsets of the data, to make predictions by voting (for classification) or averaging (for regression). A decision tree is a model that uses a tree-like structure to split the data based on feature values, leading to a decision outcome. It segments the data into branches to classify or predict outcomes by making sequential, rule-based decisions at each node. This approach is based on ensembling, which means using multiple models to generate output. By combining the predictions of multiple models, ensembling reduces the risk of overfitting and improves the model's robustness, making it effective for handling large, high-dimensional datasets. For more information on random forests, you can refer to the paper by Breiman (2001), *Random Forests* [1].

Causal forests, however, are specifically designed to estimate HTE, which means they assess how different subpopulations respond to treatments or interventions. They incorporate mechanisms from causal inference, such as using a splitting criterion that maximizes differences in treatment effects and employing honest estimation (we will cover this later in the chapter) by splitting data into parts for tree construction and effect estimation. This ensures more reliable estimates and balanced treatment and control groups within each leaf of the tree, enhancing the validity of causal claims while retaining the strengths of traditional random forests in managing complex data.

More reliable estimates and balanced treatment and control groups are achieved by employing a splitting criterion. This criterion maximizes differences in treatment effects and, using honest estimation, the algorithm creates subgroups that are more likely to have distinct treatment responses. This approach helps isolate the effect of the treatment from confounding factors, leading to more balanced treatment and control groups within each leaf. As a result, the causal forest can make more accurate and valid inferences about HTE across different subpopulations. The specific mechanisms and detailed explanations of these processes are covered in depth later in the chapter.

Evidently, causal forests combine ideas from decision trees/random forests and causal inference to estimate the **conditional average treatment effect** (**CATE**) across different subpopulations. Each tree in the forest is built by recursively partitioning the data into subsets that are more homogeneous in terms of the treatment effect. The final estimate is obtained by averaging the results from multiple trees, reducing variance and improving accuracy.

So, you can see that causal forests provide an efficient tool for understanding and estimating how treatments or interventions impact different segments of a population. They specifically provide fine-grained insights beyond what traditional predictive models can offer. Let's discuss the history of causal forests next.

Historical development and key researchers

The concept of using tree-based methods for causal inference has evolved over the past few decades, influenced by advancements in both machine learning and statistical theory.

In that spirit, we've come a long way with decision trees in machine learning since the 1980s. They started as simple tools for prediction but have evolved into powerful instruments for understanding cause and effect. In the early days, algorithms such as CART and ID3 made decision trees popular for classification and regression tasks [1, 2, 8]. Over time, they became more versatile, handling complex scenarios such as multi-label learning.

The real game-changer came in the 2000s when researchers started exploring how decision trees could uncover causal relationships. In 2016, Susan Athey and Guido Imbens introduced causal trees, which revolutionized how we partition data based on treatment effects rather than just outcomes [3, 9]. Building on this, Athey and Stefan Wager developed causal forests, further refining our ability to estimate HTE [9]. These advancements have made decision trees invaluable for applied causal inference across various fields.

The work of influential figures such as Susan Athey, Guido Imbens, and Judea Pearl has been crucial in shaping this evolution [3]. Their contributions have transformed decision trees from simple predictive tools into sophisticated instruments for causal analysis, keeping them at the forefront of machine learning as we tackle new challenges in data interpretation.

Theoretical foundations of causal forests

This section explores the theoretical foundations of causal forests, including the necessary conditions for their application, such as large sample sizes and high-dimensional covariates. It also discusses the advantages of causal forests, such as their flexibility in handling complex data and ability to capture heterogeneous effects. We also discuss their limitations, including computational complexity and challenges in interpretability. Next, we will explain when you can apply causal forests effectively.

Conditions necessary for causal forest applications

For causal forests to be appropriately applied, the following conditions are preferred:

- **Large sample size**: Causal forests, like other machine learning methods, benefit from large datasets to improve the accuracy and reliability of the treatment effect estimates.

- **High-dimensional covariates**: The method is particularly useful when dealing with high-dimensional covariates where traditional methods may struggle to handle the complexity and interactions among variables.

- **Variation in treatment assignment**: There should be sufficient variation in the treatment assignment across the covariate space to ensure that the model can learn the HTE accurately.

Now, we will study some advantages and limitations of this approach.

Advantages and limitations

Let's first list the strengths of causal forests:

- **Flexibility in handling complex data**: Causal forests can manage high-dimensional and complex data structures effectively, capturing intricate interactions between covariates and treatment effects.

- **Non-parametric nature**: Unlike traditional parametric models, causal forests do not assume a specific functional form for the relationship between covariates and treatment effects, allowing for more flexible modeling.

- **HTE**: One of the key strengths of causal forests is their ability to estimate HTE, providing insights into how different subgroups within the data respond to the treatment differently.

- **Robustness to overfitting**: The ensemble nature of causal forests, which averages over multiple trees, helps in reducing the variance of the estimates and mitigates the risk of overfitting.

Then, what could the limitations be? Let's list them:

- **Computational complexity**: Causal forests can be computationally intensive, especially with very large datasets. This can be addressed by using parallel computing techniques and optimizing the implementation of the algorithms.

- **Interpretability**: While causal forests provide precise estimates, the interpretability of the model can be challenging compared to simpler models. Methods such as feature importance metrics and partial dependence plots can help in interpreting the results.

- **Assumption sensitivity**: The validity of causal forest estimates depends on the underlying assumptions (e.g., unconfoundedness, SUTVA). Ensuring that these assumptions hold is crucial and can be challenging in observational studies. Sensitivity analyses and robustness checks are essential to validate the findings.

- **Data requirements**: Large and diverse datasets are often required for causal forests to perform well. In cases of limited data, alternative methods or additional data collection may be necessary to ensure reliable estimates.

In summary, causal forests offer a robust and flexible approach to causal inference, capable of handling complex data structures and uncovering HTE. However, attention must be paid to computational demands, model interpretability, and the validation of underlying assumptions to ensure reliable and actionable insights. Now, let's deep-dive into the mathematical aspects of causal forests.

Understanding the math behind causal forests

Causal forests are built on the framework of potential outcomes and, as such, use ensemble methods to estimate treatment effects. Now, what are ensemble methods?

An **ensemble method** is an approach that leverages the collective wisdom of a large number of models, reducing the impact of individual model biases and errors. It does so by combining different machine learning models, where the goal is to improve predictive performance and robustness.

Many such methods are available to apply such as bagging, boosting, and stacking (random forests use the bagging technique). **Bagging** involves training multiple models on random subsets of data and averaging their predictions to reduce variance (e.g., random forests). **Boosting** sequentially trains models, each correcting errors from the previous one, to reduce bias. **Stacking** combines different models by using their predictions as input for a final model, improving overall predictive power. For more details, please read these references [5, 6, 7]. Evidently, causal forests utilize ensemble methods to estimate treatment effects by combining predictions from multiple decision trees trained on different subsets of data and treatment assignments. Read more about them in this work by Dietterich et al. [2].

In causal forests, our primary goal is to estimate the CATE function $\tau(X)$, defined as follows:

$$\tau(X) = \mathbb{E}\left[Y_{i(1)} - Y_{i(0)} \mid X_i = x \right] \qquad (1)$$

Here, $Y_{i(1)}$ and $Y_{i(0)}$ are the potential outcomes for individual i under treatment and control, respectively, and X_i represents the covariates.

Propensity scores (as you must know by now) are defined as the probability of receiving treatment given the covariates and are used to balance the treatment and control groups. The propensity score $e(X)$ is given by the following:

$$e(X) = P(T_i = 1 \mid X_i = x) \qquad (2)$$

In causal forests, propensity scores help create balanced subsets of the data, ensuring fair comparisons between treated and control units. This helps isolate the effect of the treatment from confounding variables.

The treatment effect for an individual i can be estimated as follows:

$$\hat{\tau}(X_i) = 1/|N(X_i)| \sum_j \in N(X_i)\left(Y_j - \hat{m}(X_j)\right) \qquad (3)$$

Here, $N(X_i)$ represents the neighborhood of i in the causal forest, and $\hat{m}(X_j)$ is the estimate of the mean outcome conditional on the covariates.

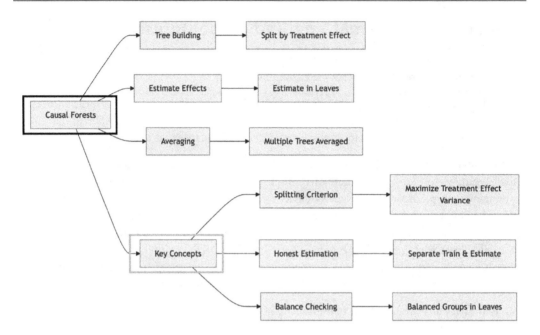

Figure 14.1 – Various concepts in causal forests and how they relate to each other

Deep-dive into causal forests

Causal forests account for the variability of treatment effects across different segments of the population by taking the following steps. An overview of the concepts can be seen in *Figure 14.1*:

1. **Tree building**: Similar to decision trees, the data is recursively split into subsets. However, the splitting criterion focuses on creating subsets that are more homogeneous regarding the treatment effect rather than the outcome. In traditional decision trees, splits are typically determined using measures such as Gini impurity or entropy. Gini impurity measures the probability of incorrectly classifying a randomly chosen element if it were randomly labeled according to the distribution of labels in the subset. Entropy, on the other hand, quantifies the amount of information or uncertainty in the data. Both measures aim to create purer subsets with respect to the target variable. In contrast, causal forests adapt these concepts to focus on treatment effect heterogeneity, aiming to create subgroups with distinct causal effects rather than just similar outcomes.

2. **Estimating treatment effects**: Each leaf of the tree contains subsets of data that are more similar in their treatment effects. The treatment effect is then estimated within these leaves.

3. **Averaging across trees**: To reduce variance and improve reliability, multiple trees are built (forming a "forest"), and the treatment effect estimates are averaged across these trees.

Some concepts that we should highlight more are the following:

- **Splitting criterion**: Traditional decision trees split data to maximize the homogeneity of the outcome variable within nodes. Causal forests, however, split the data to maximize the heterogeneity of the treatment effects across child nodes. This is achieved by finding splits where the difference in estimated treatment effects (CATE) between the resulting groups is maximized.

- **Honest estimation**: The data is split into two halves:

  - *Training half*: Used to construct the tree.

  - *Estimation half*: Used to estimate the treatment effects within the leaves of the tree.

  This approach prevents overfitting by ensuring the model is not evaluated on the same data it was trained on.

- **Balance checking**: After the splits, each leaf (subset of data) should have balanced treatment and control groups to ensure valid causal inferences. Balance checking involves verifying that the distribution of covariates, X, is similar for treated and control units within each leaf.

These steps and methodologies ensure that causal forests provide robust and reliable estimates of treatment effects, making them a very useful tool for applied causal inference [3, 4].

Now, let's understand this effect with a scenario.

Scenario – classroom cohort

Let's think of a teacher wishing to understand different treatment methods they can follow and their impact in terms of causality on students' performance. They apply the method to a subset of students (treatment group) while the rest continue with the traditional method (control group). The performance is then measured at the end of the term. So what needs to be done? Let's see:

1. **Data collection**: First, they must collect data on student performance, Y. They also need to know whether the students received the new teaching method, T, in addition to all the student characteristics (e.g., prior grades, attendance, and socioeconomic status) as covariates, X.

2. **Building the trees:** Causal forests create splits to maximize the difference in the estimated treatment effect (CATE) across subsets, grouping students based on how the teaching method's impact varies, rather than on performance alone.

3. **Estimating treatment effects**: Within each subset (leaf), the average treatment effect of the new teaching method is estimated. For instance, in one subset, the new method might show a significant improvement for students with high prior grades while, in another, it may show less improvement for students with lower prior grades.

4. **Honest estimation**: The teacher splits the student data into two halves and uses one half to build the trees and the other half to estimate the treatment effects in each subset, ensuring the reliability of the estimates.

5. **Averaging the results**: Multiple trees are built, and the treatment effect estimates from each tree are averaged to obtain a more robust estimate of the impact of the teaching method across different student subpopulations.

6. **Checking balance**: The teacher ensures that within each subset (leaf), the characteristics of students receiving the new method are similar to those continuing with the traditional method, ensuring valid comparisons.

Now, let's spin up R and try this in practice.

Using R to understand causal forests

In this code, we will begin with data generation and see, hands-on, how to build causal forests. You don't have to apply everything from scratch; we can utilize powerful R packages and start from there.

Installing and loading necessary packages

First, we need to load the `grf` package. `grf`, which stands for **generalized random forests**. It is an open source R package developed by researchers at Stanford University. It provides a powerful and flexible framework for implementing various types of random forest algorithms, including causal forests, instrumental forests, and regression forests.

One of the key strengths of the `grf` package is its ability to handle different types of data structures, such as panel data, clustered data, and data with missing values. It also supports a wide range of statistical estimation problems, including treatment effect estimation, HTE estimation, and quantile regression. Let's install and load the `grf` package:

```
# Install the grf package
install.packages("grf")
# Load the grf package
library(grf)
```

Next, let's set a seed ensuring that the random numbers generated are reproducible:

```
# Set seed for reproducibility
set.seed(42)
```

Simulating data

We simulate a dataset representing a classroom cohort with student characteristics, treatment assignments, and outcomes. First, we generate a matrix, X, with 1,000 rows and 5 columns. Each column represents a different characteristic of the students (such as prior grades, attendance, etc.). The values are randomly generated from a normal distribution, simulating how these characteristics might vary among students:

```
# Number of students
n <- 1000
# Simulate student characteristics (covariates)
X <- matrix(rnorm(n * 5), nrow = n, ncol = 5)
colnames(X) <- c(
  "PriorGrades", "Attendance", "SocioeconomicStatus",
  "InterestInSubject", "ParentalSupport")
```

Next, we assign each student to either a treatment group (1) or a control group (0). This is done randomly with a 50% chance for each student to be in either group:

```
# Simulate treatment assignment (1 if treated, 0 if control)
T <- rbinom(n, 1, 0.5)
```

Now, we should calculate the outcome, Y, for each student. The outcome is based on several factors. A base value of 5 is added. The student's prior grades, attendance, and socioeconomic status each weigh differently. We also add some random noise (from a normal distribution) to add variability. A treatment effect is computed that depends on whether the student was treated (T) and their prior grades:

```
# Simulate outcomes with a heterogeneous treatment effect
Y <- 5 + X[,1] + 2*X[,2] + 0.5*X[,3] + rnorm(n) +
  T * (2 + 0.5*X[,1])
# Combine into a data frame for convenience
data <- data.frame(Y, T, X)
```

To summarize, our data is generated for 1,000 students. For each student, we simulate five student characteristics (covariates). Each characteristic is normally distributed. Next, we simulate the treatment assignment, in which each student has a 50% chance of being treated. Here, Y simulates the outcomes with an HTE, combining the covariates and treatment effect with some random noise. `data <- data.frame(Y, T, X)` combines the outcome, treatment, and covariates into a single data frame for convenience.

Training a causal forest

Once we have the data we need, we train a causal forest model using the grf package to estimate CATE:

```
# Train a causal forest
causal_forest <- causal_forest(X, Y, T)
# Print a summary of the causal forest
summary(causal_forest)
```

The forest aims to estimate CATE using the covariates, outcomes, and treatment assignments. The summary(causal_forest) command prints a summary of the trained causal forest, providing information about the model.

Estimating treatment effects

We estimate the treatment effects for each student using the trained causal forest:

```
# Estimate treatment effects
tau_hat <- predict(causal_forest)$predictions
# Add the estimated treatment effects to the data frame
data$tau_hat <- tau_hat
# View the first few rows of the data frame
head(data)
```

We first predict the treatment effect for each student using the trained causal forest. Then, we add the predicted treatment effects to the data frame. Now that we have a model, let's learn how to validate it.

Validating the model

We start by splitting the data into training and estimation sets to perform honest estimation:

```
# Split the data into training and estimation sets
train_indices <- sample(1:n, size = n/2)
est_indices <- setdiff(1:n, train_indices)
# Train a causal forest on the training set
causal_forest_train <- causal_forest(
  X[train_indices, ], Y[train_indices], T[train_indices])
# Estimate treatment effects on the estimation set
tau_hat_est <- predict(
  causal_forest_train, newdata = X[est_indices, ]
  )$predictions
```

Next, we combine the estimation set with the estimated treatment effects:

```
data_est <- data[est_indices, ]
data_est$tau_hat_est <- tau_hat_est
head(data_est)
```

Then, we compute the following metrics to evaluate the performance of the causal forest model:

- **Average treatment effect (ATE) bias**: The difference between the estimated overall treatment effect (mean of `tau_hat_est`) and the true overall treatment effect (mean difference between treated and control outcomes) is calculated. A smaller bias indicates a more accurate estimation of the overall treatment effect.

- **CATE accuracy**: The **root mean squared error** (**RMSE**) between the estimated CATE (`tau_hat_est`) and the true CATE (difference between treated and control outcomes) is calculated. Lower RMSE values suggest a better estimation of HTE across different covariate values. This metric allows for a more intuitive interpretation of the model's performance in estimating treatment effects, as it represents the average deviation of the estimated CATE from the true CATE in the original scale of the outcome variable.

In this code, we split the data into training and estimation sets and then train a causal forest on the training set. Here, `tau_hat_est` estimates the treatment effects on the estimation set:

```
# Calculate the average treatment effect (ATE) bias
true_ate <- mean((Y[est_indices][T[est_indices] == 1]) -
                 (Y[est_indices][T[est_indices] == 0]))
est_ate <- mean(tau_hat_est)
ate_bias <- est_ate - true_ate
print(paste("Average Treatment Effect (ATE) Bias:", ate_bias))
```

Now, we are ready to calculate the CATE accuracy (we only show partial code here, for brevity. For the full code, please refer to the Git repository.):

```
# Calculate the CATE accuracy using RMSE
true_cate <- 2 + 0.5 * X[
  est_indices, 1]  # True CATE based on the data generation process
cate_rmse <- sqrt(mean((tau_hat_est - true_cate)^2))
print(paste("Conditional Average Treatment Effect (CATE) RMSE:",
            cate_rmse))
```

The plot in *Figure 14.2* visualizes the relationship between students' prior grades and the estimated treatment effect of the intervention. Each point represents a student, with their prior grades on the *x*-axis and their estimated individual treatment effect on the *y*-axis. The scatter of points reveals how the treatment effect varies across different academic performance levels, helping identify whether certain students benefit more or less from the intervention based on their prior academic achievement.

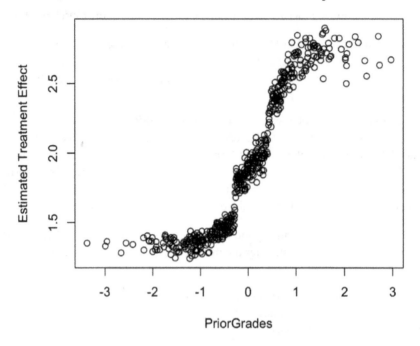

Figure 14.2 – Prior grades distribution with estimated treatment effect

Basically, this scatter plot is used to visualize the estimated treatment effects (`tau_hat_est`) against one of the covariate values (`X[est_indices, 1]`). This visualization can help identify potential heterogeneity in the treatment effects across different covariate values. In the next step, we will extract leaf indices.

Extracting leaf indices

We extract leaf indices for each observation in the causal forest. You may wonder why.

Extracting leaf indices in causal forests is essential for identifying which subset of the data each observation falls into within the forest's trees. This step allows you to group similar observations together, ensuring that treatment and control groups within each leaf are comparable in terms of their covariates. By understanding these groupings, you can accurately estimate and interpret treatment effects, as well as validate the balance of covariates within each leaf, which is crucial for making valid causal inferences. Let's extract the leaf indices in the following code snippet:

```
# Extract the leaf indices
leaf_indices <- predict(
  causal_forest, type = "leaf.index")$leaf.index
# Add the leaf indices to the data frame
data$leaf <- leaf_indices
```

Here, the `leaf_indices` variable extracts the leaf indices for each observation, indicating which leaf each observation falls into in the forest. Next, `data$leaf <- leaf_indices` adds the leaf indices to the data frame.

Now, we define a function to check the balance of covariates between the treatment and control groups within each leaf and apply this function to each leaf:

```
# Check balance within each leaf
balance_check <- function(leaf) {
  leaf_data <- data[data$leaf == leaf, ]
  treated <- leaf_data[leaf_data$T == 1, ]
  control <- leaf_data[leaf_data$T == 0, ]
  if (nrow(treated) > 1 & nrow(control) > 1) {
    balance <- sapply(leaf_data[, 3:7], function(col) {
      t.test(col ~ leaf_data$T)$p.value
    })
    return(balance)
  } else {
    return(rep(NA, 5)) # Return NA if not enough data for t-test
  }
}
# Apply balance check to each leaf
balance_results <- lapply(unique(data$leaf), balance_check)

# Print the balance results for the first few leaves
balance_results[1:5]
# Debugging: Print the first few leaves and their data
for (i in 1:5) {
  cat("Leaf:", unique(data$leaf)[i], "\n")
  print(data[data$leaf == unique(data$leaf)[i], ])
}
```

We defined `balance_check` as a function that checks the balance of covariates within each leaf. It compares the covariates between treated and control groups using t-tests and returns the p-values. Returning p-values in the `balance_check` function helps determine whether covariates are balanced between treated and control groups within each leaf. High p-values suggest the groups are similar, indicating good balance and reliable causal inferences, while low p-values indicate potential imbalance, which could bias the treatment effect estimates.

Next, `treated` and `control` extract treated and control observations within the leaf, respectively. The `sapply` function applies the t-test to each covariate and returns the p-values. If there are not enough data points for the t-test, the function returns `NA` and then we apply the `balance_check` function to each unique leaf. In this, `balance_results[1:5]` prints the balance results for the first few leaves. The final loop prints the data for the first few leaves to debug and ensure the balance check is performed correctly.

By following the aforementioned steps, we can estimate treatment effects using causal forests and check the balance of covariates within each leaf to ensure valid causal inferences. In this exercise, we learned how to use the `grf` package in R to simulate data, build a causal forest model, and estimate CATE. We also explored how to validate the model by splitting data into training and estimation sets, checking the balance of covariates within each leaf, and ensuring accurate causal inferences by analyzing the p-values from t-tests. These steps help ensure that the treatment effects estimated by the causal forest are reliable and valid.

Now, we are ready for some machine learning approaches to this topic.

Machine learning approaches to heterogeneous causal inference

Causal forests are designed to estimate HTE by adapting decision trees to focus on treatment effect variation across subpopulations. Unlike standard trees, causal forests split data to capture differences in treatment effects, estimating CATE for more targeted insights. By building multiple trees, causal forests identify how interventions impact different groups. Let's start with a recap of HTE as it is related to causal forests.

HTE refers to the variation in treatment effects across different subpopulations or individuals. Unlike ATE, which provides a single estimate of the treatment effect across the entire population, HTE recognizes that the effect of an intervention can differ based on various characteristics or contexts. For example, a medication might be more effective for younger patients compared to older ones, or a marketing campaign might work better for a particular demographic group.

HTE is formally defined as CATE, which is the expected difference in outcomes between treated and untreated groups, conditional on a set of covariates (X):

$$\tau(X) = \mathbb{E}\left[Y_i(1) - Y_i(0) \big| X_i = x\right] \tag{4}$$

Here, $\left(Y_i(1)\right)$ and $\left(Y_i(0)\right)$ represent the potential outcomes under treatment and control, respectively.

HTEs are crucial in several fields, including the following:

- **Healthcare**: Understanding HTEs can help personalize treatments, ensuring that patients receive the therapies most likely to benefit them. For instance, precision medicine relies on identifying which subgroups of patients will respond best to specific treatments based on genetic, environmental, and lifestyle factors.

- **Economics**: Policy evaluations often benefit from insights into HTEs to tailor interventions more effectively. For example, job training programs might be more beneficial to certain subpopulations, and understanding these differences can improve program design and implementation.

- **Marketing**: In marketing, HTEs help in targeting campaigns more effectively by identifying which segments of the customer base are more responsive to specific marketing strategies, thus optimizing resource allocation.

Machine learning enhances causal forests by leveraging advanced techniques to improve the estimation of HTEs:

- **Enhanced splitting criteria**: Machine learning models use sophisticated splitting criteria that can better capture the heterogeneity in treatment effects compared to traditional trees.

- **Regularization techniques**: Techniques such as **least absolute shrinkage and selection operator** (**LASSO**) and ridge regression are used to prevent overfitting, ensuring that the models generalize well to new data.

- **Honest estimation**: Causal forests often employ an "honest" approach where the data is split into training and estimation sets to avoid overfitting and provide unbiased estimates of treatment effects.

- **Feature selection and importance**: Machine learning models can identify and rank the importance of features, helping researchers understand which covariates are most influential in determining treatment effects.

By ranking the influence of different features, causal forests highlight the factors most responsible for the differences in treatment effects across subpopulations.

Impact of social media using causal forests in R

In this section, we explore the practical implementation of causal forests in R to evaluate the impact of a social media campaign on user engagement. The scenario involves creating a synthetic dataset that simulates various user demographics, engagement history, and responses to the campaign. We will use several R packages to facilitate this process, including `grf` for generalized random forests, `tidyverse` for data manipulation and visualization, and `caret` for model training and tuning. The step-by-step process includes setting up the environment, preparing and preprocessing the data, building and tuning the causal forest models, and finally, interpreting and visualizing the results. By following these steps, we aim to derive meaningful insights into the effectiveness of the campaign and inform future decision-making.

Setting up the environment

To implement causal forests in R, we need the following packages:

- `grf`: This includes causal forest functionality
- `tidyverse`: A collection of R packages for data manipulation and visualization
- `caret`: A package for model training and tuning

To install these packages, use the following commands in R:

```
install.packages("grf")
install.packages("tidyverse")
install.packages("caret")
```

Load the libraries into your R session:

```
library(grf)
library(tidyverse)
library(caret)
```

Data preparation and preprocessing

Let's create a synthetic dataset that mimics a social media campaign scenario. We'll include multiple variables such as user demographics, engagement history, and the response to the campaign:

```
set.seed(123)
n <- 1000
data <- tibble(
  user_id = 1:n,
  age = round(runif(n, 18, 65)),
  gender = sample(c("Male", "Female"), n, replace = TRUE),
  location = sample(c("Urban", "Suburban", "Rural"),
                    n, replace = TRUE),
  previous_engagement = runif(n, 0, 1),
  treatment = sample(c(0, 1), n, replace = TRUE),
  engagement_rate = NA
)
```

Now, we generate an engagement rate based on treatment and covariates, as shown here:

```
data <- data %>%
  mutate(
    engagement_rate = 0.3 * age / 100 +
      0.5 * ifelse(gender == "Female", 1, 0) +
      0.4 * ifelse(location == "Urban", 1, 0) +
      0.2 * previous_engagement +
      0.6 * treatment +
      rnorm(n, 0, 0.1)
  )


# Preview the data
head(data)
```

Next, we need to examine the dataset for any missing values and address them appropriately. Unresolved missing values can introduce bias into our analysis and lead to incomplete insights:

```
# Check for missing values
sum(is.na(data))

# Assuming no missing values since it's synthetic data
# Handling outliers (example: cap outliers at the 99th percentile)
quantiles <- quantile(
  data$engagement_rate, probs = c(0.01, 0.99))
data$engagement_rate <- pmin(pmax(
  data$engagement_rate, quantiles[1]), quantiles[2])
```

Building and tuning causal forest models

Now, we should build the model. First, let's define the covariates, treatment, and outcome:

```
# Define the covariates, treatment, and outcome
covariates <- data %>% select(
  age, gender, location, previous_engagement)
treatment <- data$treatment
outcome <- data$engagement_rate
```

Next, we convert categorical variables to one-hot encoding. We do this to transform them into a numerical format that machine learning algorithms can understand. This approach prevents the model from misinterpreting categorical data as ordinal and helps capture the relationships between different categories effectively:

```
covariates_matrix <- model.matrix(~. - 1, data = covariates)
```

As the data is ready, we can fit the causal forest model:

```
# Fit the causal forest model
causal_forest_model <- causal_forest(as.matrix(
  covariates_matrix), outcome, treatment)
```

You can see the summary of the model to understand the results:

```
summary(causal_forest_model)
```

To tune the model, we can manually perform cross-validation by iterating over a grid of hyperparameters:

```
# Define a grid of hyperparameters with sampling fraction less than
0.5
tune_grid <- expand.grid(min.node.size = c(5, 10, 15),
                         sample.fraction = c(0.3, 0.4, 0.45))
```

In this part of the code, we manually perform cross-validation to optimize our causal forest model. We start by setting up variables to track the RMSE and corresponding parameters. Using nested loops, we iterate through different combinations of min_node_size and sample_fraction from our tuning grid. For each combination, we train a causal forest model, allowing us to systematically find the best hyperparameters that enhance the model's accuracy and reliability in estimating treatment effects.

```
# Manual cross-validation
best_rmse <- Inf
best_params <- list()
```

This code snippet evaluates each model's performance and updates our best parameters. We generate predictions, calculate the RMSE, and compare it to the current best RMSE. If we find a better model, we update our best_rmse and best_params variables. After testing all combinations, we print the optimal hyperparameters that produced the lowest RMSE, completing our manual cross-validation process:

```
for (min_node_size in tune_grid$min.node.size) {
  for (sample_fraction in tune_grid$sample.fraction) {
    model <- causal_forest(
      as.matrix(covariates_matrix), outcome, treatment,
      min.node.size = min_node_size,
      sample.fraction = sample_fraction
    )
    predictions <- predict(model)$predictions
    rmse <- sqrt(mean((outcome - predictions)^2))

    if (rmse < best_rmse) {
      best_rmse <- rmse
      best_params <- list(min.node.size = min_node_size,
```

```
                              sample.fraction = sample_fraction)

      }
    }
}

print(best_params)
```

Finally, we refit the model with the best parameters found:

```
# Refit the model with the best parameters
causal_forest_model <- causal_forest(
  as.matrix(covariates_matrix), outcome, treatment,
  min.node.size = best_params$min.node.size,
  sample.fraction = best_params$sample.fraction
)
```

Interpreting results and model validation

Now, we focus on interpreting the results from our causal forest model and validating its performance. After fitting the model, we estimate the treatment effects, summarize these effects, and ensure the model's accuracy and reliability. We employ several techniques to validate the results, including checking for covariate balance and evaluating model performance using metrics such as RMSE. Additionally, we explore methods to visualize the causal inference results using tools such as ggplot2 to create interpretable visualizations that can aid decision-makers in understanding the impact of the treatment. These steps are crucial in confirming the robustness and practical relevance of our model's findings.

First, we use a causal forest model to estimate the treatment effects. Essentially, it predicts how different treatments impact the outcome for each individual in our dataset. This gives you a clearer understanding of the effect of the treatment you're analyzing:

```
# Estimate the treatment effects
treatment_effects <- predict(causal_forest_model)$predictions
```

Next, a summary will provide you with key statistics, such as ATE, which helps you understand the overall impact of the treatment across your sample:

```
# Summarize treatment effects
summary(treatment_effects)
```

The predicted treatment effects from the causal forest model (refer the summary statement in the code) range from a minimum of 0.5543 to a maximum of 0.6221, with a median of 0.5848 and a mean of 0.5866. These statistics suggest that the treatment consistently increases engagement rates by approximately 0.59 units across the population, with relatively small variation in the effect size.

We then check for covariate balance:

```
# Plot the covariate balance before and after matching
plot_balance(causal_forest_model)
```

Next, we can evaluate the model performance using the RMSE metric:

```
# Assess the model's performance using RMSE
rmse <- sqrt(mean((outcome - treatment_effects)^2))
print(rmse)
```

We can then use the `ggplot2` package for creating visualizations:

```
# Load ggplot2
library(ggplot2)

# Plot treatment effects
ggplot(data, aes(x = treatment_effects)) +
  geom_histogram(binwidth = 0.05) +
  labs(title = "Distribution of Estimated Treatment Effects",
       x = "Estimated Treatment Effect",
       y = "Frequency")
```

Here, using `ggplot2`, we create two visualizations for interpreting the results of our causal forest model. The first plot is a scatter plot (*Figure 14.3a*) of estimated treatment effects versus actual engagement rates, with a linear regression line added to show the trend. The second plot visualizes treatment effects by user demographics, showing how these effects vary with age and between genders (*Figure 14.3b*). These visualizations help in understanding the relationships and variations in treatment effects, aiding in a clearer interpretation of the model's outcomes:

```
# Scatter plot of treatment effects vs. actual engagement rate
ggplot(data, aes(
  x = treatment_effects, y = engagement_rate)) +
  geom_point(alpha = 0.5) +
  geom_smooth(method = "lm", col = "red") +
  labs(title = "Treatment Effects vs. Engagement Rate",
       x = "Estimated Treatment Effect",
       y = "Engagement Rate")

# Plot treatment effects by user demographics
ggplot(data, aes(
  x = age, y = treatment_effects, color = gender)) +
  geom_point(alpha = 0.5) +
  labs(title = "Treatment Effects by Age and Gender",
       x = "Age",
```

```
        y = "Estimated Treatment Effect",
        color = "Gender")
```

We get the following plots:

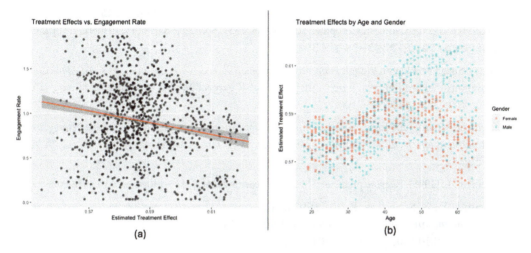

(a)   (b)

Figure 14.3: (a) Treatment effects versus engagement; (b) treatment effects by age and gender

This comprehensively covers the essential steps to implement causal forests in R, from setting up the environment to interpreting and visualizing the results. By following these steps, you can effectively analyze the impact of a social media campaign on user engagement metrics and make data-driven decisions. However, the insights gained from this exploration will empower you to apply causal forests to a wide range of problems you may encounter in the future.

Summary

In this chapter, we guided you through both the theoretical foundations and practical implementation of causal forests using R, particularly in the context of evaluating a social media campaign's impact on user engagement. We began by explaining the theory behind causal forests, including their ability to estimate HTE and the mathematical formulations involved. After setting up the necessary environment with packages such as `grf`, `tidyverse`, and `caret`, you learned how to create a synthetic dataset representing user demographics and engagement history, prepare this data, and build causal forest models. We then focused on interpreting the results, validating model performance through covariate balance checks and RMSE calculations, and creating insightful visualizations using `ggplot2`. By following these steps, you were able to understand and estimate the nuanced effects of interventions across different subpopulations, enhancing your data-driven decision-making process.

In the next chapter, we'll dive into causal discovery, an exciting new concept that might be unfamiliar to you.

References

1. Breiman, L. (2001). *Random Forests*. *Machine Learning*, 45(1), 5-32. (https://doi.org/10.1023/A:1010933404324)

2. Dietterich, T.G. (2000). *Ensemble Methods in Machine Learning*. In: Multiple Classifier Systems. MCS 2000. Lecture Notes in Computer Science, vol. 1857. Springer, Berlin, Heidelberg. https://doi.org/10.1007/3-540-45014-9_1

3. Athey, Susan and Stefan Wager. *Estimating Treatment Effects with Causal Forests: An Application*. Observational Studies, vol. 5 no. 2, 2019, p. 37-51. Project MUSE, https://muse.jhu.edu/article/793356/pdf

4. Clément Bénard, Julie Josse. *Variable importance for causal forests: breaking down the heterogeneity of treatment effects*. 2023. ffhal-04177493f, https://hal.science/hal-04177493/document

5. Breiman, L. (1996). *Bagging predictors*. Machine Learning, 24(2), 123-140. doi:10.1007/BF00058655.

6. Freund, Y., & Schapire, R. E. (1997). *A decision-theoretic generalization of on-line learning and an application to boosting*. Journal of Computer and System Sciences, 55(1), 119-139. doi:10.1006/jcss.1997.1504.

7. Wolpert, D. H. (1992). *Stacked generalization*. Neural Networks, 5(2), 241-259. doi:10.1016/S0893-6080(05)80023-1.

8. Blockeel H, Devos L, Frénay B, Nanfack G, Nijssen S. *Decision trees: from efficient prediction to responsible AI*. Front Artif Intell. 2023 Jul 26;6:1124553. doi: 10.3389/frai.2023.1124553. PMID: 37565044; PMCID: PMC10411911.

9. Athey, Susan and Guido Imbens. *Recursive partitioning for heterogeneous causal effects*. Proceedings of the National Academy of Sciences 113 (2015): 7353 - 7360.

15

Implementing Causal Discovery in R

In this chapter, we'll explore a new topic known as **causal discovery**. In causality, this methodology aims to learn the underlying causal structure between various factors. As we learn the basics and applications of this technique, we'll go back to the good old representational approaches such as **directed acyclic graphs (DAGs)** and **structural causal models (SCMs)**, which will serve as our tools to map out causal relationships. Furthermore, we plan to dive deep into various causal discovery techniques, including constraint-based methods such as the PC algorithm, score-based methods such as **Greedy Equivalence Search (GES)**, and other hybrid approaches, each offering unique strengths and challenges. Implementing this practically in R with scenario-based problem solving will illustrate the transformative potential of causal discovery across diverse fields, from public health to economics. By addressing challenges such as identifiability issues, confounding variables, and data quality, we aim to show you this robust and reliable new approach to causality.

In this chapter, we'll cover the following topics:

- Introduction to causal discovery
- Methods for causal discovery
- Implementing causal discovery with Bayesian networks in R
- A multi-algorithm comparative approach to causal discovery in R

Technical requirements

You can find the code examples for this chapter in this book's GitHub repository: `https://github.com/PacktPublishing/Causal-Inference-in-R/tree/main/chap_15`.

Introduction to causal discovery

Causal discovery isn't a new area. It's another approach within the umbrella of causal inference that goes beyond mere association to learn the true drivers of observed phenomena. Through the discussions in this chapter, you'll see that its importance in data analysis lies in its ability to provide actionable insights, inform policy and decision-making, advance scientific understanding, reduce bias, and analyze complex systems. So, let's go ahead and start peeling back the layers.

Definition and importance

When we talk about causal discovery, we're trying to understand how things work. We want to figure out what causes what in the world around us, using the data we observe. It's a method that helps us identify and understand causal relationships from observational data. Its main goal is to uncover direct causal influences, which helps us better grasp the underlying mechanisms behind what we observe. By definition, it's the process of identifying and estimating causal effects by recovering causal graphs from data. It uses graphical models to uncover underlying causal relationships, often from observational data [12].

This process involves using algorithms and statistical techniques to build a causal model to represent the underlying causal structure as a DAG. In a DAG, nodes are variables, and directed edges show causal influences. For example, if variable A causes variable B, there would be a directed edge from A to B (as you learned in *Chapters 4 and 5*). As you may recall, DAGs are a way of showing which variables influence others, kind of like a family tree of causes.

Closely related to this are SCMs. These are a bit more mathematical, but they're incredibly useful. They also help us describe causal relationships formally, using equations to show how different variables are connected. We can use these models to make predictions and understand complex systems better.

However, the true beauty of causal discovery is that it lets us go beyond just noticing that things happen together. We're not just saying, "When A happens, B often happens." Instead, we are figuring out if A causes B, or if there's something else going on. It's a powerful approach that helps us understand the world in a much deeper way and gets to the root of how things work.

So, what do we need to run a causal discovery model?

The input for causal discovery typically consists of the following:

- **Observational data**: A dataset containing measurements of various variables of interest. This data is usually collected without experimental intervention.

- **Domain knowledge (optional)**: Prior information about potential causal relationships or constraints, which can help guide the discovery process.

The output of causal discovery is usually in one of the following forms:

- **Causal graph**: A DAG representing the inferred causal structure. In this graph, nodes represent variables and directed edges represent causal relationships between variables.

- **Causal model**: A formal representation of the causal relationships, which may include the following aspects:

  - **Structural equations**: Describing the functional relationships between variables.

  - **Parameters**: Quantifying the strength of causal effects.

- **Uncertainty measures**: Estimates of the confidence or uncertainty associated with the inferred causal relationships.

- **Potential interventions**: Suggestions for experimental interventions that could confirm or refine the discovered causal relationships.

You can say that the goal of causal discovery is to move from purely observational data to a structured understanding of the causal mechanisms at play, providing insights that can guide further research, decision-making, and, potentially, interventions in the system under study. We'll discuss a few opportunities for it in various fields later.

> **What distinguishes causal discovery from causal inference?**
> While causal inference focuses on estimating the effect of a treatment or intervention given a known causal model, causal discovery is concerned with determining the structure of the causal model in terms of data.

Understanding causal relationships is often fundamental to scientific inquiry and practical decision-making. Here are several reasons why causal discovery is crucial:

- **Prediction and control**: Causal discovery allows us to make more accurate predictions by identifying the true drivers of outcomes. In a situation where we're studying the factors influencing student performance, there can be numerous factors at play. Causal discovery can reveal that parental involvement directly causes improved grades, rather than just being correlated with them. With this knowledge, we can design interventions. These can target increasing parental involvement, rather than focusing on factors that might be merely correlated with good grades.

- **Policy and decision-making**: In public health, let's say we're investigating the causes of a recent increase in childhood obesity. Causal discovery might show that increased screen time is a direct cause of obesity, rather than just being associated with it. This insight allows policymakers to design targeted interventions, such as promoting outdoor activities or limiting screen time in schools. Here, causal discovery can be a game-changer. How? By pinpointing the causal relationship, rather than just observing correlations, we can develop more effective and efficient public health strategies.

- **Optimize strategies**: Businesses leverage causal discovery to optimize marketing strategies, enhance customer satisfaction, and boost operational efficiency. By identifying the direct effects of specific marketing actions on sales, they move beyond simple correlations. For instance, understanding the causal impact of a marketing campaign on sales helps with allocating resources more effectively. This data-driven approach leads to smarter decisions, better customer engagement, and higher returns on investment.

- **Scientific understanding**: Scientists seek to learn causal mechanisms behind observed phenomena. Evidently, in fields such as epidemiology, economics, and social sciences, causal discovery helps build theories that explain how different factors interact and influence each other. In such cases, causal discovery can provide a rigorous foundation for scientific experiments and observational studies, ensuring that any conclusions that are drawn are based on true causal relationships rather than spurious correlations.

- **Bias reduction and robustness**: As you've seen throughout this book, identifying and adjusting for confounding variables in causality is a big problem. You can use causal discovery to identify confounders and bias estimates in observational studies. As an example, in healthcare studies, confounders such as age and lifestyle factors need to be accounted for. Why? To accurately estimate the effect of a treatment. The robustness of conclusions is enhanced when causal models are validated and tested against new data. This iterative process of model refinement leads to more reliable and generalizable findings.

- **Complex systems analysis**: You may or may not know this, but many real-world systems are complex and involve numerous/hundreds of interacting components. Causal discovery provides tools to better understand these interactions and understand the underlying structure of complex systems, such as ecosystems, financial markets, and social networks. By mapping out causal relationships, you can better understand feedback loops, emergent behaviors, and potential points of intervention within these systems.

Next, let's look at the historical background of causal discovery.

Historical background

Historically, causal discovery spans several disciplines, such as statistics, computer science, and epidemiology. This history reflects the evolution of ideas and techniques to extract the underlying causal structures from observational data.

In the 1920s, Sewall Wright (a geneticist and statistician) developed path analysis, which allowed correlations to be decomposed into direct and indirect effects. This method was one of the first attempts to delineate causal pathways using statistical models [1, 11], setting the stage for more sophisticated causal discovery techniques. Building on path analysis, **structural equation modeling** (**SEM**) emerged as a powerful framework in the mid-20th century. SEMs help combine theoretical knowledge with statistical data to model complex causal relationships among variables [1].

A pioneering computer scientist and philosopher, Judea Pearl, significantly advanced the representation of causal relationships in the field of causal discovery by formalizing the use of graphical models (specifically DAGs). His work in the late 20th century introduced a new, rigorous language for causality.

The late 20th and early 21st centuries saw the development of new causal discovery algorithms such as the **Peter-Clark** (**PC**) algorithm and the **Inductive Causation** (**IC**) algorithm. These algorithms, which were developed by researchers Peter Spirtes, Clark Glymour, and Richard Scheines, utilized statistical tests to infer the structure of causal graphs from data [2].

Now, let's look at some interdisciplinary applications of causal discovery (more examples can be found in the *References* section):

- **Epidemiology and public health**: Causal discovery methods have been extensively applied in epidemiology to identify risk factors and causal pathways of diseases. Techniques such as the **Fast Causal Inference** (**FCI**) algorithm have been used to tackle issues regarding confounding and latent variables [3].

- **Economics and social sciences**: Economists and social scientists have leveraged causal discovery to better understand policy impacts, social dynamics, and economic mechanisms. Tools such as GES have been developed to efficiently search for causal structures in large datasets [1, 2].

With this, we're prepared to learn the underpinnings of the theory behind causal discovery.

Theoretical foundations

In causal discovery, we aim to construct models from data, especially in situations where domain knowledge is limited, or the causal structure is unknown. This section dives into the theoretical foundations, methodologies, and practical applications of causal discovery.

Causal graphs and models

As a reminder, our primary goal is to uncover the true causal graph that represents the dependencies among variables. For this, we should learn about causal graphs, something we already discussed in *Chapters 4* and *5*. We'll summarize this to help you catch up.

Understanding direct and indirect causation

Let's take a closer look at direct and indirect causation:

- **Direct causation**: This occurs when a variable directly influences another variable without any mediating variables. In a causal graph, this is represented by a direct edge from one node to another.

- **Indirect causation**: This occurs when the effect of one variable on another is mediated through one or more intermediary variables. Indirect effects can be decomposed into sequences of direct effects, and their identification often requires more complex analytical techniques (refer to *Figure 15.1*):

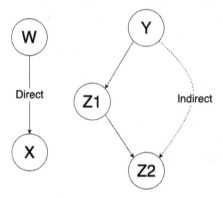

Figure 15.1 – Direct and indirect graphs

Challenges in causal discovery

Despite the potential of causal discovery to provide deep insights into data, several challenges need to be addressed to make the process robust and reliable. The key challenges include identifiability issues, confounding variables, and considerations of sample size and data quality.

Identifiability issues

Identifiability is one of the most significant challenges in causal discovery. It is the issue of uniquely determining the causal structure from the data. In many cases, different causal structures can lead to the same statistical distribution of the data, making it impossible to distinguish between them based solely on observational data. This issue is often referred to as the problem of equivalence classes of causal graphs [4, 5].

Functional causal models (FCMs) have been shown to improve identifiability by making additional assumptions about the data distribution beyond conditional independence relations. FCMs represent the effect variable, Y, as a function of the direct causes, X, and some noise term, E – that is, $Y = f(X, E)$ – where E is independent of X. This approach can help us distinguish between different DAGs in the same equivalence class [5].

In addition, several other methods have been proposed to address identifiability issues, including the use of interventional data and the incorporation of domain knowledge. Interventional data, obtained by actively manipulating some variables and observing the outcomes, can provide additional information that helps in distinguishing between equivalent causal structures. For example, methods such as do-calculus, developed by Judea Pearl, allow for the use of interventional data to improve identifiability [6, 2].

Additionally, techniques such as linear non-Gaussian models and non-linear models have been developed to address identifiability issues in causal discovery.

Integrating domain expertise can guide the causal discovery process by providing prior knowledge about the relationships between variables. It helps in narrowing down the set of plausible causal structures. Expert knowledge can be particularly useful when certain causal relationships are well-established or when experiments are infeasible.

It's worth noting that while these methods can improve identifiability, the challenge remains significant, especially in high-dimensional datasets. Scaling causal discovery so that it can work with large numbers of variables is an ongoing area of research and development in the field of causality.

Confounding variables

Confounding variables pose a significant challenge in causal discovery by potentially masking true causal relationships or creating spurious associations in observational data. This challenge is further complicated by latent confounders, which are unobserved variables affecting both the cause and outcome. In real-world scenarios, identifying and adjusting for all relevant confounders is often difficult as recognizing, measuring, and incorporating these variables into causal models can be complex. If not properly addressed, confounders can lead to biased causal estimates and incorrect conclusions. Moreover, many causal discovery algorithms assume causal sufficiency (that all common causes are observed), an assumption frequently violated in practice due to latent confounders. Collectively, these factors make confounding a critical hurdle in accurately identifying causal structures in observational data [7].

Sample size and data quality considerations

The quality and quantity of data are critical factors that influence the success of causal discovery. Small sample sizes can lead to overfitting and unstable causal estimates, while large sample sizes are generally more reliable but may still suffer from issues related to data quality.

Noise and missing data can significantly impact the accuracy of causal discovery algorithms [8]. Methods such as robust statistics [9] and imputation techniques are commonly used to handle noise and missing values, respectively. As the dimensionality of the data increases, so does the complexity of the causal discovery task. Why? Because the number of potential causal relationships grows combinatorially with the number of variables.

Moreover, the presence of mixed data types (continuous and discrete) can complicate the application of causal discovery algorithms. Some methods are specifically designed to handle mixed data types, but their applicability might be limited by the assumptions they make about the underlying data-generating process.

So, what main factors are needed to overcome the challenges of the robustness of causal discovery methods? These include advances in algorithms, the integration of domain knowledge, and the use of interventional data. Speaking of methods, we'll focus on them in the next section.

Methods for causal discovery

Prominent causal discovery methodologies can be categorized into three types: constraint-based, score-based, and hybrid methods. Each has its strengths and weaknesses and each employs distinct mechanisms for identifying causal structures. Various algorithms have been developed for causal discovery, including the PC algorithm, GES, and the IC algorithm. These algorithms utilize different principles, such as independence tests and scoring functions, to infer causal relationships. The choice of algorithm depends on the nature of the data and the specific requirements of the analysis as each approach has its strengths and weaknesses in uncovering causal structures from observational data. We'll discuss the methodologies in detail in the following sections.

Constraint-based methods

Researchers Clark Glymour, Peter Spirtes, and Richard Scheines developed constraint-based methods such as the PC algorithm, which uses conditional independence tests to uncover causal structures from observational data. These methods addressed the limitations of purely statistical approaches by incorporating graph theory and conditional independence principles.

The PC algorithm, the most well-known constraint-based method, proceeds in two main steps:

1. **Skeleton identification**: The algorithm identifies the undirected skeleton of the causal graph using conditional independence tests to determine edges between variable pairs.

2. **Edge orientation**: Once the skeleton has been identified, the algorithm applies rules to orient the edges, forming a DAG.

Starting with a fully connected graph, the PC algorithm iteratively removes edges corresponding to conditional independencies found in the data. It then uses rules such as collider detection (for example, $X \rightarrow Z \leftarrow Y$) to orient the edges.

This approach is flexible in terms of the conditional independence tests that are used and can sometimes relax causal sufficiency assumptions. However, it's prone to error propagation, where mistakes in early edge deletions can lead to incorrect final graphs.

The PC algorithm's output is typically a **partially directed acyclic graph** (**PDAG**) or a **completed PDAG** (**CPDAG**), representing the equivalence class of causal structures consistent with the observed conditional independencies. The PC algorithm was influenced by the IC algorithm, proposed by Judea Pearl and Thomas Verma. The IC algorithm introduced principles for inferring causal structures from observational data using conditional independence tests. Its three main steps are identifying the skeleton, v-structures, and orienting edges. Though conceptual, IC laid the foundation for inferring causal structures from statistical dependencies.

Now, let's discuss the advantages of the PC algorithm:

- **Soundness and completeness**: The PC algorithm is sound and complete under the assumptions of causal sufficiency (no hidden confounders) and faithfulness (all observed conditional independencies are due to the true causal structure). Let's deep dive a bit more. Soundness ensures that any causal relationships identified by the algorithm are correct, while completeness guarantees that all determinable causal relationships will be found. These properties hold when the assumptions of causal sufficiency, faithfulness, and the causal Markov condition are met. Causal sufficiency assumes there are no hidden confounders influencing the observed variables. This is a strong assumption that may not always hold in real-world scenarios. Faithfulness requires that all conditional independence relationships in the probability distribution are reflected in the causal graph structure, without accidental cancellations of causal effects. The causal Markov condition states that each variable is conditionally independent of its non-descendants given its direct causes.

- **Scalability**: The PC algorithm is relatively scalable and can handle large datasets with hundreds or even thousands of variables, making it a preferred choice for complex data analysis tasks in various fields. This algorithm achieves its scalability through a clever approach of performing conditional independence tests, starting with small conditioning sets and gradually increasing their size. This strategy allows it to quickly eliminate many potential edges in sparse graphs, making it particularly effective for high-dimensional settings with relatively few true causal connections. However, the algorithm's performance and accuracy may decrease when the sample size is limited relative to the number of variables or when the assumptions are violated.

Now, let's glean over the limitations of the PC algorithm:

- **Sensitivity to statistical errors**: The performance of constraint-based methods heavily depends on the accuracy of the conditional independence tests, which can be sensitive to statistical errors, especially in small samples.

- **Difficulty with latent variables**: The PC algorithm assumes causal sufficiency, meaning it can struggle with hidden confounders and latent variables.

Score-based methods

Score-based methods, such as GES, approach causal discovery by assigning a score to different possible graphs and selecting the graph with the highest score. This approach often employs techniques from machine learning and optimization.

One prominent example is the use of Bayesian networks, where each graph is scored based on its likelihood given the data and a prior distribution over graphs. The **Bayesian Information Criterion (BIC)** is commonly used as a scoring metric, balancing model fit and complexity:

$$BIC = \ln L - \left(\frac{k}{2}\right)\ln n \qquad (1)$$

Here, L is the log-likelihood of the model, k is the number of parameters, and n is the sample size.

There's quite a wide array of techniques for searching for the optimal graph, including greedy search, simulated annealing, and more. The search procedure can be exhaustive or heuristic, with GES being a common example. Score-based methods directly optimize a global criterion, potentially leading to more accurate models, but they're computationally intensive and sensitive to the choice of scoring function and search procedure.

Similar to what we did for constraint-based methods, let's consider the advantages and limitations of score-based approaches. Here are the advantages:

- **Handling latent variables**: Score-based methods can be more robust to the presence of latent variables and hidden confounders.

- **Flexibility in model assumptions**: These methods can incorporate different scoring functions and priors, providing flexibility in modeling assumptions.

Here are the limitations:

- **Computational complexity**: The search space of possible graphs is super-exponential, making score-based methods computationally intensive.

- **Local optima**: Greedy search strategies can get stuck in local optima, potentially missing the true causal structure.

Hybrid methods

Hybrid methods aim to leverage the strengths of both constraint-based and score-based approaches. One common strategy is to use constraint-based methods to reduce the search space, followed by score-based methods to fine-tune the causal structure.

Let's check out some examples and practical applications.

Max-Min Hill-Climbing (MMHC)

This hybrid algorithm combines the **Max-Min Parents and Children** (**MMPC**) algorithm and applies a constraint-based method to identify a candidate set of edges (relationships between variables). MMPC works in two stages:

- **Maximization step**: For each variable, the algorithm seeks the parents and children that maximize the conditional dependency with the target variable while controlling for subsets of other variables.

- **Minimization step**: It refines this set by removing variables that become conditionally independent of the target, given the found set of variables. This identifies candidate edges for the graph.

This is done to test for conditional independence. After the initial skeleton of the graph is generated using MMPC, the algorithm switches to a score-based approach to orient the edges and refine the structure of the DAG. A score-based method evaluates different DAG structures and selects the one that optimizes a scoring criterion, often based on model fit (for example, BIC). The hill-climbing part refers to a greedy search, where the algorithm incrementally modifies the graph (by adding, removing, or reversing edges) and selects the modification that improves the score the most.

Fast Causal Inference

FCI is an extension of the PC algorithm that accounts for latent variables, using both conditional independence tests and scoring methods to identify potential latent structures. This makes it more robust when hidden variables might affect the observed variables.

The original PC algorithm assumes that there are no hidden variables and uses conditional independence tests to discover the structure of the DAG. The FCI algorithm, however, relaxes this assumption by accounting for latent variables that may introduce unobserved confounding into the system. FCI extends the PC algorithm by adding a crucial step – *detecting and encoding structures that could be explained by unmeasured variables*. Specifically, it introduces bidirected edges to represent the possible presence of unmeasured confounders between two variables. The resulting structure is a **Partial Ancestral Graph (PAG)**, which encodes both the observed causal relationships and the presence of latent confounders.

First, FCI applies conditional independence tests to prune non-essential edges and then employs a scoring method to direct the edges and refine the graph. This process identifies not only the direct causal relationships but also which variables might be linked via unmeasured confounders, marking them with specific edge types (for example, bidirected edges).

FCI is especially useful in scenarios where it's unrealistic to assume that all relevant variables are observed. It provides a more accurate representation of the possible causal structure when some factors influencing the relationships aren't included in the dataset. This makes FCI a powerful tool in social sciences, economics, and epidemiology, where latent variables are common. It may help if we also briefly touch upon some other applications:

- **Biomedical research**: Hybrid methods are frequently used in genomics and epidemiology to identify causal relationships between genetic factors and diseases.

- **Economics**: These methods help in understanding causal relationships in economic systems, such as the impact of policy changes on economic indicators.

As you can see, causal discovery helps in various domains, from biomedical research to economics. By employing constraint-based, score-based, and hybrid methods, you can uncover causal relationships that inform decision-making and policy development. Each approach has its strengths and limitations, and when choosing a method, you should consider the specific context and challenges of the data at hand.

Functional Causal Models

FCMs represent a distinct approach to causal discovery, differing significantly from constraint-based, score-based, and hybrid methods. FCMs assume specific functional relationships between variables, allowing for non-linear and non-Gaussian relationships, and directly model these relationships using techniques such as **Independent Component Analysis (ICA)**. This approach contrasts with constraint-based methods, which rely on conditional independence tests, and score-based methods, which evaluate graph structures using scoring functions.

The **Linear Non-Gaussian Acyclic Model (LiNGAM)** is a well-known example of FCMs that assumes linear relationships and non-Gaussian distributions. LiNGAM can provide unique solutions even when the underlying DAG can't be identified by traditional methods, making it effective for datasets with non-Gaussian distributions. However, it assumes a specific functional form, which may not hold in all cases. FCMs excel in handling non-Gaussian data and complex scenarios but require assumptions about the functional form of relationships.

Which causal discovery method should we use?

The choice of causal discovery method depends on various factors, such as data characteristics, research goals, and computational resources. For linear relationships with non-Gaussian distributions, LiNGAM is suitable, especially when prior knowledge about functional relationships is available. Constraint-based methods are ideal for large datasets with potentially sparse causal structures and limited computational resources, while score-based methods are better for smaller datasets with latent variables and flexible modeling assumptions.

For complex scenarios involving large-scale applications, hybrid methods offer a balance between efficiency and accuracy by combining constraint-based and score-based approaches. Other factors to consider include sample size, temporal data, data types, and available computational power. Ultimately, the most appropriate method should be selected based on the specific context and data characteristics to ensure accurate and insightful results.

Next, we'll try implementing causal discovery in R.

Implementing causal discovery with Bayesian networks in R

In this section, we'll explore the practical application of causal discovery using R. We'll guide you through the step-by-step process of constructing a robust causal discovery model while leveraging various R packages to streamline the implementation, including the Bayesian-network-based approach.

Using R packages

Alright, first things first! R has many libraries that are suitable for problems in causal discovery. Each one brings something unique to the table. Let's look at some of them as we load them into our environment:

```
library(pcalg)
library(bnlearn)
library(causaleffect)
library(igraph)
library(graph)
```

The `pcalg` package, released in 2009, is a solid tool for implementing causal discovery algorithms, notably the PC algorithm, and estimating causal effects. It supports flexible conditional independence tests and handles mixed data types through the `micd` add-on. The package also addresses missing data with test-wise deletion and multiple imputation methods. There's also the `tpc` package, which extends the `pcalg` package's functionality by incorporating time-ordering constraints and context variables, both of which are crucial for longitudinal data analysis. The `micd` add-on extends the `pcalg` package's capabilities by providing functions for handling missing data through test-wise deletion and multiple imputation. For instance, `gaussCItwd`, `disCItwd`, and `mixCItwd` handle test-wise deletion for Gaussian, discrete, and mixed data, respectively. Multiple imputation can be done using the `mice` package, and the resulting imputed datasets can be analyzed using `gaussMItest`, `disMItest`, or `mixMItest`.

The `bnlearn` package, introduced in 2007, focuses on Bayesian network learning and inference. It provides built-in conditional independence tests, such as the conditional Gaussian test for mixed data, and allows users to specify edge constraints through whitelist and blacklist options. This package resolves edge orientation conflicts using a heuristic based on p-values, ensuring order-independent output. The `bnlearn` package is user-friendly and suitable for straightforward causal discovery tasks, particularly when handling mixed data and temporal constraints.

The `pcalg` package and its add-ons (`tpc` and `micd`) provide extensive customization options and flexibility in handling mixed data and missing data. On the other hand, the `bnlearn` package offers straightforward options for specifying edge constraints and handling mixed data, making it suitable for users who need a more streamlined approach.

For practical recommendations, it's advisable to use multiple packages to perform causal discovery and compare the results, something we'll do in this section. This approach can lend plausibility to your findings and help you identify potential issues.

Scenario for our problem

Here's our scenario: in a fictional study of women's involvement in drug trafficking in Colombia, researchers have collected data on various factors that might influence women's participation in this criminal activity. The study focuses on 1,000 women from different regions of Colombia, some involved in drug trafficking and others not, to identify potential risk factors and societal conditions that may contribute to women entering this illegal trade.

In this scenario, causal discovery aims to identify the underlying causal relationships between various personal, economic, and societal factors and women's involvement in drug trafficking in Colombia. By uncovering these causal links, researchers can better understand the root causes, predict risk factors, and develop more effective interventions and policies to address this complex social issue.

Some factors that are considered in the study include age, education level, employment status, income level, number of dependents, history of domestic violence, region (urban/rural), presence of organized crime in the area, local unemployment rate, access to social services, and involvement in drug trafficking (target variable).

> **Important note on sensitivity and analytical focus**
>
> In this situation, there's a complex interplay of personal, economic, and societal factors that might influence a woman's likelihood of becoming involved in drug trafficking in Colombia. By considering a wide range of variables, the study aims to identify potential risk factors and societal conditions that may contribute to women's participation in this criminal activity.
>
> Before proceeding, it's crucial to acknowledge that this fictional study involves complex and sensitive issues with profound societal implications. The primary goal here is to teach analytical methods, not to interpret results or draw conclusions about real-world situations. This book focuses on computational and methodological aspects of data analysis using synthetic data. The insights derived are meant to demonstrate statistical and computational techniques, not to provide definitive answers to social issues. You should remember that while this analysis can't fully account for the multifaceted nature of these factors in real-life scenarios, the analytical approach will empower you to use these skills in real settings.

Creating the dataset

This *fictional* dataset includes individual-level factors (such as age, education, and personal history) and broader societal and economic indicators (such as local unemployment rates and the presence of organized crime). Here's the dataset's structure. For the full code, please refer to this book's GitHub repository:

```
# Create a complex dataset
set.seed(123)
n <- 100
data <- data.frame(
   age = sample(18:60, n, replace = TRUE),
   education = sample(0:5, n, replace = TRUE), # 0: No education, 1:
Primary, 2: Secondary, 3: High school, 4: University, 5: Postgraduate
   ....
)
```

This scenario and dataset could be used to explore patterns and risk factors associated with women's involvement in drug trafficking, potentially informing policy decisions and intervention strategies.

First, we calculate a probability score for each individual based on their characteristics. Each term represents a factor's contribution:

- **Age**: (age - 30)^2 / 1,000 gives higher values for ages further from 30
- **Education**: -0.1 * education means higher education reduces the probability
- **Employment**: 0.2 * (1 - employed) increases the probability of unemployment
- **Income**: (7,000 - income) / 10,000 increases the probability of lower incomes

Other factors are weighted similarly.

As can be seen, the probability of involvement is calculated using a complex formula that takes into account all the factors, with some having positive influences (for example, unemployment and presence of organized crime) and others having negative influences (for example, education level and access to social services):

```
prob_involvement <- with(data,
  (age - 30)^2 / 1000 -
  0.1 * education +
  0.2 * (1 - employed) +
  (7000 - income) / 10000 +
  0.05 * dependents +
  0.2 * domestic_violence +
  0.1 * (region == "Rural") +
  0.1 * organized_crime +
  0.01 * local_unemployment -
  0.1 * social_services
)
```

Next, we normalize the values in each factor as computed here. This means we scale the values to be between 0 and 1, as they are probabilities. We apply the calculated values to randomly assign each individual a binary involvement status (0 or 1). Then, we convert the `region` variable from categorical (Urban/Rural) into numeric (1/2) for the causal discovery algorithm:

```
prob_involvement <- (prob_involvement - min(prob_involvement)) /
  (max(prob_involvement) - min(prob_involvement))
data$involved <- rbinom(n, 1, prob_involvement)
data$region <- as.numeric(factor(data$region))
```

Implementing PC algorithm

Moving forward, we'll be performing causal discovery. We'll use the PC algorithm to infer the causal structure from the data. The algorithm will create a list of sufficient statistics (correlation matrix and sample size) and then apply the PC algorithm:

```
# Perform causal discovery
suffStat <- list(C = cor(data), n = n)
pc.fit <- pc(suffStat, indepTest = gaussCItest,
             p = ncol(data), alpha = 0.05)
```

We now will write a function that defines how to plot the causal graph. It converts the graph to an `igraph` object (readable by the `igraph` library). Then we call this plotting function to visualize the causal graph.

```
# Call the function to plot the graph
plot_graph(pc.fit@graph)

# Print summary of the causal graph
summary(pc.fit)
```

We get the following graph:

Causal Graph: Women's Involvement in Drug Trafficking

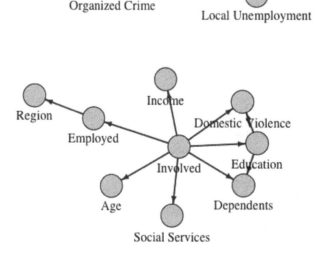

Figure 15.2 – Graph estimation by the PC algorithm

Finally, we print a summary of the causal graph, including the number of nodes, edges, and specific causal relationships discovered. Overall, the code walks you through a complex scenario, simulates data based on hypothesized relationships, applies a causal discovery algorithm to infer the causal structure from the data, and then visualizes and summarizes the results. It's an effective demonstration of how causal discovery methods can be applied to complex social science scenarios:

```
Graphical properties of skeleton:
==================================
Max. number of neighbours:  2 at node(s) 5 6
Avg. number of neighbours:  0.9090909
```

Figure 15.3 – Graph attribute numbers

Using bnlearn for Bayesian networks

Next, let's try Bayesian networks using the `bnlearn` package. The `hc` function from this library uses a Hill-Climbing algorithm; it climbs a mountain of possible network structures to find the best one that describes the underlying causal structure in our data.

First, we need to convert variables into factors (that is, make them categorical). Once we've done that, we're ready to learn the network structure. For this, we'll use the Hill-Climbing algorithm to figure out how all these factors might be related:

```
data[] <- lapply(data, factor)

# Learn the structure of the Bayesian Network
bn <- hc(data)

# Call the function to plot the Bayesian Network
plot_bn(bn)
```

Finally, we must start plotting to see what this network looks like (*Figure 15.4*). We're creating a visual map of how all our factors might be connected. We can also print the summary to check what we have:

```
print("Summary of the Bayesian Network:")
print(bn)
```

Here's the output:

Bayesian Network: Women's Involvement in Drug Trafficking

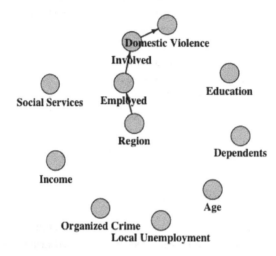

Figure 15.4 – The network structure that was learned using the Bayesian network approach

Now, we want to know what factors are most closely related to women's involvement in drug trafficking. The Markov Blanket, parents, and children give us different perspectives on this:

```
cpdag <- cpdag(bn)
print("\nMarkov Blanket of 'involved':")
print(mb(cpdag, "involved"))
fitted_bn <- bn.fit(bn, data)
```

At this point, we know how things are connected, but how strong are these connections? To find out, let's print a conditional probability table (*Figure 15.5*):

```
print("\nConditional Probability Table for 'involved':")
print(fitted_bn$involved)
```

Here's the output:

```
Conditional probability table:

          domestic_violence
involved          0          1
       0  0.4558611  0.3527508
       1  0.5441389  0.6472492
```

Figure 15.5 – Conditional probability table

Now, what's the probability of involvement for a highly educated, employed woman?

```
cpquery(fitted_bn, event = (involved == 1),
        evidence = (education == 5 & employed == 1))
```

How well does this network explain our data? The BIC score helps us understand this:

```
print("\nNetwork Score (BIC):")
print(score(bn, data, type = "bic"))
```

We can also cross-validate the network. Cross-validation helps us understand how well our network might perform on new, unseen data:

```
cv_score <- bn.cv(data, bn, k = 10, loss = "logl")
print("\nCross-validated log-likelihood loss:")
print(mean(cv_score))
```

In this example, each step builds on the previous one, taking us from raw data to a comprehensive understanding of the factors influencing women's involvement in drug trafficking.

More causal discovery methods

Do you want to try more methods in R for causal discovery? Let's start with the MMHC algorithm (a hybrid method). This algorithm infers the causal relationships between variables based on the observed data:

```
# PC algorithm (constraint-based)
pc_dag <- pc.stable(data)
# MMHC algorithm (hybrid)
mmhc_dag <- mmhc(data)
```

Next, we'll visualize the DAGs that have been learned by both the PC and MMHC algorithms. This will create a PDF file with side-by-side plots of the two DAGs. Then, it will print out the edges (connections between variables) discovered by each algorithm, which represent the inferred causal relationships:

```
par(mfrow = c(1, 2))  # Plot two graphs side by side

# PC Algorithm (Constraint-based)
plot_styled_dag(pc_dag, title = "PC Algorithm")

# MMHC Algorithm (Hybrid)
plot_styled_dag(mmhc_dag, title = "MMHC Algorithm")
```

Here's the MMHC algorithm's causal structure:

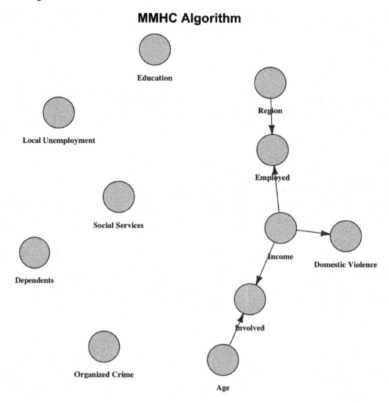

Figure 15.6 – The causal structure learned by the MMHC algorithm

Finally , we print the edges of each learned DAG:

```
cat("PC Algorithm Edges:\n")
print(pc_dag$arcs)
cat("\nMMHC Algorithm Edges:\n")
print(mmhc_dag$arcs)
```

Estimating causal effects

Next, we'll convert the learned DAGs into a format compatible with the `igraph` library. Then, we'll estimate the causal effect of education on involvement in organized crime using both the PC and MMHC algorithm results. The `tryCatch` blocks are used to handle potential errors in calculating the causal effect:

```
# Convert bnlearn graphs to igraph objects
pc_dag_igraph <- igraph::graph_from_edgelist(
```

```
  as.matrix(pc_dag$arcs), directed = TRUE)
mmhc_dag_igraph <- igraph::graph_from_edgelist(
  as.matrix(mmhc_dag$arcs), directed = TRUE)
# Estimate causal effects
# We'll estimate the causal effect of 'education' on 'involved'
# Using the PC algorithm result
tryCatch({
  effect_pc <- causal.effect(y = "involved", x = "education",
                             G = pc_dag_igraph)
  cat("\nCausal Effect (PC algorithm):\n")
  print(effect_pc)
}, error = function(e) {
  cat("\nError in calculating causal effect for PC algorithm:",
      conditionMessage(e), "\n")
})
# Using the MMHC algorithm result
tryCatch({
  effect_mmhc <- causal.effect(y = "involved", x = "education",
                               G = mmhc_dag_igraph)
  cat("\nCausal Effect (MMHC algorithm):\n")
  print(effect_mmhc)
}, error = function(e) {
  cat("\nError in calculating causal effect for MMHC algorithm:",
      conditionMessage(e), "\n")
})
```

At this point, we should bootstrap on the PC algorithm to assess the stability of the learned structure. This will create 200 bootstrap samples and apply the PC algorithm to each. The results are then averaged to create a more robust network structure. The averaged network is plotted (*Figure 15.7*) and saved as a PDF.

Finally, the edges of the bootstrapped network that have a strength greater than 0.5 are printed, indicating more stable relationships:

```
# Perform bootstrapping to assess the stability of the PC algorithm
boot_pc <- boot.strength(data, R = 200, algorithm = "pc.stable")

# Get the averaged network from the bootstrap results
avg_network <- averaged.network(boot_pc)

# Save the plot as a PDF and use the custom plotting function
plot_styled_bootstrap_dag(
  avg_network, title = "Bootstrapped PC Network", label_cex = 1.2
)
```

We get the following structure:

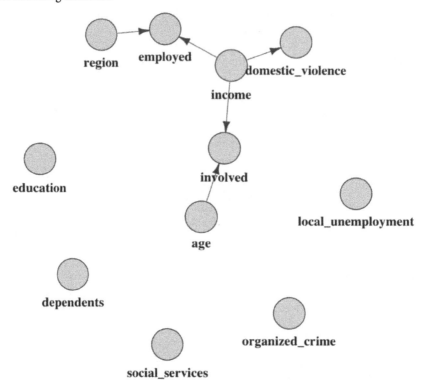

Figure 15.7 – The causal structure learned by the bootstrapped PC approach

In this section, we learned how to apply a wide range of causal discovery methods in R Next, we'll focus on another case study that provides a more comprehensive application of causal discovery algorithms.

A multi-algorithm comparative approach to causal discovery in R

Now that you have a grasp on causal discovery in R, let's consider a marketing research scenario where you can apply multiple techniques in R.

Unlike the first study, which primarily used Bayesian networks and the PC algorithm, this case study employs multiple methods, including the PC algorithm, the MMHC algorithm, and bootstrapping techniques. This multi-method approach allows for a comparative analysis of different causal discovery techniques, providing a more robust and comprehensive understanding of the causal relationships in the marketing data.

A marketing research firm is conducting a study to understand the complex relationships between demographic factors, digital engagement, brand perception, and consumer behavior in the modern marketplace. They've collected data on various aspects of consumer interaction with a brand, and they want to uncover the causal relationships between these factors to inform marketing strategies and improve customer engagement.

Let's write down what we want to discover through causal analysis:

- The impact of age on income and social media usage, and how these factors indirectly influence other variables.

- How social media usage affects brand perception and product awareness.

- The relationship between income and customer satisfaction.

- The factors that directly and indirectly influence purchase intent and repurchase rate.

- The potential causal pathways from demographic factors (age) to outcomes (purchase intent and repurchase rate).

- Are there any others that come to mind? Jot them down.

This situation is perfect for causal discovery. Here's why:

- **Complex interrelationships**: The dataset includes multiple variables with potential complex relationships. Causal discovery can help untangle these relationships and reveal direct and indirect effects.

- **Mixed variable types**: The dataset includes continuous variables (age, income, social media usage, brand perception, and customer satisfaction) and binary variables (purchase intent, product awareness, and repurchase rate). Causal discovery can handle this mix of variable types.

- **Non-experimental data**: This dataset represents observational data, which is common in marketing research as controlled experiments are often not feasible. As we discussed previously, causal discovery methods are designed to infer causal relationships from such observational data.

- **Multiple potential pathways**: There are several potential causal pathways (for example, age → social media usage → brand perception → purchase intent). Causal discovery can help identify these pathways and their relative strengths.

- **Informing strategic decisions**: Understanding the causal structure can help the marketing firm make more informed decisions about where to focus their efforts for maximum impact on key outcomes such as purchase intent and repurchase rate.

- **Hypothesis generation**: The results of causal discovery can generate new hypotheses about consumer behavior that can be further tested or explored in future studies.

- **Identifying key leverage points**: By understanding the causal structure, the firm can identify key variables that have the most influence on desired outcomes, helping to prioritize marketing efforts and resource allocation.

From this, you must be convinced that in this scenario, causal discovery can provide valuable insights beyond traditional correlation analysis. It can help the marketing firm understand not just what factors are related to important outcomes such as purchase intent and repurchase rate, but how these factors are causally related. This understanding can lead to more effective, targeted marketing strategies and a deeper understanding of the customer journey in the digital age.

Setting up and generating data

Let's begin by setting up our environment and generating our synthetic dataset:

```
library(bnlearn)
library(pcalg)
library(Rgraphviz)
set.seed(123)
n <- 10000
data <- data.frame(
  age = rnorm(n, mean = 35, sd = 10)
)
data$income <- exp(log(10) + 0.03 * data$age + rnorm(n, 0, 0.5))
data$social_media_usage <- pmax(0, round(
  10 - 0.15 * data$age + rnorm(n, 0, 2)))
data$brand_perception <- 3 + 0.5 * scale(
  data$social_media_usage) + rnorm(n, 0, 0.5)
data$customer_satisfaction <- 5 + 2 * scale(
  log(data$income)) + rnorm(n, 0, 1)
data$purchase_intent <- factor(rbinom(n, 1, plogis(
  -2 + 0.5 * data$customer_satisfaction + 0.5 *
    data$brand_perception)))
data$product_awareness <- factor(rbinom(n, 1, plogis(
  -1 + 0.3 * data$social_media_usage)))
data$repurchase_rate <- factor(rbinom(n, 1, plogis(
  -3 + 0.5 * data$customer_satisfaction)))
```

In this setup, we're creating a synthetic dataset with known causal relationships. We start with age as our root variable and build a network of dependencies from there. Income grows with age, while social media usage decreases. Brand perception is influenced by social media usage, and customer satisfaction is tied to income. The binary variables (purchase intent, product awareness, and repurchase rate) are influenced by various factors, mirroring real-world consumer behavior patterns.

This data generation process allows us to test our models against a known ground truth. We can evaluate how well each method recovers the relationships we've built into the data.

Let's take a quick look at the correlations in our data:

```
data$social_media_usage <- as.numeric(data$social_media_usage)
cor_matrix <- cor(data[sapply(data, is.numeric)])
print("Correlation matrix of numeric variables:")
print(cor_matrix)
```

This correlation matrix gives us an initial view of the linear relationships between our numeric variables. It's a starting point for understanding the data structure, but remember, correlation doesn't imply causation – that's why we need more sophisticated causal discovery methods.

Constraint-based methods in R

Now, let's apply a constraint-based method, the Semi-Interleaved HITON-PC algorithm:

```
print("\nLearning network structure using Semi-Interleaved HITON-PC:")
si_hiton_pc_result <- si.hiton.pc(data)
print(arcs(si_hiton_pc_result))
```

Here's the output:

```
            from                  to
   [1,]  "age"                 "income"
   [2,]  "age"                 "social_media_usage"
   [3,]  "age"                 "brand_perception"
   [4,]  "age"                 "customer_satisfaction"
   [5,]  "income"              "age"
   [6,]  "income"              "customer_satisfaction"
   [7,]  "income"              "product_awareness"
   [8,]  "income"              "repurchase_rate"
   [9,]  "income"              "purchase_intent"
  [10,]  "social_media_usage"  "age"
  [11,]  "social_media_usage"  "brand_perception"
  [12,]  "social_media_usage"  "product_awareness"
  [13,]  "brand_perception"    "age"
  [14,]  "brand_perception"    "social_media_usage"
  [15,]  "brand_perception"    "purchase_intent"
  [16,]  "customer_satisfaction" "age"
```

Figure 15.8 – A snapshot of the connections that were discovered

The Semi-Interleaved HITON-PC algorithm [10] is a sophisticated constraint-based method that uses conditional independence tests to infer the causal structure of the data. It's particularly effective for mixed data types and can handle high-dimensional data.

As you examine the output, each arc represents a potential causal link between two variables. The direction of the arc suggests the direction of causality, but this should be interpreted cautiously. The absence of an arc between two variables suggests potential conditional independence.

The output from the causal discovery algorithms (Hill-Climbing and semi-interleaved HITON-PC) reveals the limitations in inferring causal relationships from observational data. There are bidirectional edges, such as between income and age, which shows that the algorithms couldn't definitively determine the direction of causality. This results in PDAGs or CPDAGs, representing equivalence classes of DAGs. These bidirectional edges should be interpreted as relationships between variables where the causal direction is uncertain given the available data and assumptions.

To address these limitations, several approaches can be considered:

- **Post-processing steps** could be implemented to remove cycles, potentially using domain knowledge to decide on directions.

- **Time-series data**, if available, could better capture temporal relationships.

- **More advanced causal discovery algorithms**, such as FCI, might handle these situations better, especially with latent confounders.

Incorporating domain expertise is crucial to resolve ambiguities in causal direction. While DAGs are powerful tools for causal reasoning, the presence of bidirectional edges highlights the challenges in inferring true causal structures from observational data and emphasizes the need for careful interpretation and potentially additional analysis.

Score-based methods in R

Next, let's apply a score-based method – the Hill-Climbing algorithm:

```
print("\nLearning network structure using original mixed data:")
mixed_hc_result <- hc(data)
print(arcs(mixed_hc_result))
```

The Hill-Climbing algorithm searches for a network structure that maximizes a scoring function. It starts with an empty graph and iteratively adds, removes, or reverses edges to improve the score.

Each arc in the output represents a direct dependency between variables. The network structure represents the joint probability distribution of the variables. While the direction of arcs suggests potential causal relationships, they should be interpreted cautiously since the algorithm may reverse edge directions to achieve a better score.

Hybrid methods in R

Finally, let's try a hybrid approach by discretizing our data and then applying the Hill-Climbing algorithm:

```
data_discrete_more_bins <- as.data.frame(lapply(data, function(x) {
  if(is.numeric(x)) {
    cut(x, breaks = 5, labels = c(
      "Very Low", "Low", "Medium", "High", "Very High"))
  } else {
    x
  }
}))
print("\nLearning network structure using data with more bins:")
hc_result_more_bins <- hc(data_discrete_more_bins)
print(arcs(hc_result_more_bins))
```

This hybrid approach combines elements of constraint-based methods (through discretization) and score-based methods. Discretization can help in detecting non-linear relationships and reduce the impact of outliers. However, it may also lead to information loss.

When interpreting these results, compare them with the previous two methods. Look for relationships that appear consistently across all methods as these are likely to be robust. Also, pay attention to relationships that only appear here as they might indicate non-linear associations.

Visualizing causal relationships

To better understand our results, let's visualize the causal graphs:

```
par(mfrow = c(1, 3))  # Adjust layout for multiple plots in one row

# Plot each graph with the improved style and title
plot_graph(mixed_hc_result, "Hill-Climbing (Mixed Data)",
           label_cex = 1.2)
plot_graph(hc_result_more_bins, "Hill-Climbing (More Bins)",
           label_cex = 1.2)
plot_graph(si_hiton_pc_result, "Semi-Interleaved HITON-PC",
           label_cex = 1.2)
```

We get the following graph:

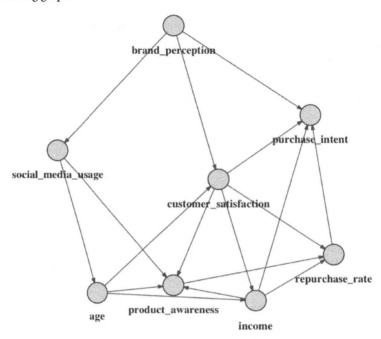

Figure 15.9 – The Hill-Climbing (Mixed Data) output's algorithm

As we discussed previously, the Hill-Climbing (Mixed Data) algorithm is a score-based method in R that's used for causal discovery that iteratively adjusts a graph to maximize a scoring criterion such as the BIC score. In this process, nodes represent variables such as customer behavior (for example, social media usage, brand perception, and customer satisfaction) and demographics (for example, age and income), while arrows indicate causal relationships between them. For instance, social media usage affects brand perception, which then influences customer satisfaction, and income has a direct causal effect on customer satisfaction.

Satisfaction drives purchase intent and repurchase rate, indicating that satisfied customers are more likely to buy and rebuy products. Demographic factors such as income and age also influence satisfaction. This model highlights how customer perceptions and behaviors interrelate, providing insights for improving customer engagement and loyalty. The algorithm adapts to mixed data types (continuous and discrete) to infer the most likely causal structure.

There are other visualizations in the code that allow you to quickly grasp the overall structure of the learned causal relationships. Compare the graphs from different methods while looking for consistent patterns and notable differences (refer to this book's GitHub repository for the full code and visualizations).

To dive deeper into the results, let's examine the Markov Blankets and check for edges in each network:

```
cat("\nMarkov Blankets for each variable (Semi-Interleaved HITON-
PC):\n")
for (var in names(data)) {
  mb <- mb(si_hiton_pc_result, node = var)
  cat(var, ": ", paste(mb, collapse = ", "), "\n")
}


check_edges <- function(network, name) {
  if(length(arcs(network)) > 0) {
    cat("\nEdges found in", name, ":\n")
    print(arcs(network))
  } else {
    cat("\nNo edges found in", name, "\n")
  }
}
check_edges(mixed_hc_result, "Hill-Climbing (Mixed Data)")
check_edges(hc_result_more_bins, "Hill-Climbing (More Bins)")
check_edges(si_hiton_pc_result, "Semi-Interleaved HITON-PC")
```

The Markov Blankets show us, for each variable, which other variables are most relevant for predicting its value. A Markov Blanket for a variable is a set of variables that, when known, renders the variable conditionally independent of all other variables in the system. It includes the variable's parents, children, and co-parents, effectively shielding it from outside influences. This concept is vital for understanding causal relationships and for feature selection in statistical and machine learning applications.

This can be crucial for understanding the immediate causal environment of each variable. By comparing the edges found in each network, you can see which relationships are identified consistently across different methods. Look for edges that appear in all or most of the networks – these are likely to represent the strongest and most reliable relationships in your data.

Interpretation from code

Our causal discovery analysis reveals several important relationships between demographic factors, digital engagement, brand perception, and consumer behavior in the modern marketplace. Let's interpret the results:

```
                             age      income social_media_usage brand_perception customer_satisfaction
age                    1.0000000  0.4724077         -0.5828863       -0.4260703             0.4637279
income                 0.4724077  1.0000000         -0.2762650       -0.2056882             0.8176985
social_media_usage    -0.5828863 -0.2762650          1.0000000        0.7058035            -0.2706503
brand_perception      -0.4260703 -0.2056882          0.7058035        1.0000000            -0.1968138
customer_satisfaction  0.4637279  0.8176985         -0.2706503       -0.1968138             1.0000000
```

Figure 15.10 – The correlation matrix in the data

Correlation matrix

The correlation matrix (*Figure 15.10*) shows strong relationships between several variables:

- **Age** is negatively correlated with social media usage (-0.58) and brand perception (-0.43), indicating that younger consumers tend to use social media more and have a more positive brand perception.

- **Income** is positively correlated with customer satisfaction (0.82), suggesting that higher-income customers are generally more satisfied.

- **Social media usage** is positively correlated with brand perception (0.71), implying that increased social media engagement is associated with a more positive brand image.

Network structures

The three different network learning algorithms Hill-Climbing (Mixed Data), Hill-Climbing with more bins, and Semi-Interleaved HITON-PC) provided insights into the causal relationships. We learned the following:

- Income strongly influences customer satisfaction across all models, indicating that higher income leads to greater satisfaction.

- Age affects social media usage in all models, confirming that younger consumers tend to use social media more frequently.

- Social media usage influences brand perception, suggesting that increased social media engagement leads to a more positive brand image.

- Customer satisfaction impacts repurchase rate and purchase intent, highlighting the importance of keeping customers satisfied to drive sales and loyalty.

- Income also influences age, product awareness, and purchase intent, indicating that income level affects various aspects of consumer behavior and demographics.

- Brand perception affects purchase intent, showing that a positive brand image can lead to increased sales.

The Markov Blankets from the Semi-Interleaved HITON-PC model provide insights into the most influential factors for each variable:

- Age is influenced by income, social media usage, brand perception, and customer satisfaction.

- Income is connected to age, customer satisfaction, purchase intent, product awareness, and repurchase rate, highlighting its central role in consumer behavior.

- Social media usage is linked to age, brand perception, and product awareness, emphasizing its importance in shaping consumer perceptions and knowledge.

- Customer satisfaction is influenced by age, income, purchase intent, and repurchase rate, underscoring its crucial role in driving consumer behavior.

So, based on what we've learned what could be the implications? Let's discuss this:

- Target younger audiences through social media to improve brand perception and product awareness.

- Focus on improving customer satisfaction, especially for higher-income segments, to drive repurchase rates and purchase intent.

- Develop age-specific marketing strategies while considering the different levels of social media usage and brand perception across age groups.

- Leverage social media to enhance brand perception, which can lead to increased purchase intent.

- Consider income levels when developing marketing campaigns since income influences various aspects of consumer behavior.

In conclusion, this causal discovery analysis provides valuable insights into the complex relationships between consumer characteristics and behaviors. By understanding these causal links, the marketing research firm can develop more targeted and effective strategies to improve customer engagement and drive business outcomes.

Future steps

Based on the results and interpretations of the analysis, there are several additional steps and actions we can take to further leverage the insights gained:

- **Conduct path analysis**: Examine the specific paths between variables to understand the direct and indirect effects. For example, analyze how age influences purchase intent through social media usage and brand perception.

- **Perform sensitivity analysis**: Test how changes in one variable affect others in the network. This can help you identify which factors have the most significant impact on key outcomes, such as purchase intent or customer satisfaction.

- **Develop targeted marketing campaigns**: Use the identified relationships to create personalized marketing strategies for different age groups and income levels. For instance, focus on social media engagement for younger audiences to improve brand perception.

- **Optimize customer satisfaction initiatives**: Given the strong link between income, customer satisfaction, and repurchase rate, develop strategies to enhance satisfaction across different income segments.

- **Improve product awareness strategies**: Utilize the connection between social media usage and product awareness to refine digital marketing efforts and increase product visibility.

- **Conduct A/B testing**: Design experiments to validate the causal relationships identified in the models. For example, test how changes in social media engagement affect brand perception and purchase intent.

- **Develop predictive models**: Use the discovered relationships to build predictive models for key outcomes such as purchase intent or repurchase rate. This can help in forecasting and decision-making.

- **Investigate feedback loops**: Analyze potential feedback loops in the network, such as the bidirectional relationship between age and income, to understand their long-term effects on consumer behavior.

- **Perform segment analysis**: Perform separate analyses for different customer segments (for example, age groups or income levels) to identify segment-specific patterns and relationships.

- **Perform temporal analysis**: If time series data is available, extend the analysis to include temporal aspects, examining how relationships evolve.

- **Integrate with other data sources**: Combine these findings with other relevant data (for example, market trends and competitor information) to gain a more comprehensive understanding of the market dynamics.

- **Develop a customer journey map**: Use the identified relationships to create a detailed customer journey map, highlighting key touchpoints and their impact on purchase decisions and customer satisfaction.

By implementing these additional steps, the marketing research firm can gain deeper insights into consumer behavior, refine their strategies, and develop more effective marketing and customer engagement initiatives.

Summary

In this chapter, you learned that causal discovery is a useful tool within causal inference that aims to identify the underlying causal structure between various factors using observational data. We implemented representational approaches such as DAGs and SCMs to map out causal relationships. We also discussed various techniques, including constraint-based methods such as the PC algorithm, score-based methods such as GES, and hybrid approaches, each with their unique strengths and challenges.

Then, we illustrated the practical implementation in R through scenario-based problem solving, demonstrating the transformative potential of causal discovery across diverse fields such as public health and economics. We addressed challenges such as identifiability issues, confounding variables, and data quality, highlighting how causal discovery provides actionable insights, informs policy and decision-making, advances scientific understanding, reduces bias, and analyzes complex systems.

By moving from purely observational data to a structured understanding of causal mechanisms, causal discovery enables more accurate predictions and targeted interventions, offering a robust approach to uncovering the true drivers of observed phenomena.

That concludes our discussions and learnings from the science of causal inference, with a complete focus on R and its application to real-world problems. Throughout this book, you explored a wide range of compelling topics, including the distinction between correlation and causation, propensity score matching, causal graphs, heterogeneity in causality, causal discovery, and others. We hope that through the theories, mathematical deep dives, and case study implementations in R, you now feel well-equipped to tackle problems related to causal inference. Specifically, you should be able to view your data through the lens of causality, thereby enhancing your interpretations and judgment calls about how things work in real life by applying the principles you've learned.

However, the field of causal inference is rapidly evolving. I encourage you to continue learning by reading new papers, blogs, and articles to stay at the forefront of this critical expertise in data science. This book serves as a foundational stepping stone for understanding how causal principles can be applied, but your learning journey doesn't end here. True expertise and confidence are built when you apply the concepts you've learned to your endeavors and make continuous improvements.

References

1. Mitra N, Roy J, Small D. *The Future of Causal Inference*. Am J Epidemiol. 2022 Sep 28;191(10):1671-1676. doi: 10.1093/aje/kwac108. PMID: 35762132; PMCID: PMC9991894.

2. Glymour C, Zhang K, Spirtes P. *Review of Causal Discovery Methods Based on Graphical Models*. Front Genet. 2019 Jun 4;10:524. doi: 10.3389/fgene.2019.00524. PMID: 31214249; PMCID: PMC6558187.

3. Broadbent A, Grote T. *Can Robots Do Epidemiology? Machine Learning, Causal Inference, and Predicting the Outcomes of Public Health Interventions*. Philos Technol. 2022;35(1):14. doi: 10.1007/s13347-022-00509-3. Epub 2022 Feb 26. PMID: 35251906; PMCID: PMC8881939.

4. Shen, X., Ma, S., Vemuri, P. et al. *Challenges and Opportunities with Causal Discovery Algorithms: Application to Alzheimer's Pathophysiology*. Sci Rep 10, 2975 (2020). https://doi.org/10.1038/s41598-020-59669-x.

5. *Identifiability of Causal Graphs using Functional Models*, by Jonas Peters, Joris Mooij, Dominik Janzing, and Bernhard Schoelkopf. Available at https://arxiv.org/pdf/1202.3757.

6. *The Do-Calculus Revisited*, by Judea Pearl. Available at https://arxiv.org/abs/1210.4852.

7. David Kaltenpoth and Jilles Vreeken. 2023. *Causal discovery with hidden confounders using the algorithmic Markov condition*. In Proceedings of the Thirty-Ninth Conference on Uncertainty in Artificial Intelligence (UAI '23), Vol. 216. JMLR.org, Article 96, 1016–1026.

8. *Sample, estimate, aggregate: A recipe for causal discovery foundation models*, by Menghua Wu, Yujia Bao, Regina Barzilay, and Tommi Jaakkola. Available at `https://arxiv.org/pdf/2402.01929`.

9. *Robust Statistics / Estimation (Robustness) & Breakdown Point*: `https://www.statisticshowto.com/robust-statistics/`.

10. Yang, Wei. *A Fast HITON_PC Algorithm*. 2010 International Conference on Computational Intelligence and Security (2010): 47-50.

11. Glymour, Clark, Kun Zhang, and Peter Spirtes. *Review of Causal Discovery Methods Based on Graphical Models*. Frontiers in Genetics 10 (2019):

Index

packtpub.com

Subscribe to our online digital library for full access to over 7,000 books and videos, as well as industry leading tools to help you plan your personal development and advance your career. For more information, please visit our website.

Why subscribe?

- Spend less time learning and more time coding with practical eBooks and Videos from over 4,000 industry professionals

- Improve your learning with Skill Plans built especially for you

- Get a free eBook or video every month

- Fully searchable for easy access to vital information

- Copy and paste, print, and bookmark content

Did you know that Packt offers eBook versions of every book published, with PDF and ePub files available? You can upgrade to the eBook version at packtpub.com and as a print book customer, you are entitled to a discount on the eBook copy. Get in touch with us at customercare@packtpub.com for more details.

At www.packtpub.com, you can also read a collection of free technical articles, sign up for a range of free newsletters, and receive exclusive discounts and offers on Packt books and eBooks.

Other Books You May Enjoy

If you enjoyed this book, you may be interested in these other books by Packt:

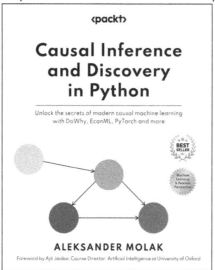

Causal Inference and Discovery in Python

Aleksander Molak

ISBN: 978-1-80461-298-9

- Master the fundamental concepts of causal inference
- Decipher the mysteries of structural causal models
- Unleash the power of the 4-step causal inference process in Python
- Explore advanced uplift modeling techniques
- Unlock the secrets of modern causal discovery using Python
- Use causal inference for social impact and community benefit

Modern Generative AI with ChatGPT and OpenAI Models

Valentina Alto

ISBN: 978-1-80512-333-0

- Understand generative AI concepts from basic to intermediate level
- Focus on the GPT architecture for generative AI models
- Maximize ChatGPT's value with an effective prompt design
- Explore applications and use cases of ChatGPT
- Use OpenAI models and features via API calls
- Build and deploy generative AI systems with Python
- Leverage Azure infrastructure for enterprise-level use cases
- Ensure responsible AI and ethics in generative AI systems

Packt is searching for authors like you

If you're interested in becoming an author for Packt, please visit `authors.packtpub.com` and apply today. We have worked with thousands of developers and tech professionals, just like you, to help them share their insight with the global tech community. You can make a general application, apply for a specific hot topic that we are recruiting an author for, or submit your own idea.

Share Your Thoughts

Now you've finished *Causal Inference in R*, we'd love to hear your thoughts! Scan the QR code below to go straight to the Amazon review page for this book and share your feedback or leave a review on the site that you purchased it from.

https://packt.link/r/1-837-63902-7

Your review is important to us and the tech community and will help us make sure we're delivering excellent quality content.

Download a free PDF copy of this book

Thanks for purchasing this book!

Do you like to read on the go but are unable to carry your print books everywhere?

Is your eBook purchase not compatible with the device of your choice?

Don't worry, now with every Packt book you get a DRM-free PDF version of that book at no cost.

Read anywhere, any place, on any device. Search, copy, and paste code from your favorite technical books directly into your application.

The perks don't stop there, you can get exclusive access to discounts, newsletters, and great free content in your inbox daily

Follow these simple steps to get the benefits:

1. Scan the QR code or visit the link below

https://packt.link/free-ebook/9781837639021

2. Submit your proof of purchase
3. That's it! We'll send your free PDF and other benefits to your email directly